THE ADDISON-WESLEY SCIENCE HANDBOOK

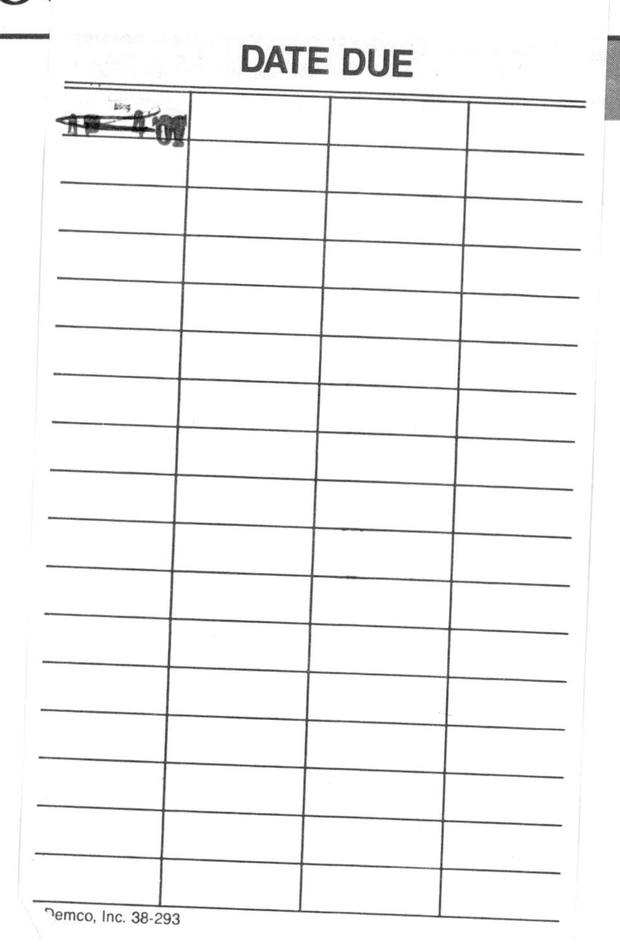

HELIX BOOKS

Addison-Wesley Publishers Limited

Don Mills, Ontario • Reading, Massachusetts • Menlo Park, California
New York • Wokingham, England • Amsterdam • Bonn
Sydney • Singapore • Tokyo • Madrid • San Juan
Paris • Seoul • Milan • Mexico City • Taipei

Publisher: Ron Doleman
Managing Editor: Linda Scott
Editors: Michael Cuddy, Suzanne Schaan
Production Coordinator: Melanie van Rensburg
Manufacturing Coordinator: Sharon Latta Paterson
Book Design and Layout: Lexigraf

Canadian Cataloguing in Publication Data

Coleman, Gordon J.

The Addison-Wesley science handbook

ISBN 0-201-76652-3

1. Science – Handbooks, manuals, etc.

I. Dewar, David (David C.). II. Title.

Q199.C65 1997 500 C95-931771-6

ISBN 0-201-76652-3

A B C D E — WC — 00 99 98 97 96

PREFACE

Today's science industry faces many challenges. There's at least one difficulty—the lack of convenient reference material—that we felt we could do something about. Science students, technical writers, and people in various science-related fields struggle through their work with inadequate tools.

The problem is not a lack of information. On the contrary, sometimes it seems that there is too much. Faced with textbooks, journals, technical libraries, databases—not to mention the Internet—how do you access the information you need in an efficient manner? In other words, how do you prevent the search for a simple definition from turning into a three-hour ordeal?

One way is to use *The Addison-Wesley Science Handbook*. It includes all the things you're expected to know but can't always memorize. Because science is interdisciplinary, we have gathered the most vital information from biology, chemistry, geology, physics, and mathematics, and placed it at your fingertips. Each discipline has its own unit containing key concepts, formulas, illustrations, and tables.

The Addison-Wesley Science Handbook is a warehouse of information in an accessible format. We designed the book to include both fundamental and advanced concepts; users will be pleased with its coverage and depth. Students won't get lost in advanced material, while technical writers and others can use the tables, glossaries, and summaries as quick memory refreshers. The book includes something useful for all levels—high school, college, university, and workforce.

The Addison-Wesley Science Handbook brings together a broad range of essential science information in one easy-to-use guide. Whether you are writing reports or cramming for an exam, this book is the reference manual you need.

Gordon J. Coleman
David Dewar

ACKNOWLEDGMENTS

Without these people this book would never have been possible:

Joe Bauche
Beverly Bertrand
Beth Bruder
Donna Dewar
Bill Dewar
Mary Dewar
Ron Doleman
Kateri Lanthier
Julie Nolan

Reviewers and Advisors:

Linda Barlow
Craig Carroll
David Caughell
Reed Kirkpatrick
Colin Morozuk
Chris Pawlowicz
Geoff Rayner-Canham

We would also like to thank the following people:

Air Canada
John Anderson
Wayne Archer
Marie-France Bertrand
Patrick Bertrand
Professor Buchanan
Canada Post
Carleton University
Data & Spooky
Marc de la Roche
Mike Dewar
Laurie Duval
Nicole Facette
Tom Facette
Kelly Fines
Noelani Laycock
Daryl Lindberg
Mike Manders
Pink Floyd
The Rouge River Group
John Showman
Al Stewart
John Strain
Molly Toman
Barbra Vajdik

Authors' photograph by Monica Royal

CONTENTS

CONSTANTS FOR SCIENCE

1

Constants for Science

Quantity	Symbol	Value	Common rounded value
Absolute zero	0 K	-273.15°C	-273°C
Average acceleration due to gravity on Earth	g	9.80665 m/s^2	9.81 m/s^2
Atomic mass unit	u	1.6605 x 10^{-27} kg 931.43 MeV/c^2	1.66 x 10^{-27} kg 931 MeV/c^2
Avogadro's number	N_A	6.0220 x 10^{23} particles/mol	6.02 x 10^{23} particles/mol
Biot-Savart constant	k_m	10^{-7} T m/A	10^{-7} T m/A
Bohr magneton	μ_B	9.2740 x 10^{-24} J/T	9.27 x 10^{-24} J/T
Bohr radius	a_0	5.2918 x 10^{-11} m	5.29 x 10^{-11} m
Boltzmann's constant	k_B	1.3807 x 10^{-23} J/K	1.38 x 10^{-23} J/K
Compton wavelength	λ_C	2.4263 x 10^{-12} m	2.43 x 10^{-12} m
Coulomb constant	k	8.9875 x 10^{9} N m^2/C^2	9 x 10^{9} N m^2/C^2
Deuteron mass	m_d	3.3436 x 10^{-27} kg 2.0136 u	3.34 x 10^{-27} kg 2.01 u
Electron charge	e	-1.6022 x 10^{-19} C	-1.6 x 10^{-19} C
Electron mass	m_e	9.1094 x 10^{-31} kg 5.4858 x 10^{-4} u 0.51100 MeV/c^2	9.11 x 10^{-31} kg 5.486 x 10^{-4} u 0.511 MeV/c^2
Electron volt	eV	1.6022 x 10^{-19} J	1.6 x 10^{-19} J
Elementary charge	e	1.6022 x 10^{-19} C	1.6 x 10^{-19} C
Faraday's constant	F	96 485 C/mol	96 500 C/mol
Gas constant	R	8.3145 Pa m^3/K mol 8.3145 J/K mol 8.2057 x 10^{-2} L atm/K mol 8.2057 x 10^{-5} m^3 atm/K mol	8.31 Pa m^3/K mol 8.31 J/K mol 0.0821 L atm/K mol 8.21 x 10^{-5} m^3 atm/K mol
Gravitational constant	G	6.672 x 10^{-11} N m^2/kg^2	6.67 x 10^{-11} N m^2/kg^2

(For additional physical data, please refer to Physics, Section 7.10.)

Constants for Science

Quantity	Symbol	Value	Common rounded value
Hydrogen ground state	E_0	13.606 eV	13.6 eV
Ideal gas molar volume at SATP 298.15 K and 100 kPa	V_m, $\overline{V}$	24.790 L/mol 2.4790×10^{-2} m^3/mol	24.8 L/mol 0.0248 m^3/mol
at STP 273.15 K and 101.3 kPa	V_m, $\overline{V}$	22.414 L/mol 2.2414×10^{-2} m^3/mol	22.4 L/mol 0.0224 m^3/mol
Josephson frequency-voltage ratio	2e/h	4.8360×10^{14} Hz/V	4.84×10^{14} Hz/V
Magnetic flux quantum	ϕ_0	2.0678×10^{-15} Wb	2.07×10^{-15} Wb
Neutron charge		0	0
Neutron mass	m_n	1.6749×10^{-27} kg 1.0087 u 939.57 MeV/c^2	1.68×10^{-27} kg 1.009 u 939.6 MeV/c^2
Nuclear magneton	μ_N	5.0508×10^{-27} J/T	5.05×10^{-27} J/T
Permeability of free space	μ_0	$4\pi \times 10^{-27}$ N/A^2	1.26×10^{-26} N/A^2
Permittivity of free space	$\in_0$	8.8542×10^{-12} C^2/N m^2	8.85×10^{-12} C^2/N m^2
Planck's constant	h	6.6261×10^{-34} J s	6.63×10^{-34} J s
Proton charge	e	1.6022×10^{-19} C	1.6×10^{-19} C
Proton mass	m_p	1.6726×10^{-27} kg 1.0073 u 938.27 MeV/c^2	1.67×10^{-27} kg 1.007 u 938.3 MeV/c^2
Quantized hall resistance	h/e^2	25812.8 Ω	25800Ω
Rydberg's constant	R_H	1.0974×10^7 m^{-1} 2.1799×10^{-18} J	$1.10 \times 10^7 m^{-1}$ 2.18×10^{-18} J
Speed of light in a vacuum	c	2.9979×10^8 m/s	3.00×10^8 m/s
Stefan's constant	σ	5.6696×10^{-8} W/ m^2 K^4	5.67×10^{-8} W/ m^2 K^4

UNITS & CONVERSIONS

2

2.0 Units & Conversions

Life, as we know it, would be impossible without systems of measurement.

"I have three."

"Three what?"

"Just three."

Numbers are meaningless without units; units are meaningless without numbers. This chapter introduces you to the world of measurement. You'll see how the numbers and symbols we use today are related to the classical Greek alphabet and Roman numbering systems. We also look at the most widely used scientific measuring system in the world, the Système international (SI). You'll see how to save time writing numbers by using numerical prefixes, and use conversion tables to speed up calculations between different measurement systems. Rules for determining significant figures are also covered. Because the methods of scientific proof are based upon standard measuring systems, it's important to understand the principles in this chapter. The information is useful in both the academic and "real world."

Suppose your car tire is getting flat. You know that it should be inflated to 35 psi. At the service station however, the pump measures in kPa. Now the pressure is on you! If you have a calculator and *The Addison-Wesley Science Handbook* in the glove compartment, you've solved the problem.

2.1 The Greek Alphabet

Greek Alphabet		
Α	α	alpha
Β	β	beta
Γ	γ	gamma
Δ	δ	delta
Ε	ε	epsilon
Ζ	ζ	zeta
Η	η	eta
Θ	θ	theta
Ι	ι	iota
Κ	κ	kappa
Λ	λ	lambda
Μ	μ	mu
Ν	ν	nu
Ξ	ξ	xi
Ο	ο	omicron
Π	π	pi
Ρ	ρ	rho
Σ	σ	sigma
Τ	τ	tau
Υ	υ	upsilon
Φ	φ	phi
Χ	χ	chi
Ψ	ψ	psi
Ω	ω	omega

The various branches of science use so many variables and constants that the English alphabet does not have enough letters to cover them. The English alphabet is essentially the same as the Roman alphabet, which has been in use for over 700 000 days! Because of the English alphabet's limitations, scientists also use the Greek alphabet to signify variables and constants.

In mathematical equations, the Greek capital sigma, Σ, is used almost universally to indicate "the sum of." The small sigma, σ, is used to indicate more than 15 different variables and constants, depending on the context, such as area per molecule, conductivity, NMR shielding constant, population standard deviation, and wave number in a medium.

2.2 Numerical Prefixes

Scientists study an incredibly wide range of phenomena, from the infinitesimally small (the average mass of a hydrogen atom is about 0.000000000000000000000000001674 kg) to the immensely huge (the mass of the Earth is about 5 980 000 000 000 000 000 000 000 kg). Obviously, scientists (and you) have better things to do than waste time writing out all those zeros. There are two ways to handle large and small numbers: scientific notation and numerical prefixes.

Scientific notation. To change a large number to scientific notation, move the decimal point to the left and stop at the first digit. Count the number of spaces you moved. Write this as a positive power of ten. Represent the number in the following form:

Mass of the Earth = 5.98×10^{24} kg

To change a small number to scientific notation, move the decimal point to the right, but stop to the right of the first non-zero digit. Count the number of spaces you moved. Write this as a negative power of ten. Represent the number in the following form, but don't forget the minus sign:

Average mass of a hydrogen atom = 1.674×10^{-27} kg

Numerical prefixes. These save even more time than scientific notation. Prefix symbols can be used to replace certain powers of ten. Instead of writing 1 000 000 000 m (one billion meters), or 1×10^{9} m, it's much easier to write 1 Gm (one gigameter). Similarly, 0.000000001 m (one billionth of a meter), or 1×10^{-9} m, can be written as 1 nm (one nanometer).

Powers of Ten

Power	10^{1}	10^{2}	10^{3}	10^{6}	10^{9}	10^{12}	10^{15}	10^{18}
Prefix	deca	hecto	kilo	mega	giga	tera	peta	exa
Symbol	da	h	k	M	G	T	P	E

Subpowers of Ten

Power	10^{-1}	10^{-2}	10^{-3}	10^{-6}	10^{-9}	10^{-12}	10^{-15}	10^{-18}
Prefix	deci	centi	milli	micro	nano	pico	femto	atto
Symbol	d	c	m	μ	n	p	f	a

The prefixes in the following table are used less formally to modify words. These prefixes represent common multiples and fractions. For example, a person who speaks two languages is *bilingual*.

Common Multiples

Multiple	Prefix	Multiple	Prefix	Multiple	Prefix
½	hemi-	12	dodeca-	26	hexacosa-
1	mono-	13	trideca-	27	heptacosa-
1½	sesqui-	14	tetradeca-	28	octacosa-
2	bi-, di-	15	pentadeca-	29	nonacosa-
2½	hemipenta-	16	hexadeca-	30	triaconta-
3	tri-	17	heptadeca-	40	tetraconta-
4	tetra	18	octadeca-	50	pentaconta-
5	penta-	19	nonadeca-	60	hexaconta-
6	hexa-	20	eicosa-*	70	heptaconta-
7	hepta-	21	heneicosa-*	80	octaconta-
8	octa-	22	docosa-	90	nonaconta-
9	nona-	23	tricosa-	100	hecta-
10	deca-	24	tetracosa-	110	decahecta-
11	hendeca-	25	pentacosa-	200	dicta-

*Variant spellings: icosa, henicosa.

No people do so much harm as those who go about doing good. Mandell Creighton

2.3 Roman Numerals

Roman numerals are sometimes used in place of numbers. As numbers increase in size, Roman numerals get very cumbersome. Using Roman numerals, simple arithmetic can become a nightmare.

Roman Numerals					
Number	**Capital**	**Small**	**Number**	**Capital**	**Small**
1	I	i	30	XXX	xxx
2	II	ii	40	XL	xl
3	III	iii	50	L	l
4	IV	iv	60	LX	lx
5	V	v	70	LXX	lxx
6	VI	vi	80	LXXX	lxxx
7	VII	vii	90	XC	xc
8	VIII	viii	100	C	c
9	IX	ix	110	CX	cx
10	X	x	150	CL	cl
11	XI	xi	200	CC	cc
12	XII	xii	300	CCC	ccc
13	XIII	xiii	400	CD	cd
14	XIV	xiv	500	D	d
15	XV	xv	600	DC	dc
16	XVI	xvi	700	DCC	dcc
17	XVII	xvii	800	DCCC	dccc
18	XVIII	xviii	900	CM	cm
19	XIX	xix	1000	M	m
20	XX	xx	1500	MD	md
21	XXI	xxi	1900	MCM	mcm
22	XXII	xxii	2000	MM	mm

You raise your voice when you should reinforce your argument. Dr. Samuel Johnson

2.4 SI Units

Système international (SI). This is the standard system of measurement used by most scientists throughout the world. It's made up of seven base units and two supplementary units, from which all other scientific units can be derived.

Système International (SI) Base Units			
Quantity	**Symbol**	**Unit name**	**Unit abbreviation**
Amount of substance	n	mole	mol
Electric current	I	ampere	A
Length	L, l, h, d, δ	meter	m
Luminous intensity	I_V	candela	cd
Mass	m	kilogram	kg
Temperature	T	kelvin	K
Time	t	second	s

SI supplementary units. The radian and the steradian are used to measure angles. Their units cancel each other out, therefore they have a dimension of one. They may be used in derived units where they clarify the meaning of the unit.

Système International (SI) Supplementary Units			
Quantity	**Symbol**	**SI unit basis**	**Dimension**
Radian, plane angle	rad	m/m	1 (dimensionless)
Steradian, solid angle	sr	m^2/m^2	1 (dimensionless)

To believe only possibilities is not faith, but mere Philosophy. Sir Thomas Browne

Dimensional analysis. Why is this important to you? Even if you forget a formula on a test there's still hope. If you know what units your answer should be in, you can quickly use dimensional analysis to check the correctness of your answer.

For example: Find the acceleration due to gravity on the Moon's surface. Given: the mass of the Moon is $M = 7.35 \times 10^{22}$ kg, its radius is $r = 1.74 \times 10^{6}$ m, and the Universal Gravitational Constant is $G = 6.67 \times 10^{-11}$ N m^2/kg^2.

1) Start at the value which has the most complex units. In this case it's $[G]$* = (N m^2/kg^2). Express newtons (N) in terms of SI base units (kg m/s^2), then combine them with the remaining units from $[G]$. Units cancel when they are divided by themselves. After cancellation, $[G]$ becomes (m^3/kg s^2).
2) Using trial and error, determine what combination of arithmetic yields the units your answer requires. In this example, finding acceleration (m/s^2) is the goal. It might take several attempts, but you should end up with $[G] \times [M]/[r]^2$ = (m^3/kg s^2) × (kg) × (1/m^2) = m/s^2.
3) The above form gives you an answer with the correct units, so it's probably the equation you need. Dimensional analysis won't reveal the answer every time, but with practice it becomes a powerful tool. In any case, partial marks are better than no marks.

* Use brackets [ ] when you are working with units, but not numbers.

Système International (SI) Derived Units

Quantity	Symbol	Unit	SI unit basis
Acceleration	a, g	m/s^2	m/s^2
Angular acceleration	α	rad/s^2	rad/s^2
Angular frequency	ω	rad/s	rad/s
Angular momentum	L	J s	kg m^2/s
Angular velocity	ω	rad/s	s^{-1} or ^{1}/s
Area	A	m^2	m^2
Density	ρ	kg/m^3	kg/m^3
Displacement	s	meter, m	m
Dose equivalent index	H	sievert, Sv, J/kg	m^2/s^2

Système International (SI) Derived Units

Quantity	Symbol	Unit	SI unit basis
Electric capacitance	C	Farad, F	$A^2\ s^4/kg\ m^2$
Electric charge (E.C.)	Q, q, e	Coulomb, C	A s
E.C. density, line	λ	C/m	A s/m
E.C. density, surface	σ	C/m^2	$A\ s/m^2$
E.C. density, volume	ρ	C/m^3	$A\ s/m^3$
Electric conductivity	σ	$1/\Omega$ m	$A^2\ s^3/kg\ m^3$
Electric current density	J, j	A/m^2	A/m^2
Electric dipole moment	p, μ	C m	A s m
Electric field	E	V/m	$kg\ m/A\ s^3$
Electric flux	ϕ, ψ	V m	$kg\ m^3/A\ s^3$
Electric potential	V, ϕ	Volt, V	$kg\ m^2/A\ s^3$
Electrical resistance	R	ohm, Ω	$kg\ m^2/A^2\ s^3$
Electromotive force	$\mathscr{E}_{mf}, E$	Volt, V	$kg\ m^2/A\ s^3$
Energy	E, U, K	Joule, J	$kg\ m^2/s^2$
Enthalpy	H	Joule, J	$kg\ m^2/s^2$
Entropy	S	J/K	$kg\ m^2/s^2\ K$
Fluidity	ϕ	m s/kg	m s/kg
Force	F	Newton, N	$kg\ m/s^2$
Frequency	ν, f	Hertz, Hz	s^{-1}
Friction coefficients	μ_k, μ_s	dimensionless	dimensionless
Heat	Q, q	Joule, J	$kg\ m^2/s^2$
Heat capacity at constant pressure	C_p	J/K	$kg\ m^2/s^2\ K$
Heat capacity at constant volume	C_v	J/K	$kg\ m^2/s^2\ K$

Système International (SI) Derived Units

Quantity	Symbol	Unit	SI unit basis
Illuminance	F	lux, lx	cd sr/m^2
Inductance	L	Henry,H	kg m^2/A^2 s^2
Kinetic energy	E_k, K, T	Joule, J	kg m^2/s^2
Luminous flux	E	lumen, lm	cd sr
Magnetic dipole moment	μ, m	N m/T	A m^2
Magnetic field or magnetic flux density	B	Tesla, T	kg/A s^2
Magnetic flux	ϕ_m	Weber, Wb	kg m^2/A s^2
Mass number	A		
Molality	m	mol/kg	mol/kg
Molar mass	M	g/mol	g/mol
Molar volume	V_m	m^3/mol	m^3/mol
Molar specific heat	C	J/mol K	kg m^2 /s^2 K mol
Molarity	M	mol/L	mol/dm^3
Moment of inertia	I, J	kg m^2	kg m^2
Momentum	p	kg m/s	kg m/s
Osmotic pressure	Π, π	Pascal, Pa	kg/m s^2
Period	T	s	s
Permeability of space	μ_0	H/m	kg m/A^2 s^2
Permittivity of space	ε_0	F/m	A^2 s^4/kg m^3
Potential energy	E_p,U,V,ϕ	Joule, J	kg m^2/s^2
Power	P	Watt, W	kg m^2/s^3
Pressure	P, p	Pascal, Pa	kg/m s^2
Radiation absorbed dose	D	Gray, Gy, J/kg	m^2/s^2

In this world nothing is certain but death and taxes. Benjamin Franklin

Système International (SI) Derived Units

Quantity	Symbol	Unit	SI unit basis
Radioactivity	A	Becquerel, Bq	s^{-1} or 1/s
Solubility	s	g/L	kg/m^3
Specific heat	c	J/kg K	m^2/s^2 K
Speed	s	m/s	m/s
Torque	T, τ	N m	kg m^2/s^2
Velocity	v	m/s	m/s
Viscosity, dynamic	η, μ	Pa s	kg/m s
Viscosity, kinematic	υ	m^2/s	m^2/s
Volume	V	m^3	m^3
Wavelength	λ	m	m
Weight	W, P, G	Newton ,N	kg m/s^2
Work	W	Joule, J	kg m^2/s^2

Common Units Used with Système International (SI)

Quantity	Unit name	Symbol	Relation to SI
Area, nuclear cross-sectional	barn	b	10^{-28} m^2
Energy	electron volt	eV	1.6022 x 10^{-19} J
Length	angstrom	Å	10^{-10} m
Mass	metric tonne	t	1000 kg
Mass, atomic	unified atomic mass unit, dalton	u, Da	1.6605 x 10^{-27} kg
Plane angle	degree	°	(π/180) rad
Plane angle	minute	′	(π/10800) rad
Plane angle	second	″	(π/648000) rad
Pressure	bar	bar	10^5 Pa
Time	day	d	86400 s
Time	hour	h	3600 s
Time	minute	min	60 s
Volume	liter	ℓ, L	10^{-3} m^3

The only certainty is that nothing is certain. Roman scholar

2.5 Unit Conversion Factors

Professors love to introduce students to new types of units. They often do this by having the student perform endless conversions. The units yielded are often found in obscure and archaic measurement systems. By all means, practice a few conversions by hand until you understand them. The following shows a single conversion factor between units.

Find the unit you're converting from in the left hand column; find the unit you're converting to in the top row. The conversion factor is found where the row and column intersect. Multiply the quantity of your unit by the conversion factor. *Presto!* You've just converted your unit.

1. m = 100. cm = 3.2808399 feet = 39.3700787 inches = 0.001 km = ...

The above conversion starts off with exactly 1 meter. This is rarely the case. When using conversion factors, be aware of how many significant digits your answer should have. (Section 2.6 covers significant figures.) Don't assume that all the numbers a calculator gives you are significant. The table below illustrates the proper use of conversion factors.

Significant Figures in Unit Conversions

Starting unit	Number of significant figures	Multiply by	Converted unit
1 m	1	39.3700787 in/m	40 in
1.0 m	2	39.3700787 in/m	39 in
1.00 m	3	39.3700787 in/m	39.4 in
1.000 m	4	39.3700787 in/m	39.37 in
1.0000 m	5	39.3700787 in/m	39.370 in
1.00000 m	6	39.3700787 in/m	39.3701 in
1.000000 m	7	39.3700787 in/m	39.37008 in
1.0000000 m	8	39.3700787 in/m	39.370079 in
1.00000000 m	9	39.3700787 in/m	39.3700787 in
1.000000000 m	10	39.3700787 in/m	39.3700787 in

In the following conversion factors, there is a degree of uncertainty. They are exact only when printed in **bold** characters.

Otherwise, they are significant according to the rules governing significant figures.

Speak the truth, but leave immediately after. Slovenian Proverb

Your professor is not being perverse by asking you to toil away at mindless calculations. In performing conversions you learn to appreciate units of measurement. Using the following tables will take the pain out of performing conversions.

How to use the tables

Let's assume you have to convert 50. degrees to radians.

1) First, find the relevant table. In this case, use the table with the heading, **Angle**.
2) Start at the left column (From) until you find *degrees*, then read to the right until you reach the *radians* column. (Conversion factor 0.0174533, or π/180.)
3) To get your conversion, multiply 50 by either of these numbers. You should end up with 0.872665 radians.

Acceleration

↓ From/To→	cm/s^2	ft/s^2	km/h s	km/s^2	m/s^2	mi/h s
cm/s^2	**1**	0.0328084	**0.036**	$\mathbf{10^{-5}}$	**0.01**	0.0223694
ft/s^2	**30.48**	**1**	1.09728	$\mathbf{3.048 \times 10^{-4}}$	**0.3048**	0.681818
km/h s	27.7778	0.911344	**1**	2.77778×10^{-4}	0.277778	0.621371
km/s^2	**100 000**	3 280.84	**3 600**	**1**	**1000**	2 236.94
m/s^2	**100**	3.28084	3.6	**0.001**	**1**	2.23694
mi/h s	44.7040	1.46667	1.60934	4.47040×10^{-4}	0.447040	**1**

Angle

↓ From/To→	Circum-ference	Degrees	Grade	Minutes	Quadrant	Radians	Revolutions	Seconds
circum-ference	**1**	**360**	**400**	**21 600**	**4**	6.28319 (2π)	**1**	**1 296 000**
degrees	2.77778×10^{-3}	**1**	1.11111	**60**	0.0111111	0.0174533 **(π/180)**	2.77778×10^{-3}	**3 600**
grade	**0.0025**	**0.9**	**1**	**54**	**0.01**	0.0157080 **(π/200)**	**0.0025**	**3 240**
minutes	4.6296×10^{-5}	0.01667 **(1/60)**	0.0185185 **(1/54)**	**1**	1.82519×10^{-4} **(1/5400)**	2.9089×10^{-4} **(π/10800)**	4.6296×10^{-5}	**60**
quadrant	**0.25**	**90**	**100**	**5 400**	**1**	1.570796 **(π/2)**	**0.25**	**324 000**
radians	0.159155	57.29578	63.66198	3 437.747	0.6366198	**1**	0.159155	206 264.8
revolutions	**1**	**360**	**400**	**21 600**	**4**	6.28319 (2π)	**1**	**1 296 000**
seconds	7.71605×10^{-7}	2.77778×10^{-4}	3.08642×10^{-4}	0.016667 **(1/60)**	3.08642×10^{-6}	4.848137×10^{-6}	7.71605×10^{-7}	**1**

Tell your boss what you think of him, and the truth will set you free. Anonymous

Area

↓ From/To→	Acre	Are	Circular mil	Hectare	Square centimeter	Square foot
acre	1	40.4686	7.98657×10^{12}	0.404686	4.04686×10^{7}	43 560
are	0.0247105	1	1.97353×10^{11}	0.01	10^{6}	1 076.39
circular mil	1.25210×10^{-13}	5.06707×10^{-12}	1	5.06707×10^{-14}	5.06707×10^{-6}	5.45415×10^{-9}
hectare	2.47105	100	1.97353×10^{13}	1	10^{8}	107 639
square centimeter	2.47105×10^{-8}	10^{-6}	1.97353×10^{5}	10^{-8}	1	1.07639×10^{-3}
square foot	2.29568×10^{-5}	9.29030×10^{-4}	1.83346×10^{8}	9.29030×10^{-6}	929.030	1
square inch	1.59423×10^{-7}	6.45160×10^{-6}	1.27324×10^{6}	6.4516×10^{-8}	6.4516	6.94444×10^{-3}
square kilometer	247.105	10000	1.97353×10^{15}	100	10^{10}	1.07639×10^{7}
square meter	2.47105×10^{-4}	0.01	1.97353×10^{9}	10^{-4}	10 000	10.7639
square mile	640	25 899.9	5.11142×10^{15}	258.999	2.58999×10^{10}	2.78784×10^{7}
square millimeter	2.47105×10^{-10}	10^{-8}	1 973.53	10^{-10}	0.01	1.07639×10^{-5}
square mil	1.59423×10^{-13}	6.44925×10^{-12}	1.27324	6.4516×10^{-14}	6.4516×10^{-6}	6.94444×10^{-9}
square yard	2.06612×10^{-4}	8.36127×10^{-3}	1.65012×10^{9}	8.36127×10^{-5}	8361.27	9

Density

↓ From/To→	g/cm^3	g/L	g/mL	kg/dm^3	kg/m^3
g/cm^3	1	1 000	1	1	1 000
g/L	0.001	1	0.001	0.001	1
g/mL	1	1 000	1	1	1 000
kg/dm^3	1	1 000	1	1	1 000
kg/m^3	0.001	1	0.001	0.001	1
lb/ft^3	0.001601846	16.01846	0.01601846	0.01601846	16.01846
lb/gallon (Brit)	0.09977637	99.77637	0.09977637	0.09977637	99.77637
lb/gallon (US)	0.1198264	119.8264	0.1198264	0.1198264	119.8264
lb/in^3	27.679905	27 679.9	27.679905	27.679905	27 679.9
lb/yd^3	5.932764×10^{-4}	0.5932764	5.932764×10^{-4}	5.932764×10^{-4}	0.5932764
$slug/ft^3$	0.515379	515.379	0.515379	0.515379	515.379

Notes: Density

$g/mL = g/cm^3 = g/cc$ $kg/L = kg/dm^3$

The passion for the infinite is precisely subjectivity, and thus subjectivity becomes truth. Søren Kierkegaard

Area

Square inch	Square kilometer	Square meter	Square mile	Square millimeter	Square mil	Square yard
6.27264×10^{6}	4.04686×10^{-3}	4 046.86	**1.5625×10^{-3}**	4.04686×10^{9}	6.27264×10^{12}	**4 840**
1.55000×10^{5}	**10^{-4}**	**100**	3.86102×10^{-5}	**10^{8}**	1.55000×10^{11}	119.599
7.85398×10^{-7}	5.06708×10^{-16}	5.06708×10^{-10}	1.95534×10^{-16}	5.06707×10^{-4}	0.785398	6.06171×10^{-10}
1.55000×10^{7}	**0.01**	**10 000**	3.86102×10^{-3}	**10^{10}**	1.55000×10^{13}	11 959.9
0.155000	**10^{-10}**	**10^{-4}**	3.86102×10^{-11}	**100**	1.55000×10^{5}	1.19599×10^{-4}
144	9.29030×10^{-8}	0.0929030	3.58701×10^{-8}	**92 903.04**	**1.44×10^{8}**	0.111111
1	**6.45160×10^{-10}**	**6.45160×10^{-4}**	2.49098×10^{-10}	**645.16**	**10^{6}**	7.716×10^{-4}
1.55000×10^{9}	**1**	**10^{6}**	0.386102	**10^{12}**	1.55000×10^{15}	1.19599×10^{6}
1 550.00	**10^{-6}**	**1**	3.86102×10^{-7}	**10^{6}**	1.55000×10^{9}	1.19599
4.01449×10^{9}	2.58999	2.58999×10^{6}	**1**	2.58999×10^{12}	4.01449×10^{15}	**3.0976×10^{6}**
1.55000×10^{-3}	**10^{-12}**	**10^{-6}**	3.86102×10^{-13}	**1**	1 550.00	1.19599×10^{-6}
10^{-6}	**6.4516×10^{-16}**	**6.4516×10^{-10}**	2.49098×10^{-16}	**6.4516×10^{-4}**	**1**	7.71605×10^{-10}
1 296	8.36127×10^{-7}	0.836127	3.22831×10^{-7}	8.36127×10^{5}	**1.296×10^{9}**	**1**

Density

lb/ft^3	lb/gallon (Brit)	lb/gallon (US)	lb/in^3	lb/yd^3	$slug/ft^3$
62.42796	10.02241	8.34540	0.03612729	1 685.55	1.94032
0.06242796	0.01002241	8.34540×10^{-3}	3.612729×10^{-5}	1.68555	1.94032×10^{-3}
62.42796	10.02241	8.34540	0.03612729	1 685.55	1.94032
62.42796	10.02241	8.34540	0.03612729	1 685.55	1.94032
0.06242796	0.01002241	8.34540×10^{-3}	3.612729×10^{-5}	1.68555	1.94032×10^{-3}
1	0.160544	0.133681	5.78704×10^{-4}	**27**	0.0310809
6.228835	**1**	0.832674	3.60465×10^{-3}	168.179	0.193598
7.480519	1.20095	**1**	4.32900×10^{-3}	201.973	0.232502
1 728	277.419	231.000	**1**	46 656	53.7079
0.0370370	5.94606×10^{-3}	4.95113×10^{-3}	2.14335	**1**	1.15115×10^{-3}
32.1741	5.16534	4.30105	0.0186192	868.700	**1**

Everyone is entitled to my opinion. Madonna

Energy & Work

↓ From/To→	BTU	Calorie, thermal (cal)	Cubic foot atmosphere	Erg	Foot-poundal	Foot-pound force
BTU	1	252.164	0.367718	1.05506×10^{10}	25 036.9	778.169
calorie, thermal (cal)	3.96576×10^{-3}	1	1.45825×10^{-3}	$\mathbf{4.184 \times 10^{7}}$	99.2880	3.08596
cubic foot atmosphere	2.71948	685.756	1	2.86921×10^{10}	68 087.4	2 116.22
erg	9.47817×10^{-11}	2.39006×10^{-8}	3.48529×10^{-11}	1	2.37304×10^{-6}	7.37562×10^{-8}
foot-poundal	3.99411×10^{-5}	0.0100717	1.46870×10^{-5}	421 401	1	0.0310809
foot-pound force	1.28507×10^{-3}	0.324049	4.72543×10^{-4}	1.35582×10^{7}	32.1742	1
horsepower hour	2 544.43	641 616	935.633	2.68452×10^{13}	6.37047×10^{7}	$\mathbf{1.98 \times 10^{6}}$
horsepower hour metric	2 509.63	632 839	922.835	2.64780×10^{13}	6.28334×10^{7}	1.95292×10^{6}
joule	9.47817×10^{-4}	0.239006	3.48529×10^{-4}	$\mathbf{10^{7}}$	23.7304	0.737562
kilocalorie (Cal or Kcal)	3.96567	**1 000**	1.45825	$\mathbf{4.184 \times 10^{10}}$	99 288.0	3 085.96
kilogram-force-meter	9.29491×10^{-3}	2.34385	3.41790×10^{-3}	$\mathbf{9.80665 \times 10^{7}}$	232.716	7.23301
kilowatt-hour	3 412.14	860 421	1 254.70	$\mathbf{3.6 \times 10^{13}}$	8.54294×10^{7}	2.65522×10^{6}
liter atmosphere	0.0960376	24.2173	0.0353147	$\mathbf{1.01325 \times 10^{9}}$	2 404.48	74.7335
watt-hour	3.41214	860.421	1.25470	$\mathbf{3.6 \times 10^{10}}$	85 429.4	2 655.22

Notes: Energy and Work
Atomic mass unit (u) = 1.4924×10^{-10} J
Electron volt (eV) = 1.6022×10^{-19} J
Wave number (cm^{-1}) = 1.9864×10^{-23} J

There are a few definitions for the calorie floating around.
The thermal calorie (used above): 1 cal = **4.184** J (common in chemistry)
1 Calorie = 1 kcal = 1000 calories
Erg = $g\ cm^2/s^2$
Joule = A V s = Pa m^3 = N m = kg m^2/s^2

Force
Dyne = $g\ cm/s^2$
Newton = $kg\ m/s^2$

People who think they know everything are very irritating to those of us who do. Anonymous

Energy & Work

Horsepower hour	Horsepower hour (metric)	Joule	Kilocalorie (Cal or kcal)	Kilogram force meter	Kilowatt hour	Liter atmosphere	Watt hour
3.93015×10^{-4}	3.98466×10^{-4}	1 055.06	0.252164	107.586	2.93071×10^{-4}	10.4126	0.293071
1.55857×10^{-6}	1.58018×10^{-6}	**4.184**	**0.001**	0.426651	1.16222×10^{-6}	0.0412929	1.16222×10^{-3}
1.06880×10^{-3}	1.08362×10^{-3}	2 869.21	0.685754	292.579	7.97001×10^{-4}	28.3168	0.797001
3.72506×10^{-14}	3.77673×10^{-14}	**10^{-7}**	2.39005×10^{-11}	1.01972×10^{-8}	2.77778×10^{-14}	9.86923×10^{-10}	2.77778×10^{-11}
1.56974×10^{-8}	1.59152×10^{-8}	0.0421401	1.00717×10^{-5}	4.29711×10^{-3}	1.17056×10^{-8}	4.15891×10^{-4}	1.17056×10^{-5}
505051×10^{-7}	5.12057×10^{-7}	1.35582	3.24048×10^{-4}	0.138256	3.76617×10^{-7}	0.0133809	3.76617×10^{-4}
1	1.01387	2.68452×10^{6}	641.614	273 746	0.745700	26 494.1	7.45700×10^{6}
0.986321	**1**	2.64780×10^{6}	632.837	**270 000**	0.735500	26 131.7	7.35500×10^{6}
3.72506×10^{-7}	3.77673×10^{-7}	**1**	2.39005×10^{-4}	0.101972	2.77778×10^{-7}	9.86923×10^{-3}	2.77778×10^{-4}
1.55857×10^{-3}	1.58018×10^{-3}	**4 184**	**1**	426.651	1.16222×10^{-3}	41.2929	1.16222
3.65304×10^{-6}	3.70371×10^{-6}	**9.80665**	2.34384×10^{-3}	**1**	2.72407×10^{-6}	0.0967841	2.72407×10^{-3}
1.34102	1.35962	**3.6×10^{6}**	860.418	367 099	**1**	35 529.2	**1000**
3.77442×10^{-5}	3.82677×10^{-5}	**101.325**	0.0242172	10.3323	2.81458×10^{-5}	**1**	0.0281458
1.34102×10^{-3}	1.35962×10^{-3}	**3 600**	0.860418	367.099	**0.001**	35.5292	**1**

Force

↓ From/To→	Dyne	Gram force	Joules/cm	Kilogram force	Newton	Poundal	Pound force
dyne	**1**	1.01972×10^{-3}	**10^{-7}**	1.01972×10^{-6}	**10^{-5}**	7.23301×10^{-5}	2.24809×10^{-6}
gram force	**980.665**	**1**	**9.80665×10^{-5}**	**0.001**	**9.80665×10^{-3}**	0.0709316	2.20462×10^{-3}
joules/cm	**10^{7}**	10 197.2	**1**	10.1972	**100**	723.301	22.4809
kilogram force	**980 665**	**1 000**	**0.0980665**	**1**	**9.80665**	70.9316	2.20462
newton	**100 000**	101.972	**0.01**	0.101972	**1**	7.23301	0.224809
poundal	13 825.5	14.0981	1.38255×10^{-3}	0.0140981	0.138255	**1**	0.0310810
pound force	44 482.2	453.592	0.0444822	0.453592	4.44822	32.1740	**1**

HA! HA! HA! But you know that there is no choice in reality, so, say what you like. Feodor Mikhailovich Dostoevsky

Length

↓ From/To→	Angstrom	Astronomical unit	Bolt, cloth	Centimeter	Fathom	Foot	Inch	Kilometer
angstrom	1	6.68459×10^{-22}	2.73403×10^{-12}	10^{-8}	5.46807×10^{-11}	3.28084×10^{-10}	3.93701×10^{-9}	10^{-13}
astronomical unit	1.49598×10^{21}	1	4.09006×10^{9}	1.49598×10^{13}	8.18011×10^{10}	4.90807×10^{11}	5.88968×10^{13}	1.49598×10^{8}
bolt, cloth	3.6576×10^{11}	2.44495×10^{-10}	1	3 657.6	20	120	1 440	0.036576
centimeter	10^{8}	6.68459×10^{-14}	2.73403×10^{-4}	1	5.46807×10^{-3}	0.0328084	0.393701	10^{-5}
fathom	1.82880×10^{10}	1.22248×10^{-11}	0.05	182.88	1	6	72	1.8288×10^{-3}
foot	3.048×10^{9}	2.03746×10^{-12}	8.33333×10^{-3}	30.48	0.166667	1	12	3.048×10^{-4}
inch	2.54×10^{8}	1.69789×10^{-13}	6.94444×10^{-4}	2.54	0.0138889	0.0833333	1	2.54×10^{-5}
kilometer	10^{13}	6.68459×10^{-9}	27.3403	100 000	546.807	3 280.84	39 370.1	1
light-year	9.46053×10^{25}	63 239.7	2.58654×10^{14}	9.46053×10^{17}	5.17308×10^{15}	3.10385×10^{16}	3.72462×10^{17}	9.46053×10^{12}
meter	10^{10}	6.68459×10^{-12}	0.0273403	100	0.546807	3.28084	39.3701	0.001
micrometer	10 000	6.68459×10^{-18}	2.73403×10^{-8}	0.0001	5.46807×10^{-7}	3.28084×10^{-6}	3.93701×10^{-5}	10^{-9}
mil	254 000	1.69789×10^{-16}	6.94444×10^{-7}	2.54×10^{-3}	1.38889×10^{-5}	8.33333×10^{-5}	0.001	2.54×10^{-8}
mile (nautical)	1.852×10^{13}	1.23794×10^{-8}	50.6343	185 200	1 012.69	6 076.12	72 913.4	1.852
mile (statute)	1.60934×10^{13}	1.07578×10^{-8}	44	160 934	880	5 280	63 360	1.60934
millimeter	10^{7}	6.68459×10^{-15}	2.73403×10^{-5}	0.1	5.46807×10^{-4}	3.28084×10^{-3}	0.0393701	10^{-6}
nanometer	10	6.68459×10^{-21}	2.73403×10^{-11}	10^{-7}	5.46807×10^{-10}	3.28084×10^{-9}	3.93701×10^{-9}	10^{-12}
parsec	3.08568×10^{26}	206 265	8.43635×10^{14}	3.08568×10^{18}	1.68727×10^{16}	1.01236×10^{17}	1.21483×10^{18}	3.08568×10^{13}
yard	9.144×10^{9}	6.11239×10^{-12}	0.025	91.44	0.5	3	36	9.144×10^{-4}

Length

Light year	Meter	Micro-meter	Mil	Mile (nautical)	Mile (statute)	Milli-meter	Nano-meter	Parsec	Yard
1.05702 x 10^{-26}	**10^{-10}**	**10^{-4}**	3.93701 x 10^{-6}	5.39957 x 10^{-14}	6.21371 x 10^{-14}	**10^{-7}**	**0.1**	3.24078 x 10^{-27}	1.09361 x 10^{-10}
1.58128 x 10^{-5}	1.49598 x 10^{11}	1.49598 x 10^{17}	5.88968 x 10^{15}	8.07764 x 10^{7}	9.29558 x 10^{7}	1.49598 x 10^{14}	1.49598 x 10^{20}	4.84814 x 10^{-6}	1.63602 x 10^{11}
3.86616 x 10^{-15}	**36.576**	**3.6576 x 10^{7}**	**1.440 x 10^{6}**	0.0197495	0.0227273	**36 576**	3.6576 x 10^{10}	1.18535 x 10^{-15}	**40**
1.05702 x 10^{-18}	**0.01**	**10 000**	393.701	5.39957 x 10^{-6}	6.21371 x 10^{-6}	**10**	**10^{7}**	3.24078 x 10^{-19}	0.01093613
1.93308 x 10^{-16}	**1.8288**	**1.8288 x 10^{6}**	**72 000**	9.87473 x 10^{-4}	1.13636 x 10^{-3}	**1 828.8**	1.8288x 10^{9}	5.92674 x 10^{-17}	**2**
3.22180 x 10^{-17}	**0.3048**	**304 800**	**12 000**	1.64579 x 10^{-4}	1.89394 x 10^{-4}	**304.8**	**3.048 x 10^{8}**	9.87790 x 10^{-18}	0.333333
2.68483 x 10^{-18}	**0.0254**	**25 400**	**1 000**	1.37149 x 10^{-5}	1.57828 x 10^{-5}	**25.4**	**2.54 x 10^{7}**	8.23158 x 10^{-19}	0.0277778
1.05702 x 10^{-13}	**1 000**	**10^{9}**	3.93701 x 10^{7}	0.539957	0.621371	**10^{6}**	**10^{12}**	3.24078 x 10^{-14}	1 093.613
1	9.46053 x 10^{15}	9.46053 x 10^{21}	3.72462 x 10^{20}	5.10828 x 10^{12}	5.87850 x 10^{12}	9.46053 x 10^{18}	9.46053 x 10^{24}	0.306595	1.03462 x 10^{16}
1.05702 x 10^{-16}	**1**	**10^{6}**	39 370.1	5.39957 x 10^{-4}	6.21371 x 10^{-4}	**1 000**	**10^{9}**	3.24078 x 10^{-17}	1.09361
1.05702 x 10^{-22}	**10^{-6}**	**1**	0.0393701	5.39957 x 10^{-10}	6.21371 x 10^{-10}	**0.001**	**1 000**	3.24078 x 10^{-23}	1.09361 x 10^{-6}
2.68483 x 10^{-21}	**2.54 x 10^{-5}**	**25.4**	**1**	1.37149 x 10^{-8}	1.57828 x 10^{-8}	**0.0254**	**25 400**	8.23158 x 10^{-22}	2.77778 x 10^{-5}
1.95760 x 10^{-13}	**1 852**	**1.852 x 10^{9}**	7.29134 x 10^{7}	**1**	1.15078	**1.852 x 10^{6}**	**1.852 x 10^{12}**	6.00192 x 10^{-14}	2 025.37
1.70111 x 10^{-13}	1 609.34	1.60934 x 10^{9}	**6.336 x 10^{7}**	0.868976	**1**	1.60934 x 10^{6}	1.60934 x 10^{12}	5.21552 x 10^{-14}	**1 760**
1.05702 x 10^{-19}	**0.001**	**1 000**	39.3701	5.39957 x 10^{-7}	6.21371 x 10^{-7}	**1**	**10^{6}**	3.24078 x 10^{-20}	1.09361 x 10^{-3}
1.05702 x 10^{-25}	**10^{-9}**	**0.001**	3.93701 x 10^{-5}	5.39957 x 10^{-13}	6.21371 x 10^{-13}	**10^{-6}**	**1**	3.24078 x 10^{-26}	1.09361 x 10^{-9}
3.26164	3.08568 x 10^{16}	3.08568 x 10^{22}	1.21483 x 10^{21}	1.66613 x 10^{13}	1.91735 x 10^{13}	3.08568 x 10^{19}	3.08568 x 10^{25}	1	3.37454 x 10^{16}
9.66539 x 10^{-17}	**0.9144**	**914 400**	**36 000**	4.93736 x 10^{-4}	5.68182 x 10^{-4}	**914.4**	**9.144 x 10^{8}**	2.96337 x 10^{-17}	**1**

Magnetic Field

↓ From/To→	Gauss	Lines/cm^2	Lines/in^2	Tesla	Weber/cm^2	Weber/in^2	Weber/m^2
gauss	**1**	**1**	6.45161	**0.0001**	**10^{-8}**	6.45161 x 10^{-8}	**0.0001**
lines/cm^2	**1**	**1**	6.45161	**0.0001**	**10^{-8}**	6.45161 x 10^{-8}	**0.0001**
lines/in^2	**0.155**	**0.155**	**1**	**1.55 x 10^{-5}**	**1.55 x 10^{-9}**	**10^{-8}**	**1.55 x 10^{-5}**
tesla	**10 000**	**10 000**	64 516.1	**1**	**0.0001**	6.45161 x 10^{-4}	**1**
weber/cm^2	**10^{8}**	**10^{8}**	6.45161 x 10^{8}	**10 000**	**1**	6.45161	**10 000**
weber/in^2	**1.55 x 10^{7}**	**1.55 x 10^{7}**	**10^{8}**	**1 550**	**0.155**	**1**	**1 550**
weber/m^2	**10 000**	**10 000**	64 516.1	**1**	**0.0001**	6.45161 x 10^{-4}	**1**

Mass

↓ From/To→	Atomic mass unit	Carat (metric)	Dram (avdp.)	Grain	Gram	Hundred-wgt. (long)	Hundred-wgt. (short)	Kilogram
atomic mass unit	1	8.3028×10^{-24}	9.3719×10^{-25}	2.5626×10^{-23}	1.6605×10^{-24}	3.2686×10^{-29}	3.6609×10^{-29}	1.6605×10^{-27}
carat (metric)	1.2044×10^{23}	1	0.112877	3.08647	**0.2**	3.93683×10^{-6}	4.40925×10^{-6}	**0.0002**
dram(avdp.)	1.0670×10^{24}	8.85823	1	27.3437	1.77185	3.48772×10^{-5}	3.90625×10^{-5}	1.77185×10^{-3}
grain	3.9022×10^{22}	0.323995	0.0365714	1	0.0647989	1.27551×10^{-6}	1.42857×10^{-6}	6.47989×10^{-5}
gram	6.0220×10^{23}	**5**	0.564383	15.4324	1	1.96841×10^{-5}	2.20462×10^{-5}	**0.001**
hundred-weight (long)	3.0593×10^{28}	254 012	**28 672**	**784 000**	50 802.3	1	**1.12**	50.8023
hundred-weight (short)	2.7315×10^{28}	226 796	**25 600**	**700 000**	45 359.2	0.892857	1	45.3592
kilogram	6.0220×10^{26}	**5 000**	564.383	15 432.4	**1 000**	0.0196841	0.0220462	1
ounce (avdp.)	1.7072×10^{25}	141.748	**16**	**437.5**	28.3495	5.58036×10^{-4}	$\mathbf{6.25 \times 10^{-4}}$	0.0283495
ounce (troy)	1.8730×10^{25}	155.517	17.5543	**480**	31.1035	6.12232×10^{-4}	6.857×10^{-4}	0.0311035
penny weight	9.3653×10^{23}	7.77587	0.877714	**24**	1.55517	3.06122×10^{-5}	3.42857×10^{-5}	1.55517×10^{-3}
pound (avdp.)	2.7315×10^{26}	2 267.96	**256**	**7 000**	453.592	8.92857×10^{-3}	**0.01**	0.453592
scruple	7.8044×10^{23}	6.47989	0.731428	**20**	1.29598	2.55102×10^{-5}	2.85714×10^{-5}	1.29598×10^{-3}
slug	8.7885×10^{27}	72 969.5	8 236.55	225 218	14 593.9	0.28727	0.321740	14.5939
stone	3.8241×10^{27}	31 751.5	**3 584**	**98 000**	6 350.29	**0.125**	**0.14**	6.35029
ton (long)	6.1186×10^{29}	5.08023×10^{6}	**573 440**	$\mathbf{1.568 \times 10^{7}}$	1.01605×10^{6}	**20**	**22.4**	1 016.05
tonne (metric)	6.0220×10^{29}	$\mathbf{5 \times 10^{6}}$	564 383	1.54323×10^{7}	$\mathbf{10^{6}}$	19.6841	22.0462	**1 000**
ton (short)	5.4631×10^{29}	4.53592×10^{6}	**512 000**	$\mathbf{1.4 \times 10^{7}}$	907 185	17.8571	**20**	907.185

Notes: Mass
1 mg = 0.001 g = 1000 µg

If Noah had been truly wise, he would have swatted those two flies. H. Castle

Mass

Ounce (avdp.)	Ounce (troy)	Penny weight	Pound (avdp.)	Scruple	Slug	Stone	Ton (long)	Tonne (metric)	Ton (short)
5.8573×10^{-26}	5.3388×10^{-26}	1.0677×10^{-24}	3.6608×10^{-27}	1.2813×10^{-24}	1.1378×10^{-28}	2.6149×10^{-28}	1.6343×10^{-30}	1.6605×10^{-30}	1.8304×10^{-30}
7.0548×10^{-3}	6.43014×10^{-3}	0.12860	4.40925×10^{-4}	0.15432	1.3704×10^{-5}	3.14948×10^{-5}	1.9684×10^{-7}	**2×10^{-7}**	2.2046×10^{-7}
0.0625	0.056966	1.13932	3.90625×10^{-3}	1.36719	1.21410×10^{-7}	2.7902×10^{-4}	1.74386×10^{-6}	1.77184×10^{-6}	1.9531×10^{-6}
2.28571×10^{-3}	2.08333×10^{-3}	0.0416667	1.42857×10^{-4}	**0.05**	4.44014×10^{-6}	1.02041×10^{-5}	6.37755×10^{-8}	6.47988×10^{-8}	7.14285×10^{-8}
0.0352740	0.0321507	0.643015	2.20462×10^{-3}	0.771618	6.8521×10^{-5}	1.57474×10^{-4}	9.84206	**10^{-6}**	1.10231×10^{-6}
1 792.00	1 633.33	32 666.7	**112**	**39 200**	3.4811	**8**	**0.05**	0.050802	**0.056**
1 600	1 458.33	29 166.7	**100**	**35 000**	3.1081	7.14290	0.044643	0.045359	**0.05**
35.2940	32.1507	6.4301	2.20462	771.617	0.0685218	0.157474	9.84206×10^{-4}	**0.001**	1.10231×10^{-3}
1	0.911458	18.2292	**0.0625**	**21.875**	1.64256×10^{-3}	4.46431×10^{-3}	2.79018×10^{-5}	2.83495×10^{-5}	3.125×10^{-5}
1.09714	**1**	**20**	0.06857	**24**	2.13122×10^{-3}	4.89789×10^{-3}	3.06116×10^{-5}	3.11028×10^{-5}	3.4285×10^{5}
0.054857	**0.05**	**1**	3.42857×10^{-3}	**1.2**	1.06563×10^{-4}	2.47899×10^{-4}	1.53061×10^{-6}	1.55517×10^{-6}	1.71429×10^{-6}
16	14.5833	291.667	**1**	**350**	0.031081	0.071429	4.46429×10^{-4}	4.53592×10^{-4}	**0.0005**
0.045714	0.0416667	0.833333	2.85714×10^{-3}	**1**	8.88028×10^{-5}	2.04083×10^{-4}	1.27551×10^{-6}	1.29598×10^{-6}	1.42857×10^{-6}
514.784	469.203	9 384.09	32.1740	11 260.9	**1**	2.29816	0.0143634	0.0145939	0.0160870
224	20 416.6	4 083.34	**14**	**4900**	0.435134	**1**	**6.25×10^{-3}**	6.3503×10^{-3}	**0.007**
35 840	32 666.6	653 334	2 240	784 000	69.1214	160	**1**	1.016047	**1.12**
35 273.9	32 150.6	643 015	2 204.62	771 617	68.5218	157.474	0.984207	**1**	1.10231
32 000	29 166.6	583 334	**2000**	**700 000**	62.162	142.858	0.892857	0.907185	**1**

Power

↓ From/To→	BTU/hr	BTU/min	Calorie/min	Calorie/sec	Ergs/sec	Footpound force/min
BTU/hr	**1**	0.0166667	4.19994	0.0699988	2.93071×10^{6}	12.9695
BTU/min	**60**	**1**	251.997	4.19994	1.75843×10^{8}	778.170
calorie/min	0.238099	3.96832×10^{-3}	**1**	0.0166667	6.97800×10^{5}	3.08802
calorie/sec	14.2859	0.238099	**60**	**1**	**4.1868×10^{7}**	185.281
ergs/sec	3.41214×10^{-7}	5.68690×10^{-9}	1.43308×10^{-6}	2.38846×10^{-8}	**1**	4.42537×10^{-6}
footpound force/min	0.0771041	1.28507×10^{-3}	0.323833	5.39720×10^{-3}	2.25970×10^{5}	**1**
footpound force/sec	4.62624	7.71040×10^{-2}	19.4300	0.323832	1.35582×10^{7}	**60**
horsepower	2 544.43	42.4072	10 686.5	178.107	7.45700×10^{9}	**33 000**
horsepower (metric)	2 509.63	41.8271	10 540.3	175.671	7.35500×10^{9}	32 548.6
joule/hr	9.47817×10^{-4}	1.57969×10^{-5}	3.98078×10^{-3}	6.63461×10^{-5}	2 777.78	0.0122927
joule/min	0.0568690	9.47819×10^{-4}	0.238847	3.98077×10^{-3}	166 667	0.737562
kilowatt	3 412.14	56.8690	14 330.8	238.846	**10^{10}**	44 253.7
watt	3.41214	0.0568690	14.3308	0.238846	**10^{7}**	44.2537

Notes: Power
Watt (W) = J/s = A V = kg m^2/s^3
The value used for calorie here is 1 cal = **4.1868** J (sometimes used in physics).

Pressure

↓ From/To→	Atmosphere	Bar	cm of Hg	dynes/cm²	ft of water	in of Hg	in of water
atmosphere	**1**	**1.01325**	**76**	1.01325×10^{6}	33.8985	29.9213	406.782
bar	0.986923	**1**	75.0062	**10^{6}**	33.4552	**29.5300**	401.463
cm of Hg	0.0131579	0.0133322	**1**	13 332.2	0.446033	0.393701	5.35240
dynes/cm²	9.86923×10^{-7}	**10^{-6}**	7.50062×10^{-5}	**1**	3.34552×10^{-5}	2.95300×10^{-5}	4.01463×10^{-4}
ft of water	0.0294998	0.0298907	2.24199	29 890.7	**1**	0.882672	**12**
in of Hg	0.0334211	0.0338639	**2.54**	33 863.9	1.13292	**1**	13.5951
in of water	2.45832×10^{-3}	2.49089×10^{-3}	0.186832	2 490.89	0.0833332	0.0735560	**1**
kg-force /cm²	0.967841	0.980665	73.5560	980 665	32.8083	28.9590	393.701
kg-force/m²	9.67841×10^{-5}	9.80665×10^{-5}	7.35560×10^{-3}	98.0665	3.28083×10^{-3}	2.89590×10^{-3}	0.0393701
kilopascal	9.86923×10^{-3}	**0.01**	0.750062	**10 000**	0.334552	0.295300	4.01463
lb-force/ft²	4.72542×10^{-4}	4.78803×10^{-4}	0.0359132	478.803	0.0160185	0.0141391	0.192222
lb-force/in²(psi)	0.0680460	0.0689476	5.17150	68 947.6	2.30666	2.03602	27.6799
mm of Hg	1.315789×10^{-3}	1.33322×10^{-3}	**0.1**	1 333.22	0.0446033	0.0393701	0.535240
newton/cm²	0.0986923	**0.1**	7.50062	**100 000**	3.34552	2.95300	40.1463
pascal	9.86923×10^{-6}	**10^{-5}**	7.50062×10^{-4}	**10**	3.34552×10^{-4}	2.95300×10^{-4}	4.01463×10^{-3}
torr	1.315789×10^{-3}	1.33322×10^{-3}	**0.1**	1 333.22	0.0446033	0.0393701	5.35240

Power

Footpound force/sec	Horsepower	Horsepower (metric)	Joule/hr	Joule/min	Kilowatt	Watt
0.216158	3.93014×10^{-4}	3.98465×10^{-4}	1 055.06	17.5843	2.93071×10^{-4}	0.293071
12.9695	0.0235809	0.0239080	63 303.5	1 055.06	0.0175843	17.5843
0.0514671	9.35764×10^{-5}	9.48743×10^{-5}	251.208	**4.1868**	6.97800×10^{-5}	0.0697800
3.08802	5.61458×10^{-3}	5.69246×10^{-3}	15 072.5	251.208	**4.1868×10^{-3}**	**4.1868**
7.37562×10^{-8}	1.34102×10^{-10}	1.35962×10^{-10}	**3.6×10^{-4}**	**6×10^{-6}**	**10^{-10}**	**10^{-7}**
0.0166667	3.03030×10^{-5}	3.07233×10^{-5}	81.3492	1.35582	2.25970×10^{-5}	0.0225970
1	1.81818×10^{-3}	1.84340×10^{-3}	4 880.94	81.3491	1.35582×10^{-3}	1.35582
550	**1**	1.01387	2.68452×10^{6}	44 742.0	0.745700	745.700
542.476	0.986319	**1**	2.64780×10^{6}	44 129.9	0.735500	735.500
2.04878×10^{-4}	3.72506×10^{-7}	3.77672×10^{-7}	**1**	0.0166667	2.77778×10^{-7}	2.77778×10^{-4}
0.0122927	2.23504×10^{-5}	2.26603×10^{-5}	**60**	**1**	1.66667×10^{-5}	0.0166667
737.562	1.34102	1.35962	**3.6×10^{6}**	**60 000**	**1**	**1 000**
0.737562	1.34102×10^{-3}	1.35962×10^{-3}	**3 600**	**60**	**0.001**	**1**

Pressure

kg-force / cm^2	kg-force $/m^2$	kilo-pascal	lb-force / ft^2	lb-force $/in^2$ (psi)	mm of Hg	Newton / cm^2	Pascal	Torr
1.03323	10 332.3	**101.325**	2 116.21	14.6959	**760**	**10.1325**	**101 325**	**760**
1.01972	10 197.2	**100**	2 088.54	14.5038	750.062	**10**	**100 000**	750.062
0.0135952	135.952	1.33322	27.8449	0.193368	**10**	0.133322	1 333.22	**10**
1.01972×10^{-6}	0.0101972	**0.0001**	2.08854×10^{-3}	1.45038×10^{-5}	7.50062×10^{-4}	**10^{-5}**	**0.1**	7.50062×10^{-4}
0.0304801	304.801	2.98907	62.4279	0.433528	22.4199	0.298907	2 989.07	22.4199
0.0345317	345.317	3.38639	70.7261	0.491154	**25.4**	0.338639	3 386.39	**25.4**
2.54×10^{-3}	**25.4**	0.249089	5.20232	0.0361273	1.86832	0.0249089	249.089	1.86832
1	**10 000**	98.0665	2 048.16	14.2233	735.560	9.80665	98 066.5	735.559
0.0001	**1**	9.80665×10^{-3}	0.204816	1.42233×10^{-3}	0.0735560	9.80665×10^{-4}	9.80665	0.0735560
0.0101972	101.972	**1**	20.8854	0.145038	7.50062	**0.1**	**1 000**	7.50062
4.88245×10^{-4}	4.88245	0.0478803	**1**	6.94445×10^{-3}	0.359132	4.78803×10^{-3}	47.8803	0.359132
0.0703072	703.072	6.89476	**144**	**1**	51.7150	0.689476	6 894.76	51.7150
1.35952×10^{-3}	13.5952	0.133322	2.78449	0.0193368	**1**	0.0133322	133.322	**1**
0.101972	1 019.72	**10**	208.854	1.45038	75.0062	**1**	**10 000**	75.0062
1.01972×10^{-5}	0.101972	**0.001**	0.0208854	1.45038×10^{-4}	7.50062×10^{-3}	**10^{-4}**	**1**	7.50062×10^{-3}
1.35952×10^{-3}	13.5952	0.133322	2.78449	0.0193368	**1**	0.0133322	133.322	**1**

Temperature

↓ From/To→	°Celsius	°Fahrenheit	Kelvin	°Rankine
°Celsius	1	(°C x 1.8) + 32	°C + 273.15	(°C + 273.15) x 1.8
°Fahrenheit	(°F – 32)/1.8	1	((°F–32)/1.8) + 273.15	°F + 459.67
Kelvin	K – 273.15	(K x 1.8) – 459.67	1	K x 1.8
°Rankine	(°Rank/1.8) – 273.15	°Rank - 459.67	°Rank/1.8	1

Converting temperatures is not as easy as other conversions. To convert between temperatures, you add (+), subtract (–), divide (/), or multiply (×).

Velocity

↓ From/To→	cm/h	cm/min	cm/s	ft/h	ft/min	ft/s	km/h
cm/h	1	0.0166667	2.77778×10^{-4}	0.0328084	5.46807×10^{-4}	9.11345×10^{-6}	10^{-5}
cm/min	60	1	0.0166667	1.96850	0.0328085	5.46808×10^{-4}	0.0006
cm/s	3 600	60	1	118.110	1.19685	0.0328084	0.036
ft/h	30.48	0.508000	8.46667×10^{-3}	1	0.0166667	2.77778×10^{-4}	3.048×10^{-4}
ft/min	1 828.80	30.48	0.50800	60	1	0.0166667	0.0182880
ft/s	109 728	1 828.80	30.48	3 600	60	1	1.09728
km/h	100 000	1 666.67	27.7778	3 280.84	54.6807	0.911344	1
km/min	6×10^{6}	100 000	1 666.67	196 850	3 280.85	54.6808	60
km/s	3.6×10^{8}	6×10^{6}	100 000	1.18110×10^{7}	196 850	3 280.84	3 600
knots	185 200	3 086.66	51.4444	6 076.12	101.267	1.68781	1.852
m/h	100	1.66667	0.0277778	3.28084	0.0546810	9.11344×10^{-4}	0.001
m/min	6 000	100	1.66667	196.840	3.28085	0.0546810	0.06
m/sec	360 000	6 000	100	11 811.0	196.850	3.28084	3.6
mi/h	160 934	2 682.24	44.7040	5 280	88	1.46700	1.60934
mi/min	9.65606×10^{6}	160 934	2 682.24	316 800	5 280	88	96.5606
mi/s	5.79364×10^{8}	9.65606×10^{6}	160 934	1.90080×10^{7}	316 800	5280	5 793.64

Time

↓ From/To→	Day	Hour	Minute	Second	Week	Leap year	Mean year	Standard year
day	1	24	1 440	86 400	0.142857	2.73224×10^{-3}	2.73786×10^{-3}	2.73973×10^{-3}
hour	0.041667	1	60	3 600	5.95238×10^{-3}	1.13843×10^{-4}	1.14077×10^{-4}	1.14155×10^{-4}
minute	6.94444×10^{-4}	0.0166667	1	60	9.92064×10^{-5}	1.89739×10^{-6}	1.90129×10^{-6}	1.90259×10^{-6}
second	1.15746×10^{-5}	2.77778×10^{-4}	0.0166667	1	1.65344×10^{-6}	3.16232×10^{-8}	3.16882×10^{-8}	3.17098×10^{-8}
week	7	168	10 080	604 800	1	0.0191257	0.0191650	0.0191781
leap year	366	8 784	527 040	3.16224×10^{7}	52.2857	1	1.00206	1.00274
mean year	365.25	8 766	525 960	3.15576×10^{7}	52.1786	0.997951	1	1.00069
standard year	365	8 760	525 600	3.1536×10^{7}	52.1429	0.997268	0.999318	1

Velocity

km/min	km/s	knots	m/h	m/min	m/s	mi/h	mi/min	mi/s
1.66667×10^{-7}	2.77778×10^{-9}	5.39957×10^{-6}	0.01	1.66667×10^{-4}	2.77778×10^{-6}	6.21373×10^{-6}	1.03561×10^{-7}	1.72603×10^{-9}
0.00001	1.66667×10^{-7}	3.23975×10^{-4}	0.6	0.01	1.66667×10^{-4}	3.72824×10^{-4}	6.21368×10^{-6}	1.03562×10^{-7}
0.0006	0.00001	0.0194384	36	0.6	0.01	0.022369	3.72820×10^{-4}	6.21371×10^{-6}
5.08×10^{-6}	8.46667×10^{-8}	1.64579×10^{-4}	0.3048	5.08×10^{-3}	8.46667×10^{-5}	1.89394×10^{-4}	3.15654×10^{-6}	5.26094×10^{-8}
3.048×10^{-4}	5.08×10^{-6}	9.87489×10^{-3}	18.2880	0.3048	5.08×10^{-3}	0.011364	1.89393×10^{-4}	3.15656×10^{-6}
0.0182880	3.048×10^{-4}	0.592484	1 097.28	18.288	0.3048	0.681663	0.0113636	1.89394×10^{-4}
0.0166667	2.77778×10^{-4}	0.539957	1 000	16.6667	0.277778	0.621371	0.0103561	1.72603×10^{-4}
1	0.0166667	32.3975	60 000	1 000	16.6667	37.2824	0.621368	0.0103562
60	1	1 943.84	3.6×10^{6}	60 000	1 000	2 236.94	37.2823	0.621372
0.0308666	5.14444×10^{-4}	1	1 852	30.8667	0.514444	1.150779	0.019179	3.19661×10^{-4}
1.66667×10^{-5}	2.77778×10^{-7}	5.39957×10^{-4}	1	0.0166667	2.77778×10^{-4}	6.21373×10^{-4}	1.03561×10^{-5}	1.72603×10^{-7}
0.001	1.66667×10^{-5}	0.032397	60	1	0.0166667	0.037282	6.21368×10^{-4}	1.03562×10^{-5}
0.06	0.001	1.94384	3 600	60	1	2.23694	0.037282	6.21371×10^{-4}
0.0268224	4.47040×10^{-4}	0.868976	1 609.34	26.8224	0.44704	1	0.0166667	2.77778×10^{-4}
1.60934	0.0268224	52.1386	96 560.6	1 609.344	26.8224	60	1	0.0166667
96.5606	1.60934	3 128.31	5.79364×10^{6}	96 560.7	1 609.34	3 600	60	1

Education is what remains when we have forgotten all that we have been taught. George Savile

Volume

↓ From/To→	Acre foot	Acre inch	Barrel, petroleum	Cubic centimeter (cc)	Cubic foot
acre foot	**1**	**12**	7 758.37	1.23348×10^{9}	**43 560**
acre inch	0.0833334	**1**	646.531	1.02790×10^{8}	**3 630**
barrel, petroleum	1.28893×10^{-4}	1.54672×10^{-3}	**1**	158 987	5.61458
cubic centimeter (cc)	8.10713×10^{-10}	9.72855×10^{-9}	6.28981×10^{-6}	**1**	3.53147×10^{-5}
cubic foot	2.29568×10^{-5}	2.75482×10^{-4}	0.178108	28 316.8	**1**
cubic inch	1.32852×10^{-8}	1.59422×10^{-7}	1.03072×10^{-4}	16.3871	5.78704×10^{-4}
cubic meter	8.10713×10^{-4}	9.72855×10^{-3}	6.28981	$\mathbf{10^{6}}$	35.3147
cubic millemeter	8.10713×10^{-13}	9.72855×10^{-12}	6.28981×10^{-9}	**0.001**	3.53147×10^{-8}
cubic yard	6.19835×10^{-4}	7.43801×10^{-3}	4.80891	764 555	**27**
cup, metric	1.62143×10^{-7}	1.94571×10^{-6}	1.25796×10^{-3}	**200**	7.06293×10^{-3}
cup, US	1.91805×10^{-7}	2.30166×10^{-6}	1.48809×10^{-3}	236.588	8.35503×10^{-3}
gallon, Brit	3.68557×10^{-6}	4.42269×10^{-5}	0.0285940	**4 546.09**	0.160544
gallon, US	3.06888×10^{-6}	3.68266×10^{-5}	0.0238095	3 785.41	0.133681
liter	8.10713×10^{-7}	9.72855×10^{-6}	6.28981×10^{-3}	**1 000**	0.0353147
milliliter	8.10713×10^{-10}	9.72855×10^{-9}	6.28981×10^{-6}	**1**	3.53147×10^{-5}
ounce, Brit fluid	2.30348×10^{-8}	2.76418×10^{-7}	1.78713×10^{-4}	28.4131	1.00340×10^{-3}
ounce, US fluid	2.39756×10^{-8}	2.87708×10^{-7}	1.86012×10^{-4}	29.5735	1.04438×10^{-3}
pint, Brit	4.60697×10^{-7}	5.52836×10^{-6}	3.57426×10^{-3}	568.261	0.0200680
pint, US liquid	3.83610×10^{-7}	4.60332×10^{-6}	2.97619×10^{-3}	473.177	0.0167101
quart, Brit	9.21394×10^{-7}	1.10567×10^{-5}	7.14851×10^{-3}	1 136.52	0.0401359
quart, US liquid	7.67221×10^{-7}	9.20665×10^{-6}	5.95238×10^{-3}	946.353	0.0334201
tablespoon, metric	1.21607×10^{-8}	1.45928×10^{-7}	9.43472×10^{-5}	**15**	5.29720×10^{-4}
tablespoon, US	1.19878×10^{-8}	1.43854×10^{-7}	9.30060×10^{-5}	14.7868	5.22190×10^{-4}
teaspoon, metric	4.05357×10^{-9}	4.86428×10^{-8}	3.14491×10^{-5}	**5**	1.76573×10^{-4}
teaspoon, US	3.99594×10^{-9}	4.79513×10^{-8}	3.10020×10^{-5}	4.92892	1.74063×10^{-4}

Notes: Volume
cm^3 = cc = mL
dm^3 = L

Volume

Cubic inch	Cubic meter	Cubic millimeter	Cubic yard	Cup, metric	Cup, US	Gallon, Brit
7.52717×10^{7}	1 233.48	1.23348×10^{12}	1 613.33	6.16741×10^{6}	5.21363×10^{6}	271 328
6.27264×10^{6}	102.790	1.02790×10^{11}	134.445	513 951	434 470	22 610.7
9 702.00	0.158987	1.58987×10^{8}	0.207948	794.934	**672**	34.9723
0.0610237	**10^{-6}**	**1 000**	1.30795×10^{-6}	**0.005**	4.22676×10^{-3}	2.19969×10^{-4}
1 728	0.0283168	2.83168×10^{7}	0.037037	141.584	119.689	6.22883
1	1.63871×10^{-5}	16 387.1	2.14335×10^{-5}	0.0819353	0.0692642	3.60465×10^{-3}
61 023.7	**1**	**10^{9}**	1.30795	**5 000**	4 226.76	219.969
6.10237×10^{-5}	**10^{-9}**	**1**	1.30795×10^{-9}	**5×10^{-6}**	4.22676×10^{-6}	2.19969×10^{-7}
46 656	0.764555	7.64555×10^{8}	**1**	3 822.77	3 213.59	168.179
12.2047	**2×10^{-4}**	**200 000**	2.61590×10^{-4}	**1**	0.845352	0.0439938
14.4375	2.36588×10^{-4}	236 588	3.09446×10^{-4}	1.18294	**1**	0.0520421
277.419	4.54609×10^{-3}	4.54609×10^{6}	5.94606×10^{-3}	22.7305	19.2152	**1**
231	3.78541×10^{-3}	3.78541×10^{6}	4.95113×10^{-3}	18.9271	**16**	0.832674
61.0237	**0.001**	**10^{6}**	1.30795×10^{-3}	**5**	4.22676	0.219969
0.0610237	**10^{-6}**	**1 000**	1.30795×10^{-6}	**0.005**	4.22676×10^{-3}	2.19969×10^{-4}
1.73387	2.84131×10^{-5}	28 413.1	3.71629×10^{-5}	0.142065	0.120095	**0.00625**
1.80469	2.95735×10^{-5}	29 573.5	3.86807×10^{-5}	0.147868	**0.125**	6.50527×10^{-3}
34.6774	5.68261×10^{-4}	568 261	7.43258×10^{-4}	2.84131	2.40190	**0.125**
28.8750	4.73177×10^{-4}	473 177	6.18892×10^{-4}	2.36588	**2**	0.104084
69.3549	1.13652×10^{-3}	1.13652×10^{6}	1.48652×10^{-3}	5.68261	4.80381	**0.25**
57.75	9.46353×10^{-4}	9.46353×10^{5}	1.23778×10^{-3}	4.73176	**4**	0.208169
0.915356	**1.5×10^{-5}**	**15 000**	1.96193×10^{-5}	**0.075**	0.0634014	3.29954×10^{-3}
0.902344	1.47868×10^{-5}	14 786.8	1.93404×10^{-5}	0.0739339	**0.0625**	3.25263×10^{-3}
0.305119	**5×10^{-6}**	**5 000**	6.53976×10^{-6}	**0.025**	0.0211338	1.09985×10^{-3}
0.300781	4.92892×10^{-6}	4 928.92	6.44679×10^{-6}	0.0246446	0.0208334	1.08421×10^{-3}

Volume

↓ From/To→	Gallon, US	Liter	Milliliter	Ounce, Brit fluid	Ounce, US fluid	Pint, Brit
acre foot	325 852	1.23348×10^{6}	1.23348×10^{9}	4.34125×10^{7}	4.17090×10^{7}	2.17062×10^{6}
acre inch	27 154.3	102 790	1.02790×10^{8}	3.61769×10^{6}	3.47575×10^{6}	180 885
barrel, petroleum	**42**	158.987	158 987	5 595.57	**5 376**	279.779
cubic cm (cc)	2.64172×10^{-4}	**0.001**	**1**	0.0351951	0.0338140	1.75975×10^{-3}
cubic foot	7.48052	28.3168	28 316.8	996.614	957.506	49.8307
cubic inch	4.32901×10^{-3}	0.0163871	16.3871	0.576744	0.554113	0.0288372
cubic meter	264.172	**1000**	**10^{6}**	35 195.1	33 814.0	1 759.75
cubic mm	2.64172×10^{-7}	**10^{-6}**	**0.001**	3.51951×10^{-5}	3.38140×10^{-5}	1.75975×10^{-6}
cubic yard	201.974	764.555	764 555	26 908.6	25 852.7	1 345.43
cup, metric	0.0528344	**0.2**	**200**	7.03902	6.76280	0.351951
cup, US	**0.0625**	0.236588	236.588	8.32673	**8**	0.416337
gallon, Brit.	1.20095	**4.54609**	**4 546.09**	**160**	153.722	**8**
gallon, US	**1**	3.78541	3 785.41	133.228	**128**	6.66139
liter	0.264172	**1**	**1 000**	35.1951	33.8140	1.75975
milliliter	2.64172×10^{-4}	**0.001**	**1**	0.0351951	0.0338140	1.75975×10^{-3}
ounce, Brit fluid	7.50594×10^{-3}	0.0284131	28.4131	**1**	0.960760	**0.05**
ounce, US fluid	**7.8125×10^{-3}**	0.0295735	29.5735	1.04084	**1**	0.0520421
pint, Brit	0.150119	0.568261	568.261	**20**	19.2152	**1**
pint, US liquid	**0.125**	0.473177	473.177	16.6535	**16**	0.832674
quart, Brit	0.300238	1.13652	1 136.52	**40**	38.4304	**2**
quart, US liquid	**0.25**	0.946353	946.353	33.3070	**32**	1.66535
tablespoon, metric	3.96258×10^{-3}	**0.015**	**15**	0.527926	0.507210	0.0263963
tablespoon, US	3.90625×10^{-3}	0.0147868	14.7868	0.520422	**0.5**	0.0260211
teaspoon, metric	1.32086×10^{-3}	**0.005**	**5**	0.175975	0.169070	8.79877×10^{-3}
teaspoon, US	1.30208×10^{-3}	4.92892×10^{-3}	4.92892	0.173474	0.166667	8.67369×10^{-3}

There is always some way of understanding an idiot, a child, a primitive man or a foreigner if one has sufficient information.
Jean Paul Sartre

Volume

Pint, US liquid	Quart, Brit	Quart, US liquid	Tablespoon, metric	Tablespoon, US	Teaspoon, metric	Teaspoon, US
2.60681×10^{6}	1.08531×10^{6}	1.30341×10^{6}	8.22321×10^{7}	8.34174×10^{7}	2.46696×10^{8}	2.50254×10^{8}
217 234	90 442.7	108 617	6.85268×10^{6}	6.95150×10^{6}	2.05580×10^{7}	2.08545×10^{7}
336	139.889	**168**	10 599.2	**10 752**	31 797.5	**32 256**
2.11338×10^{-3}	8.79877×10^{-4}	1.05669×10^{-3}	0.0666667	0.0676280	**0.2**	0.202884
59.8441	24.9153	29.9221	1 887.79	1 915.01	5 663.37	5 745.04
0.0346320	0.0144186	0.0173160	1.09247	1.10822	3.27741	3.32467
2 113.38	879.877	1 056.69	66 666.7	67 628.0	**200 000**	202 884
2.11338×10^{-6}	8.79877×10^{-7}	1.05669×10^{-6}	6.66667×10^{-5}	6.76280×10^{-5}	**0.0002**	2.02884×10^{-4}
1 615.77	672.714	807.896	50 970.3	51 705.3	152 911	155 116
0.422675	0.175975	0.211338	13.3333	13.5256	**40**	40.5768
0.5	0.208168	**0.25**	15.7725	**16**	47.3176	**48**
9.60760	**4**	4.80380	303.073	307.443	909.218	922.329
8	3.33070	**4**	252.361	**256**	757.082	**768**
2.11338	0.879877	1.05669	66.6667	67.6280	**200**	202.882
2.11338×10^{-3}	8.79877×10^{-4}	1.05669×10^{-3}	0.0666667	0.0676280	**0.2**	0.202884
0.0600475	**0.025**	0.0300237	1.89420	1.92152	5.68261	5.76456
0.0625	0.0260211	**0.03125**	1.97157	**2**	5.91471	**6**
1.20095	**0.5**	0.600475	37.8841	38.4304	113.652	115.291
1	0.416339	**0.5**	31.5451	**32**	94.6353	**96**
2.40190	**1**	1.20095	75.7682	76.8607	227.305	230.584
2	0.832674	**1**	63.0902	**64**	189.271	**192**
0.0317006	0.0131982	0.0158503	**1**	1.01442	**3**	3.04326
0.03125	0.0130105	**0.015625**	0.985785	**1**	2.95735	**3**
0.0105669	4.39939×10^{-3}	5.28344×10^{-3}	0.333333	0.338140	**1**	1.01442
0.0104167	4.33685×10^{-3}	5.20833×10^{-3}	0.328595	0.333333	0.985784	**1**

What is the nature of all our reasonings concerning matters of fact? The proper answer seems to be, that they are founded on the relation of cause and effect...in one word, EXPERIENCE. David Hume

2.6 Significant Figures

Once you make a mathematical computation, you can use significant figures to standardize the precision of your answer.*

Rules for counting significant digits

1) All non-zero numbers are significant.
2) Zeros are significant only when they are in the middle of a number, or at the end of a number that includes a decimal point.

* A number is only as precise as the number of significant digits (*s.d.*).

1.01 (*3 s.d.*) 2.020 (*4 s.d.*) 0.013 (*2 s.d.*)
200 (*1 s.d.*) 2.00×10^2 (*3 s.d.*) 200. (***exact***)

Addition and Subtraction

1) Line up decimal points in the following form:

$$\begin{array}{r} 2500 \\ +13.5 \\ \hline \end{array} \qquad \begin{array}{r} 35.689 \\ -25.3 \\ \hline \end{array}$$

2) Draw a line to the right of the last significant digit in each number.

$$\begin{array}{r} 25|00 \\ +13.5| \\ \hline \end{array} \qquad \begin{array}{r} 35.689| \\ -25.3| \\ \hline \end{array}$$

3) Pick the line farthest to the left and extend. Disregard the other lines.

$$\begin{array}{r} 25|00 \\ +\ |13.5 \\ \hline | \end{array} \qquad \begin{array}{r} 35.6|89 \\ -25.3| \\ \hline | \end{array}$$

4) Add or subtract.

$$\begin{array}{r} 25|00 \\ +\ |13.5 \\ \hline 2500 \end{array} \qquad \begin{array}{r} 35.6|89 \\ -25.3| \\ \hline 10.3|89 \end{array}$$

5) If the first digit to the right of the line falls between 0 and 4, discard it. If the first digit to the right of the line falls between 5 and 9, discard it and round up the number to the left (add one). Where necessary, fill spaces from line to decimal point with zeros.

$$\begin{array}{r} 25|00 \\ +\ |13.5 \\ \hline 2500 \end{array} \qquad \begin{array}{r} 35.6|89 \\ -25.3| \\ \hline 10.4 \end{array}$$

Multiplication and Division

1) Count the number of significant digits (*s.d.*) in each number. Your answer will have the same number of significant digits as the least precise number, so remember this number.

$$\begin{array}{r} 2.000\ (4\ s.d.) \\ \times\ 3.11\ (3\ s.d.) \leftarrow \\ \hline \end{array} \qquad \frac{27\ (2\ s.d.)}{3.\ (\textbf{exact})}$$

2) Multiply or divide.

$$\begin{array}{r} 2000 \\ \times 3.11 \\ \hline 6.220 \end{array} \qquad \frac{27}{3.} = 9$$

3) Starting from the left, count the significant digits. Draw a line after the last one.

$$\begin{array}{r} 2.000 \\ \times\ 3.11 \\ \hline 6.22|0 \end{array} \qquad \frac{27}{3.} = 9.0|$$

4) If the first digit to the right of the line falls between 0 and 4, discard it. If the first digit to the right of the line falls between 5 and 9, discard it and round up the number to the left (add one). Where necessary, fill spaces from line to decimal point with zeros.

$$\begin{array}{r} 2.000 \\ \times 3.11 \\ \hline 6.22 \end{array} \qquad \frac{27}{3.} = 9.0$$

TRANSLATING SCIENCE INTO ENGLISH

3

3 Translating Science into English

No doubt you've heard some students complain, "I'm being blown away by all this complex scientific terminology!" The key to learning scientific terminology lies in understanding common prefixes and suffixes. With this knowledge, you can translate all kinds of scientific terms.

For example, take the term **anaerobic bacteria**. Look up the prefixes and suffixes in the following tables (*an*, without; *aero*, air). When you know the word's roots you can deduce that anaerobic bacteria do not require oxygen to live. Or take the biology term, **autotroph** (*auto*, self; *troph*, nourishment). Autotrophic organisms manufacture their own food—but you probably figured this out by now.

Keep in mind that some prefixes and suffixes have more than one meaning. In this section, we include the most useful and important meanings.

The great tragedy of science—the slaying of a beautiful theory by an ugly fact. T.H. Huxley

Translating Prefixes

Prefix	Translation	Origin	Prefix	Translation	Origin
a-	not, without	Greek *a-*	archae- archaeo-	ancient	Greek *archaios*
ab-	away from	Latin *ab*	archi-	primitive	Greek *arche-*
abd-	led away	Latin *abductum*	asco-	sac	Greek *askos*
ablat-	carry away	Latin *ablatum*	astr- astro-	star	Greek *astron*
acro-	end, tip	Greek *akros*	aut- auto-	self	Greek *autos*
actin- actino-	ray	Greek *aktinos*	auxo-	increase	Greek *auxe*
add-	brought forward	Latin *adductus*	baro-	weight (pressure)	Greek *baros*
adip-	fat	Latin *adipis*	bathy-	deep	Greek *bathys*
aer- aero-	air	Greek *aeros*	bi-	twice	Latin *bi-*
agg-	to clump	Latin *agglutinatum*	bio-	life	Greek *bios*
agro-	land	Greek *agros*	blast- blasto-	sprout (budding)	Greek *blastos*
alb-	white	Latin *albus*	calat-	inserted	
allo-	other	Greek *allos*	calci-	lime	Latin *calcem*
ameb-	change	Greek *amoibe*	carcin-	cancer	Greek *karkinos*
amphi-	around, both	Greek *amphi-*	cardio- cardia-	heart	Greek *kardia*
amyl-	starch	Greek *amylon*	carp-	fruit, wrist	Greek *karpos*
an-	without	Greek *an-*	caseo-	cheese	Latin *caseus*
ana-	up	Greek *ana-*	cata-	down	Greek *kata*
andro-	man	Greek *andros*	caul-	stalk	Latin *caulis*
ant- anti-	opposite	Greek *anti-*	centro-	center	Latin *centrum*
anth-	flower	Greek *anthos*	cera-	horn	Greek *keras*
ap- apo-	off	Greek *apo*	chalc- chalco-	bronze, copper	Greek *chalkos*

What is the use of running when you are on the wrong road? Proverb

Translating Prefixes

Prefix	Translation	Origin	Prefix	Translation	Origin
chlor- chloro-	green	Greek *chloros*	dent- denti-	tooth	Latin *dentem*
chondr-	lump	Greek *chondros*	derm-	skin	Greek *derma*
chrom- chromo-	colour	Greek *chroma*	dextr- dextro-	right	Latin *dexter*
chrys- chryso-	gold	Greek *chrysos*	di-	two	Latin
circ- circum-	around	Latin *circum*	dia-	apart (across)	Greek *dia*
cis-	same side	Latin *cis*	diastol-	dilation, expansion	Greek *diastrole*
co-	with	Latin *co-*	dors-	back	Latin *dorsum*
copro-	dung	Greek *kopros*	dys-	bad	Greek *dys-*
cosmo-	order, world	Greek *kosmos*	ec- ecto-	outside	Greek *ektos*
cran-	helmet	Latin *cranium*	em-	inside	Greek *en-*
crist-	ridge, crest	Latin *crista*	en-	in	Greek *en-*
cryo-	cold	Greek *kryos*	end- endo-	within	Greek *endon*
crypto-	hidden	Greek *kryptos*	entero-	intestine	Greek *enteron*
cut-	skin	Latin *cutis*	epi-	at, on, over	Greek *epi*
cyan- cyano-	blue	Greek *kyanos*	equi-	equal	Latin *aequus*
cyclo-	ring, wheel	Greek *kyklos*	erg-	work	Greek *ergon*
cyst-	bladder, pouch	Greek *kystis*	erythro-	red	Greek *erythros*
cyt- cyto-	cell	Greek *kytos*	eu-	good, true	Greek *eus*
de-	undo	Latin	eury-	wide	Greek *eurys*
dendr- dendri- dendro-	tree	Greek *dendron*	ex- exo-	away, out	Greek *exo-*

Tools are more important than toys, yet he with the most toys wins! Gordon Coleman

Translating Prefixes

Prefix	Translation	Origin	Prefix	Translation	Origin
extra-	outside	Latin *extra*	hygro-	wet	Greek *hygros*
fibro-	fiber	Latin *fibra*	hyper-	above	Greek *hyper-*
flag-	whip	Latin *flagellum*	hypha-	web	Greek *hyphe-*
flav-	yellow	Latin *flavus*	hypo-	below	Greek *hypo-*
gamet- gamo-	marriage, united	Greek *gamos*	im-	not	Latin *im-*
gastr-	stomach	Greek *gastros*	infra-	under	Latin *infra*
geo-	earth	Greek *geo-*	inter-	between	Latin *inter*
glyc-	sweet	Greek *glykys*	intra-	inside of, within	Latin *intra*
gyro-	circle	Greek *gyros*	intro-	inward	Latin *intro*
halo-	salt	Greek *halos*	iso-	equal	Greek *isos*
haplo-	single	Greek *haplous*	karyo-	nut	Greek *karyon*
hem- hema- hemato-	blood	Greek *haimatos*	kerat- kerato-	horn	Greek *keratos*
hemi-	half	Greek *hemi-*	kin-	motion	Greek *kinesis*
hemo-	blood	Greek *haima*	labia- labio-	lip	Latin labium
hepat- hepa-	liver	Greek *hepatos*	lact- lacti- lacto-	milk	Latin *lactis*
hetero-	different	Greek *heteros*	leuc- leuco-	white	Greek *leukos*
histo-	web	Greek *histos*	lev- levo-	left	Latin *laevus*
holo-	whole	Greek *holos*	lig-	to bind	Latin *ligare*
homeo-	the same	Greek *homoios*	lim-	edge	Latin *limbus*
homo-	same	Greek *homos*	lip- lipo-	fat	Greek *lipos*
hydro-	water	Greek *hydor*	lith- litho-	stone	Greek *lithos*

Translating Prefixes

Prefix	Translation	Origin	Prefix	Translation	Origin
lymph- lympho-	clear water	Latin *lympha*	necro-	corpse	Greek *nekros*
lys- lyso-	break up	Greek *lyein*	neo-	new	Greek *neos*
macr- macro-	large	Greek *makros*	neur- neuro-	nerve	Greek *neuron*
mamm-	breast	Latin *mamma*	noct-	night	Latin *nox*
meg- mega-	great	Greek *megas*	nucleo-	kernel	Latin *nucleus*
meio-	less	Greek *meion*	ob-	against	Latin *ob-*
melan-	black	Greek *melanos*	octa-	eight	Greek *okta*
meningo-	membrane	Greek *meningos*	odont- odonto-	tooth	Greek *odontos*
mes- meso-	middle	Greek *mesos*	oligo-	few	Greek *oligos*
met- meta-	after, transition	Greek *meta*	onco-	bulk	Greek *onkos*
micr- micro-	small	Greek *mikros*	oo-	egg	Greek *oion*
mit-	thread	Greek *mitos*	orb-	circle	Latin *orbis*
mon- mono-	one	Greek *monos*	orni-	bird	Greek *ornis*
morpho-	form, shape	Greek *morphe*	ortho-	straight	Greek *orthos*
muc- muco-	slime	Latin *mucus*	osmo-	thrust	Greek *osmos*
multi-	many	Latin *multus*	oss- osseo- osteo-	bone	Latin *osseus* Greek *osteon*
mut-	change	Latin *mutare*	ovi- ovu-	egg	Latin *ovum*
myo-	muscle	Greek *myos*	pale- paleo-	ancient	Greek *palaios*
nas-	nose	Latin *nasus*	pan-	all	Greek *pan-*

Translating Prefixes

Prefix	Translation	Origin	Prefix	Translation	Origin
para- para-	beside, near	Greek *para*	pod-	foot	Greek *podos*
pariet-	partition, wall	Latin *paries*	poie-	making	Greek *poiein*
patho-	disease	Greek *pathos*	poly-	many	Greek *polys*
pelv-	basin	Latin *pelvis*	post-	after	Latin *post*
pent- penta-	five	Greek *pente*	pre-	before	Latin *prae*
peri-	around	Greek *peri*	pro-	forward	Greek *pro*
petro-	rock	Greek *petra*	prot- proto-	first	Greek *protos*
phag- phago-	eat	Greek *phagein*	pseud- pseudo-	false	Greek *pseudes*
pharmaco-	drug	Greek *pharmakon*	psychro-	cold	Greek *psychros*
phen- pheno-	appear	Greek *phainein*	pyo-	pus	Greek *pyon*
phil- philo-	love	Greek *philos*	pyr- pyro-	fire	Greek *pyr*
phono-	sound	Greek *phone*	quadri-	four	Latin *quattro*
photo-	light	Greek *phos*	radio-	ray	Latin *radius*
physio-	nature	Greek *physis*	ren-	kidney	Latin *renalis*
pico-	small number	Spanish *pico*	reticul-	network	Latin *reticulum*
piezo-	squeeze	Greek *piezein*	rheo-	flowing	Greek *rheos*
pino-	to drink	Greek *pinein*	rhin- rhino-	nose	Greek *rhinos*
plani- plano-	flat	Latin *planus*	rhizo-	root	Greek *rhiza*
pleur-	rib, side	Greek *pleura*	sacchar- saccharo-	sugar	Latin *saccharum*
pneum-	air	Greek *pneuma*	sapr- sapro-	rotten	Greek *sapros*

Translating Prefixes

Prefix	Translation	Origin	Prefix	Translation	Origin
sarco-	flesh	Greek *sarkos*	telo-	end	Greek *telos*
schizo-	split	Greek *schizein*	tetra-	four	Greek *tettares*
scler-	hard	Greek *skleros*	therm- thermo-	temperature, heat	Greek *therme*
seb-	oil	Latin *sebum*	thio-	sulfur	Greek *theion*
seism-	to shake	Greek *seiein*	tox-	poison	Greek *toxikon*
semi-	half	Latin *semi-*	trans-	across	Latin *trans*
sept-	rotting	Greek *sepein*	tri-	three	Greek *tria*
sero-	serum	Latin *serum*	tribo-	rubbing	Greek *tribos*
sesqui-	one and a half	Latin *sesqui-*	trich-	hair	Greek *trichos*
sidero-	iron	Greek *sideros*	tropo-	turn	Greek *tropos*
soma-	body	Greek *soma*	ultra-	beyond	Latin *ultra*
spermato-	seed	Greek *sperma*	uni-	one	Latin *unus*
sporo-	seed	Greek *sporos*	uro-	tail, urine	Greek *oura*
squam-	scale	Latin *squama*	vas- vaso-	vessel	Latin *vas*
stereo-	solid	Greek *stereos*	ves-	bladder, blister	Latin *vesica*
strato-	spreading out	Latin *stratus*	vita-	life	Latin *vita*
strepto-	curved	Greek *streptos*	vitro-	glass	Latin *vitrum*
stria-	groove	Latin *stria*	vivi-	alive	Latin *vivus*
sub-	beneath	Latin *sub*	xanth- xantho-	yellow	Greek *xantos*
super- supra-	above	Latin *super* Latin *supra*	xeno-	strange	Greek *xenos*
sym- syn-	with, together	Greek *syn*	xer- xero-	dry	Greek *xeros*
systol-	contraction	Greek *systole*	xyl-	wood	Greek *xylon*
tauto-	same	Greek *tauto*	zoo-	animal	Greek *zoion*
taxis-	arrangement	Greek *taxis*	zygo-	yoke	Greek *zygon*

Be nice to people on your way up because you will need them on your way down. Wilson Mizner

Translating Suffixes

Suffix	Translation	Origin	Suffix	Translation	Origin
-aceous	like	Latin *-aceus*	-logy	the study of	Greek *logos*
-ane	saturated hydrocarbon	English	-lysis	loosening	Greek *lyein*
-ase	enzyme	English	-lyt	dissolvable	
-ate	salt of an acid or ester whose name ended in -ic	English	-mel	black	Greek *melas*
-blast	budding	Greek *blastos*	-mere	share	Greek *meros*
-carp	fruit	Greek *karpos*	-metry	measure	Greek *-metria*
-caryo	nut	Greek *karyon*	-mnesia	memory	Greek
-cide	kill	Latin *-cida*	-oid	like	Greek *eidos*
-clase	cleavage	Greek *klasis*	-ol	alcohol	English
-crin	secrete		-ole	oil	Latin *oleum*
-cul	small part		-oma	tumor	Greek *-oma*
-cut	skin	Latin *cutis*	-osis	a condition	Greek *-osis*
-cyte	cell	Greek *kytos*	-pathy	suffering	Greek *pathos*
-emia	blood	Greek *haima*	-ped	foot	Greek
-gamy	marriage	Greek *gamos*	-petal	seek	Latin *petere*
-gen	born, (agent)	Greek *-genes*	-phage	eat	Greek *phagein*
-genesis	formation	Greek *genesis*	-phil -plile	loving	Greek *philos*
-gony	reproduction	Greek *gonos*	-phore	bear, carry	Greek *pherein*
-gram	written	Greek *gramma*	-phyll	leaf	Greek *phyllon*
-graph -graphy	to write	Greek *graphein*	-phyte	plant	Greek *phyton*
-gynous	woman	Greek *gyne*	-plast -plasty	a small body	Greek *plastos*
-hema	thread	Greek *hema*	-pnoea	breathing	Greek *pneuma*
-itis	inflammation	Greek	-pod	foot	Greek *podos*
-lite	stone	Greek *lithos*	-sis	a condition	Greek *-tikos*

There is no such thing as bravery; only degrees of fear. John Wainwright

Translating Suffixes

Suffix	Translation	Origin	Suffix	Translation	Origin
-some	body	Greek *soma*	-trope	turning	Greek *trope*
-stas -stasis	halt	Greek	-tropic	influence	Greek *trope*
-stat	to stand, stabilize	Greek *-states*	-troph	nourishment	Greek *trophe*
-stome	mouth	Greek *stoma*	-ty	state of	
-taxi	touch	Greek *taxis*	-valent	strength	Latin *valentem*
-tome -tomy	cutting	Greek *-tomia*	-vorous	eat	Latin *vorare*
-tone -tonic	strength	Greek *tonos*	-yl	wood	Greek *hyle*
-tron	device	Greek *-tron*	-zyme	ferment	Greek *zyme*

It is not the meaning of life that is important, but what you can do with it. Gordon Coleman

BIOLOGY

4

4.1 Classifications (Taxonomy)

The web of life, which surrounds and includes us, consists of an immensely large number of different types of organisms. How do we begin to understand so many different life forms? How do we make sense out of such divergent information? We need a system. Taxonomy is a classification system based on the structure and function of organisms. By grouping complex networks into manageable groups, taxonomy attempts to impose order upon apparent disorder. Possibly no other branch of science has caused more disagreement and debate in the scientific community than classical taxonomy.

Taxonomy is more than just controversial. Though the subject itself is rather dull, many scientists would like to name a new bug or decide where to draw the line dividing a group of organisms. After years of scientific infighting, there is at least some agreement on how to classify the broader categories of life. Classifications are grouped into Kingdom, Phylum, Class, Order, Family, Genus and Species.* (Remember the phrase, King Philip Came Over For Good Sex.)

We include inconsistencies such as the two places where algae and molds exist. As well, we divide groups that are related but different. For example, euglenophyta (the protista kingdom) is both an autotroph and heterotroph. It is not pure algae or pure protozoa. It lies somewhere between the lines.

Keep in mind that microbiologists normally use a different classification method: *Bergey's Manual of Systematic Bacteriology*. Bergey's is organized by Volume, Section, Family and Genus. Botanists also vary from the traditional nomenclature. They use the term "division" instead of "phylum" for plants and fungi. This information is extremely variable, so every attempt is made to offer a quick solution to this confusing subject. It should be noted that the classification systems we include here do not exclusively follow any single philosophy of taxonomy.

* An organism's genus and species are underlined or (italicized), genus is capitalized: *Yersinis pestis*, or <u>Yersinis</u> <u>pestis</u>.

An Aid to Understanding

Aerobes	Oxygen-requiring organisms.
Anaerobes	Organisms that don't require atmospheric oxygen.
Autotrophs	Organisms that manufacture food from environmental compounds, such as carbon dioxide and water.
Bacteria	Unicellular prokaryotes that cannot be seen with the naked eye. Bacteria have no nuclei.
Chlorophyll	A colored pigment that absorbs light energy to fuel sugar production in phototrophic organisms (see phototrophs).
Eukaryotes	Cells that are typically 10-100 μm in size and contain at least one nucleus.
Heterotrophs	Organisms that obtain nourishment by consuming members of the food web or other organic matter.
Hyphae	Thread-like structures of fungi. Many branching hyphae make up a network called mycelium.
Morphology	A branch of biology dealing with the shape and structure of organisms.
Pathogen	A disease-causing body. Some species of bacteria are pathogenic (disease-causing).
Phototrophs	Organisms that obtain energy from light-harnessing chemical reactions (found in the monera, protista and plant kingdoms).
Physiology	A branch of biology dealing with the function of organisms and their parts.
Prokaryotes	The simplest type of cells (1-2 μm long), also referred to as bacteria. They are unicellular with no distinct nucleus.
Species	A group of organisms that normally interbreed and share many common characteristics.
Spore	A reproductive cell or body that can become an individual organism without being fertilized.
Symbiosis	A relationship between organisms of different species. In the past, this term was exclusively used when both organisms benefited from the relationship.
Taxonomy	The classification of organisms.

Cells are classified as prokaryotic or eukaryotic. Major physiological and structural differences determine whether a cell is prokaryotic or eukaryotic. The following table shows some of the criteria used in classifying cells. (See Section 4.4 for more differences between cells.)

We all know that Prime Ministers are wedded to the truth, but like other wedded couples they sometimes live apart.
H.H. Munro

Prokaryote and Eukaryote Characteristics

Characteristics	Prokaryotes	Eukaryotes
Membrane-bound organelles	Absent	Present
Nucleus	Absent	Present
Type of ribosomes	70s	80s
Kingdom	Monera	All except monera
Genetic differences	Circular DNA	Histone-bound DNA condensed to form linear chromosomes
Selfish cell division	Binary fission	Mitosis
Typical size	1-2 μm	10-100 μm

The Five Kingdoms of Life

The Monera Kingdom*

Volume	General characteristics
1 Gracilicutes (Sections 1-11)	Prokaryotes, Gram-negative, thin cell walls. Useful in medicine and commerce.
2 Fermicutes (Sections 12-17)	Prokaryotes, Gram-positive, thick cell wall. Useful in medicine and commerce.
3 Tenericutes (Sections 18-25)	Prokaryotes with pliant structures, cells lack rigidity. Includes phototrophic, lithotrophic, gliding, budding, and appendaged bacteria.
4 Mendosicutes (Sections 26-33)	Prokaryotes with "porous" cell walls lacking peptidoglycan (archaebacteria). Also includes filamentous actinomycetes.

*For a comprehensive look at how monera are classified, refer to *Bergey's Manual of Systematic Bacteriology*.

All prokaryotes belong to this extremely diverse and adaptive kingdom. The monera kingdom is divided into two groups: archaebacteria and eubacteria. Archaebacteria evolved from ancient prokaryotes. They are found living in extreme environmental conditions such as hot springs and deep sea vents. They differ significantly in structure from eubacteria. Archaebacteria have ribosomal RNA and RNA polymerase (similar to eukaryotes), cell walls that lack peptidoglycan, and lipids that are different from all other organisms. Lipids in archaebacteria have ether linkages, whereas all other organisms have ester linkages.

Two types of eubacteria exhibit differences in their cell walls. These are characterized by their reaction to Gram's staining procedures. There are two reactions to Gram's stain: purple and red. Bacteria that turn red when stained are Gram-negative; if they turn purple, they're Gram-positive.

Bacteria reproduce through binary fission (splitting down the middle) into two separate cells. Generally, bacteria come in three different shapes: rods, spheres and spirals (bacilli, cocci and spirilli). Some species show multicellular behavior by forming long filaments as the "daughter cell" remains attached to the "parent" after binary fission.

Classical taxonomy for prokaryotes is outdated and cannot facilitate grouping all microbes. For this reason we refer to *Bergey's Manual* of classifying monera. *Bergey's Manual* consists of 4 Volumes. Within each Volume are several Sections; within each Section are related families, genera and species. For review purposes, we offer a summary of Bergey's classification system in the table on the previous page.

The Protista Kingdom

Special group	Phylum or division	Common name	Example	Extra facts
Algae	Chlorophyta	Green algae	*Volvox*	Chlorophyll a+b Cellulose cell wall
	Phaeophyta	Brown algae	*Laminaria*	Chlorophyll a+c
	Rhodophyta	Red algae	*Polisiphonia*	Chlorophyll a+d Cellulose cell wall
	Pyrrophyta Dinoflagellata	Dinoflagellates	*Gonyaulax*	Chlorophyll a+c Cellulose cell wall
	Chrysophyta	Diatoms Golden algae	*Isthmix*	Chlorophyll a+c Opaline silica wall
	Euglenophyta*	Euglenoids	*Euglena*	Chlorophyll a+b Two flagella, heterotrophic, if raised in the dark; no cell wall
Protozoa	Zoomastigina Mastigophora	Flagellates	*Trypanosoma*	Many flagella
	Sarcodina Rhizopoda	Amoebas	*Amoeba*	Includes the order foraminifera
	Ciliophora	Ciliates	*Paramecium*	Two types of nuclei
	Apicomplexa Sporozoa	Sporozoans	*Plasmodium*	Parasitic
Molds	Biologists tend to classify molds as fungi.			

*Euglenophyta occupy a fine line between algae and protozoa.

Members of the protista kingdom are divided into two groups: algae and protozoa; both are mainly unicellular. The major difference between these two groups is that algae are mainly photoautotrophs whereas protozoa are chemoheterotrophs. All algae have chlorophyll (a), and use light, carbon dioxide (CO_2) and water (H_2O) to maintain their nourishment and energy requirements.

Protozoa are motile and lack a cell wall. Heterotrophic protozoa obtain their food by ingesting organic matter and other organisms. Scientists believe that the plant, animal, and fungus kingdoms all originated from the protista kingdom.

It's a recession when your neighbor loses his job; it's a depression when you lose yours. Harry S. Truman

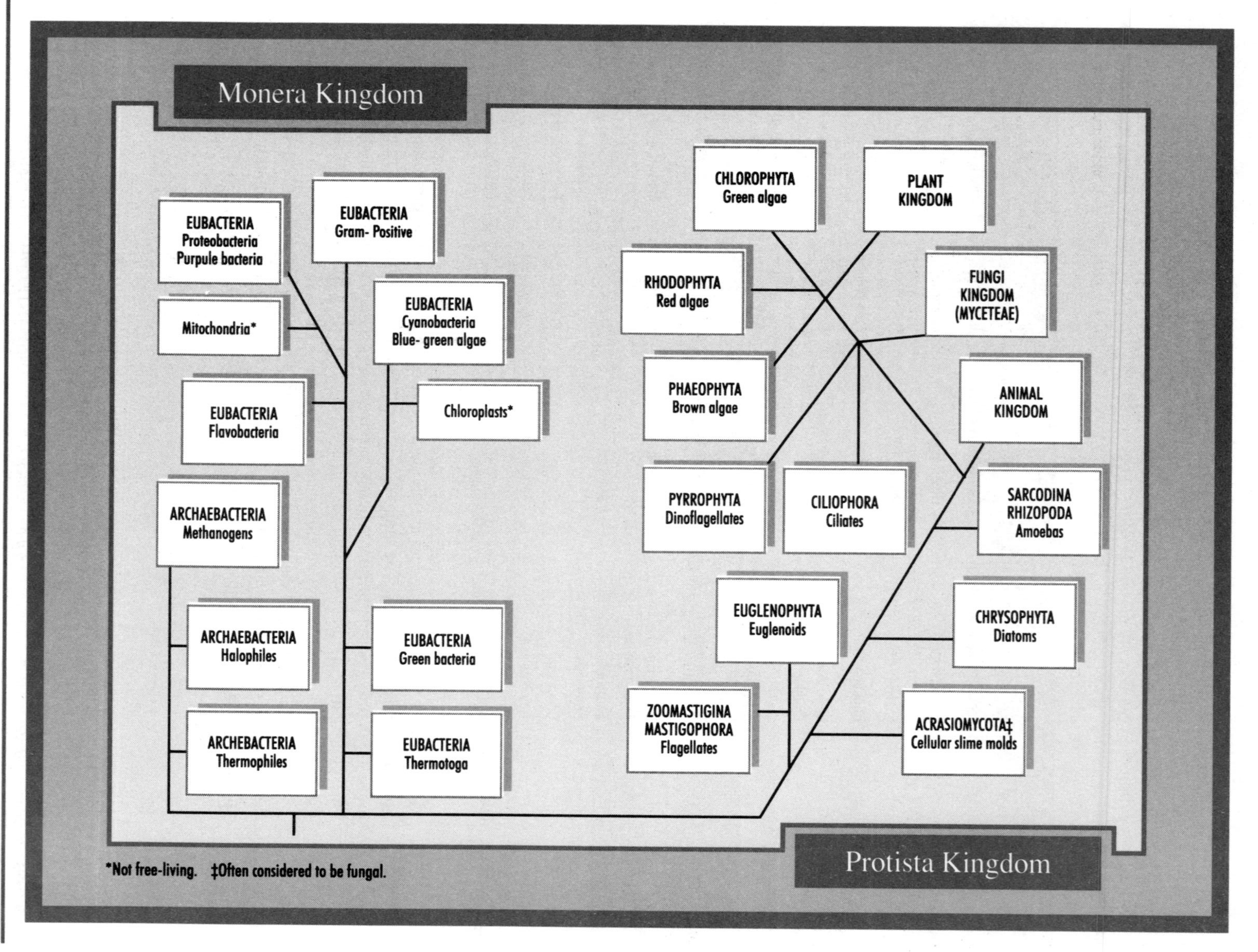

Monera Kingdom
EUBACTERIA Proteobacteria Purpule bacteria
EUBACTERIA Gram- Positive
Mitochondria*
EUBACTERIA Cyanobacteria Blue- green algae
EUBACTERIA Flavobacteria
Chloroplasts*
ARCHAEBACTERIA Methanogens
ARCHAEBACTERIA Halophiles
EUBACTERIA Green bacteria
ARCHEBACTERIA Thermophiles
EUBACTERIA Thermotoga
CHLOROPHYTA Green algae
PLANT KINGDOM
RHODOPHYTA Red algae
FUNGI KINGDOM (MYCETEAE)
PHAEOPHYTA Brown algae
ANIMAL KINGDOM
PYRROPHYTA Dinoflagellates
CILIOPHORA Ciliates
SARCODINA RHIZOPODA Amoebas
EUGLENOPHYTA Euglenoids
CHRYSOPHYTA Diatoms
ZOOMASTIGINA MASTIGOPHORA Flagellates
ACRASIOMYCOTA‡ Cellular slime molds
*Not free-living. ‡Often considered to be fungal.
Protista Kingdom

The Fungus (Mycetae) Kingdom

Special group	Phylum or division	Common name	Example	Extra facts
Eumycota	Zygomycota	Bread molds	*Rhizopus*	Sexual zygospore, coenocytic hyphae
(*Eu*-"True" fungi)	Ascomycota	Sac fungi	*Penicillium**	Sexual ascospore, septate hyphae
	Basidiomycota	Club fungi Mushrooms	*Agaricus*	Sexual basidiospore, septate hyphae
Fungi imperfecti	Deuteromycota	Imperfect fungi		Artificial group, no sexual spores, septate hyphae
Molds	Oomycota Phycomycota	Water molds	Allomyces	Sexual oospore, coenocytic hyphae
(This group is sometimes classified with the protista kingdom.)	Acrasiomycota	Cellular slime molds	*Dictyostelium discoideum*	
	Myxomycota	Plasmodial slime molds		
Lichens	A symbiotic relationship between cyanobacteria or green algae and ascomycete or basidiomycete fungi.			

******Penicillium* **was the imperfect fungi name. Once sexual spores were observed it was reclassified and named** ***Talaromyces*** **and** ***Carpenteles*****.**

The fungus kingdom (also known as mycetae), consists of divisions: zygomycota (bread molds), ascomycota (sac fungi, yeasts) and basidiomycota (mushrooms). Fungi are heterotrophic eukaryotes that almost always have a cell wall; often, they are multicellular. They reproduce by spores formed either sexually or asexually. When a spore germinates, a thread-like structure (hypha), extends and branches into many hyphae. This collective mass of tangled hyphae is called mycelium. Often, a hypha has many nuclei in the same cytoplasm (coenocytic), and is not divided into separate cells.

Yeast is a unicellular member of the ascomycota division. Yeast reproduces mainly through budding. Slime molds (cellular, plasmodial) and water molds are close to fungi in structure and function. These similarities appear to be a result of convergent evolution. Some scientists consider them to be more closely related to protists.

The deuteromycota division (fungi imperfecti) is a collection of fungi that do not appear to reproduce sexually. If sexual reproduction is ever observed among a member of deuteromycota, then it is promptly reclassified into an appropriate eumycota division.

Being a woman is terribly difficult trade, since it consists principally of dealing with men. Joseph Conrad

The Plant Kingdom

Special group	Phylum or division	Common name	Class
Vascular plants	Lycophyta	Club mosses (Microphylls)	
Seedless	Arthrophyta Sphenophyta	Horsetails	
	Pterophyta	Ferns	
	Psilophyta	Whisk ferns	
Seed-bearing Gymnosperms	Cycadophyta	Cycads	
	Gnetophyta	Vines	
	Coniferophyta	Conifers	
	Ginkgophyta	Ginkgo	
Angiosperms	Magnoliophyta	Flowering plants	Monocotyledons
	Anthophyta		Dicotyledons
Nonvascular plants	Bryophyta	Mosses	
	Hepaticophyta	Liverworts	
	Anthocerophyta	Hornworts	
Algae	Botanists tend to group green, brown and red algae in the plant kingdom as opposed to the protista kingdom.		

The plant kingdom is arranged in divisions rather than phyla. This kingdom includes plants that are multicellular eukaryotic organisms; the exception is algae, whose morphology is unicellular to leafy. The relationship between plants and algae is seen in biochemical and morphological similarities. Even though green algae is considered to be the probable origin of the plant kingdom, we side with those biologists who place algae in the protista kingdom instead of the plant kingdom.

A plant cell contains a nucleus, mitochondria, chloroplasts, chlorophyll (a+b) and a cell wall. Plants make their own food through photosynthesis. Plants made the move from water to land about 400 million years ago. Because they have the ability to prevent loss of water to the environment, they are well-suited to life on land.

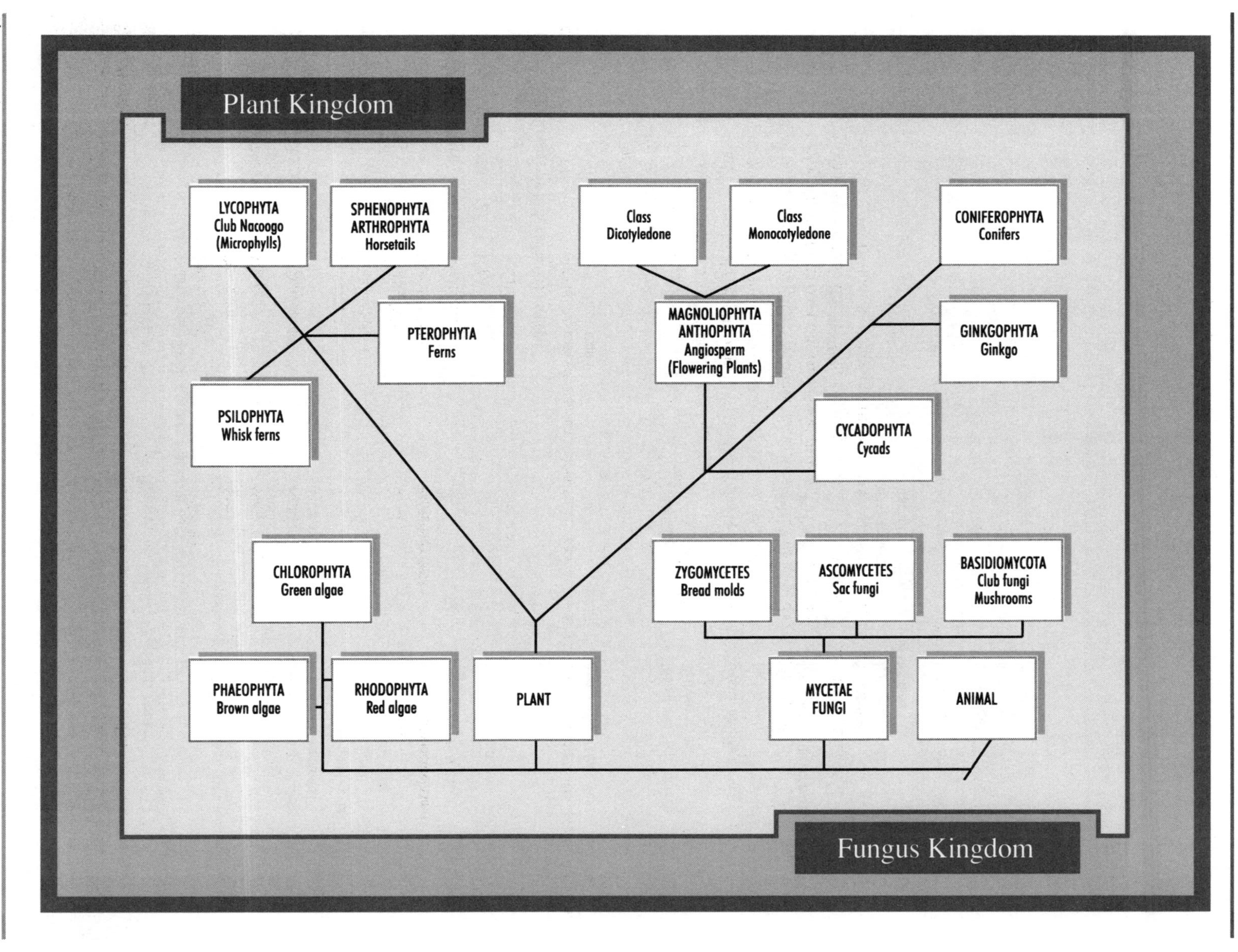

Plant Kingdom
LYCOPHYTA
Club Nacoogo
(Microphylls)
SPHENOPHYTA
ARTHROPHYTA
Horsetails
Class
Dicotyledone
Class
Monocotyledone
CONIFEROPHYTA
Conifers
PTEROPHYTA
Ferns
MAGNOLIOPHYTA
ANTHOPHYTA
Angiosperm
(Flowering Plants)
GINKGOPHYTA
Ginkgo
PSILOPHYTA
Whisk ferns
CYCADOPHYTA
Cycads
CHLOROPHYTA
Green algae
ZYGOMYCETES
Bread molds
ASCOMYCETES
Sac fungi
BASIDIOMYCOTA
Club fungi
Mushrooms
PHAEOPHYTA
Brown algae
RHODOPHYTA
Red algae
PLANT
MYCETAE
FUNGI
ANIMAL
Fungus Kingdom

The Animal Kingdom

Special name	Phylum	Class	Examples
	Subkingdom Parazoa (lack specialization)		
	Porifera		Sponges
	Subkingdom Eumetazoa (definite shape and symmetry)		
Branch Radiata (radial symmetry)	Cnidaria	Hydrozoa Scyphozoa Anthozoa	Hydras Jellyfish Corals, sea anemones
	Ctenophora		Sea walnuts, comb jellies
Branch Bilateria (bilateral symmetry)			
Acoelomates (no true body cavity)	Platyhelminthes	Turbellaria Trematoda Cestoda	Flatworms Flukes Tapeworms
	Nemertea		Probiscus worms
Pseudocoelomates (similar to coelomates but body cavity is not lined)	Rotifera		Rotifers
	Nematoda		Nematodes, round worms, pinworms
Coelomates (lined body cavity)			
Protostomes	Mollusca	Polyplacophora Cephalopoda Gastropoda Bivalvia	Marine mollusks, chitons Octopuses, squids Slugs, snails Clams, oysters
	Annelida	Oligochaeta Hirundinea Polychaeta	Earthworms Leeches Clamworms, polychaetes
	Arthropoda		
	Subphylum Chelicerata	Pycnogonida Merostomata Arachnida	Sea spiders Horseshoe crabs Spiders
	Subphylum Crustacea	Crustacea	Lobsters, shrimps
	Subphylum Uniramia	Chilopoda Diplopoda Insecta	Centipedes Millipedes Insects

The Animal Kingdom			
Special name	**Phylum**	**Class**	**Examples**
Deuterostomes	Echinodermata	Crinoidea Asteroidea Ophiuroidea Echinoidea Holothuroidea	Sea lillies Sea stars Brittle stars Sea urchins Sea cucumbers
	Chordata		
	Subphylum Urochordata		Tunicates
	Subphylum Cephalochordata		Lancelets
	Subphylum Vertebrata	Mammalia Reptilia Amphibia Chondrichthyes Aves Osteichthyes Agnatha	Humans Lizards, snakes Frogs Sharks Birds Bony fish Jawless fish
Lophophorate Animals	Phoronida		Marine worms
	Bryozoa (Ectoprocta)		Sea mosses
	Brachiopoda		Lamp shells

The animal kingdom is incredibly diverse. It is not yet clear how big this kingdom really is, but it certainly covers millions of species. All its members are multicellular eukaryotic organisms that ingest their food. Animal cells have mitochondria and a nuclear envelope, but lack a cell wall and chloroplasts. Most animals reproduce sexually.

There are two subkingdoms in the animal kingdom: parazoa and eumetazoa. Parazoa include the phylum porifera (sponges), organisms that lack internal specialization. Eumetazoa are more specialized. Organisms in this subkingdom have cells which are organized in a way that gives them a definite shape and symmetry. Some even have a developed nervous system.

In case you're wondering, humans are classified as follows: animal kingdom, phylum chordata, class mammalia, order primate, family hominidae, genus *Homo*, species *Homo sapiens*.

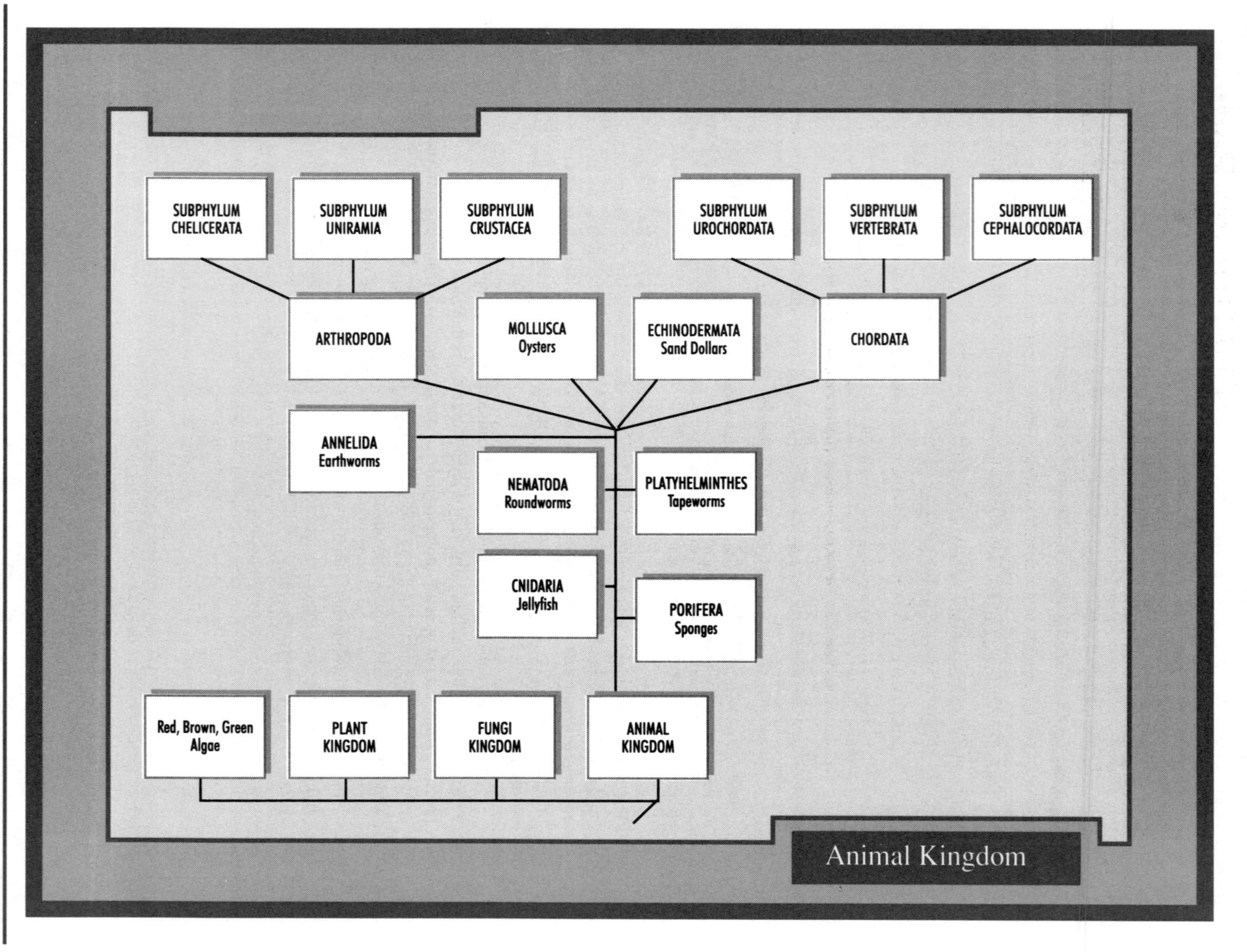
SUBPHYLUM CHELICERATA
SUBPHYLUM UNIRAMIA
SUBPHYLUM CRUSTACEA
SUBPHYLUM UROCHORDATA
SUBPHYLUM VERTEBRATA
SUBPHYLUM CEPHALOCORDATA
ARTHROPODA
MOLLUSCA Oysters
ECHINODERMATA Sand Dollars
CHORDATA
ANNELIDA Earthworms
NEMATODA Roundworms
PLATYHELMINTHES Tapeworms
CNIDARIA Jellyfish
PORIFERA Sponges
Red, Brown, Green Algae
PLANT KINGDOM
FUNGI KINGDOM
ANIMAL KINGDOM
Animal Kingdom

4.2 The Human Body

An Aid to Understanding

Anatomy	The study of structure and arrangements of body parts in organisms.
Anterior	Situated toward the front of the body. (The lips are anterior to the teeth.)
Caudal	Interchangeable with inferior. Toward the lower part of a structure. (In animals, toward the tail.)
Cranial	Interchangeable with superior. Pertaining to the skull, or toward it.
Deep	Toward the interior, away from the surface. (Bones are deep to the skin.)
Distal	Farther away from the point of origin or attachment. (The toes are distal to the ankle.)
Dorsal	Toward the back or behind in humans; in animals, the back. (The spine is dorsal to the chest.)
Inferior	Toward the lower part of, or below a structure. (The neck is inferior to the head.)
Intermediate	Between a medial and a lateral structure. (The middle ear is intermediate to the inner ear and outer ear.)
Lateral	Away from the midline, toward the outside. (The ears are lateral to the skull.)
Medial	Toward the midline, inward. (The brain is medial to the eyes.)
Midline	An imaginary vertical line, at the center of the body.
Physiology	The study of bodily functions.
Posterior	Toward the back or behind. (The throat is posterior to the tongue.)
Proximal	Closer to the point of origin or attachment. (The ankles are proximal to the toes.)
Superficial	Near, or at the surface. (The skin is the superficial tissue layer of the body.)
Superior	Toward the upper part of, or above a structure. (The head is superior to the feet.)
Ventral	Toward the front, or in front of, in humans; in animals, the belly. (The chest is ventral to the spine.)

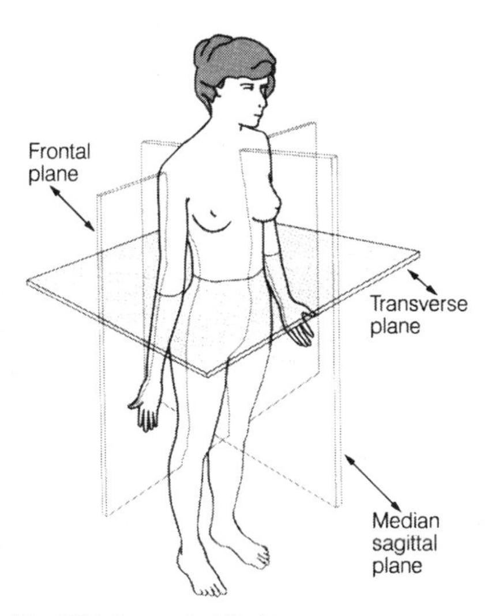

Fig. 4.2.1 Anatomical Position

When using the above terms, it's important to visualize the body standing toward you with arms to the sides, palms facing you (thumbs out). This is known as the **anatomical position**. When using the above terms, remember that directions are quoted from the anatomical position. Anterior is the front of the body, posterior is the back. This is true whether the person is facing you or not.

Ask students how many senses we have and they'll probably say "five." There are many more than this, and they are grouped into two broad types: **special senses** and **general senses**. Special senses include taste, smell, hearing, equilibrium and vision. Sensory organs (such as the eye and ear), or localized sensory tissue (responsible for taste and smell) enable us to perceive sensations.

Taste cell receptors on the tongue (tastebuds) sense each taste as a combination of sweet, sour, salty and bitter.

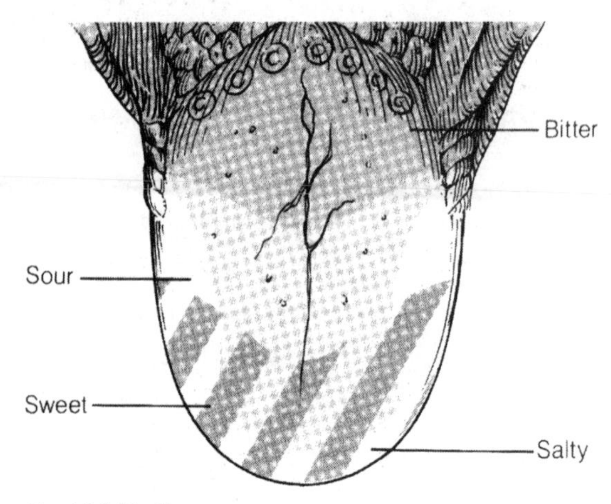

Fig. 4.2.2 The Tongue

High up in each nostril, the **olfactory epithelial membrane** contains olfactory receptor cells that endow us with the sense of smell.

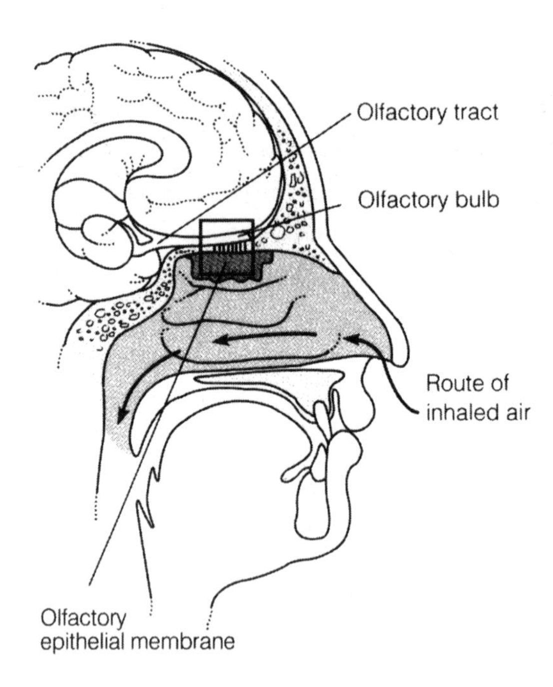

Fig. 4.2.3 The Nose

The ear receives physical vibrations (moving airwaves) which the brain interprets as sound. Our sense of hearing lies in the perception of sound. The ear is divided into three areas: outer, middle and inner. The outer ear traps sound and conducts it through the auditory canal to the **tympanic membrane** (ear drum). The middle ear consists of a series of three tiny bones: **malleus** (hammer), **incus** (anvil), and **stapes** (stirrup). These bones conduct vibrations to the **cochlea** in the inner ear. Here, vibrations are translated into neural messages that are passed on to the brain through the **auditory nerve**.

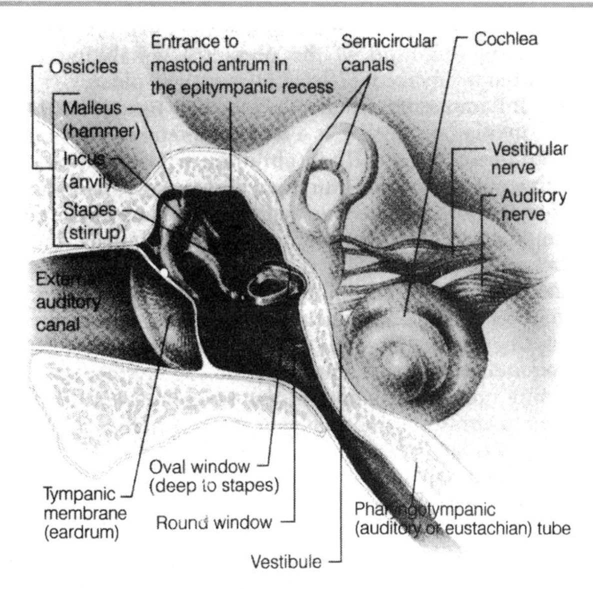

Fig 4.2.4 The Ear

Equilibrium (the sense of balance and changes in motion) originates in the **semicircular canals** within the inner ear, attached to the cochlea.

The **eye** is a complex structure that focuses light through the **lens** onto **photoreceptors** in the **retina**, a light-sensitive layer of cells.Two types of photoreceptors are found in the eye: **rods** and **cones**. Rods detect light intensity, aiding in night vision. Three types of cones densely packed at the **fovea** detect color. The **optic disk** is the spot where the optic nerve exits the eye. Because no photoreceptors are found there, it is also known as the **blind spot**.

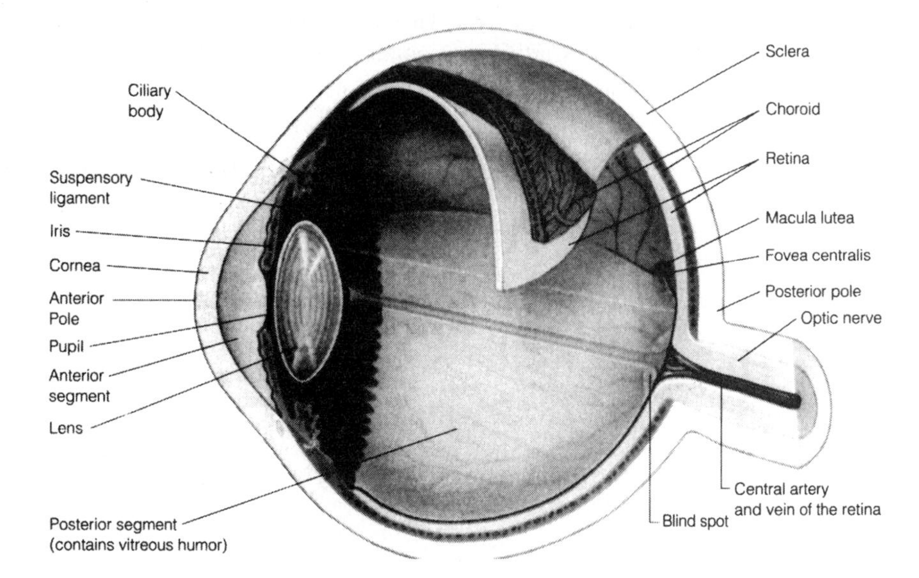

Fig. 4.2.5 The Eye

General sensory receptors are either **free** (naked) nerve endings, or **corpuscular** (covered) nerve endings. Free nerve endings detect pain, touch, heat, and cold. Corpuscular receptors detect a wide variety of stimuli such as pressure, vibration, temperature, muscle stretch, tendon stretch, joint position and joint motion.

The **human skeleton** is made up of more than **bones**. It's also made up of **cartilage**, **joints** and **ligaments**. Without a skeleton, you'd be a helpless blob of jelly, unable to get out of bed. (Be alarmed if this describes you.)

There are 206 bones in the human body. Contrary to popular belief, males and females have the same number of bones, although the pelvis differs between them. Along with providing support and protection, bones allow us to move. Cavities within the bone, filled with **marrow**, both store fat and serve as the source of the body's blood cells. Spaces in bone material contain cells that maintain calcium levels in bodily fluids.

Cartilage is a tough, flexible tissue resistant to wear and tear. There are three types of cartilage: **hyaline**, **fibrocartilage** and **elastic cartilage**.

Hyaline cartilage is found in the larynx, trachea and broncheal tubes; it supports the nose, connects the ribs to the sternum, and covers the tips of bones in joints. It is white in color. Fibrocartilage makes up invertebral disks and is found in the knee; it is grainy and resists tension. As the name implies, elastic cartilage stretches. It is yellow in color and found in the ear and epiglottis.

A **joint** is the juncture between bones. Joints secure bones, and allow us to move. There are two ways to describe joints: **functionally** or **structurally**. Functional classification includes **synarthrotic** (immovable), **amphiarthrotic** (slightly movable) and **diarthrotic** (freely movable) joints. Structural classification includes: **fibrous**, **cartilaginous** and **synovial** joints.

Fibrous joints lack a joint cavity and allow little or no movement; they connect bones with fibrous tissue. Cartilaginous joints also lack a joint cavity; they join bones with cartilage. Synovial joints are separated by a joint cavity filled with **synovial fluid** which allows them to move freely.

Ligaments are fibrous tissues that connect the bones. **Tendons** attach skeletal muscle to bones.

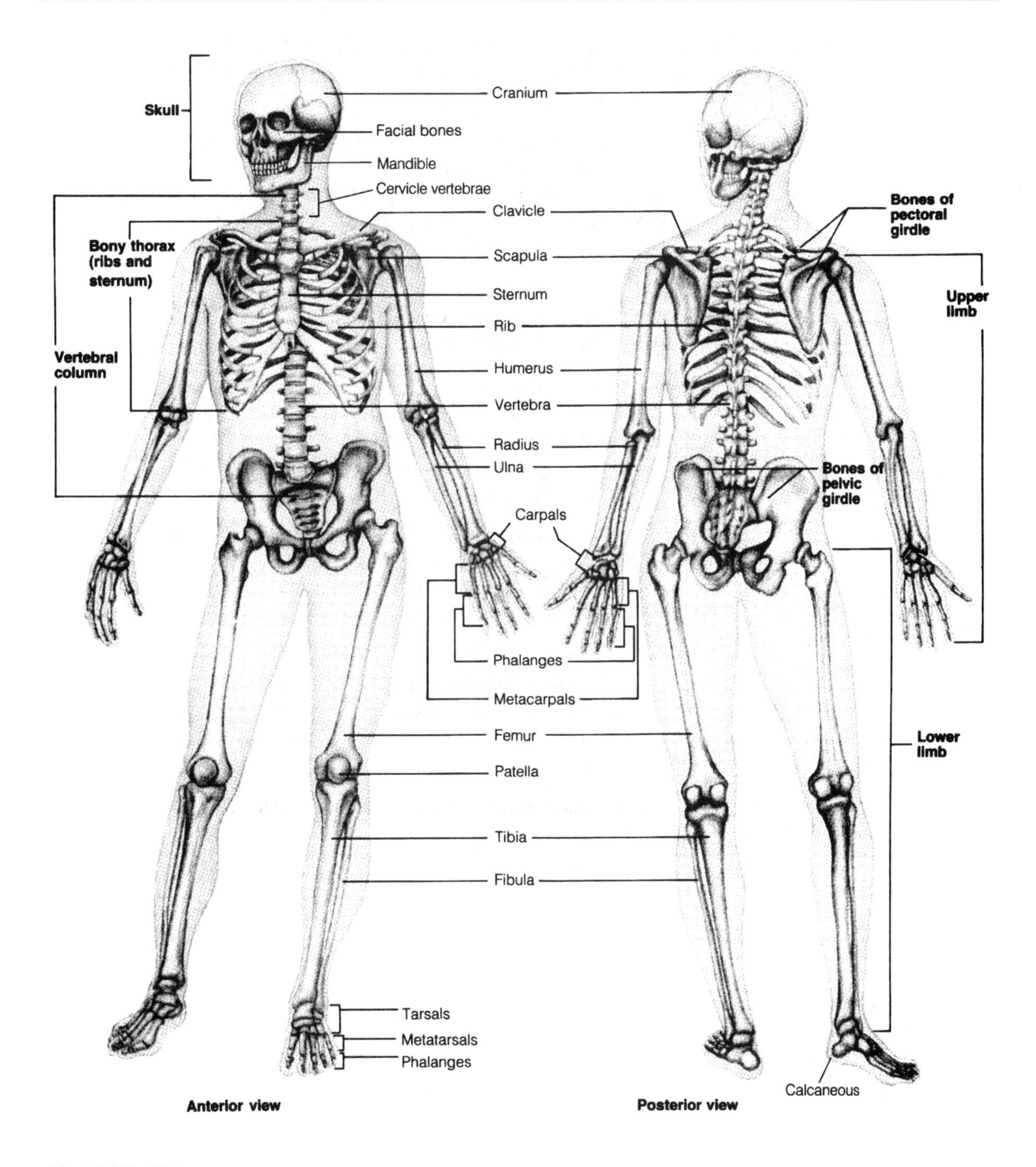

Fig. 4.2.6 The Skeleton

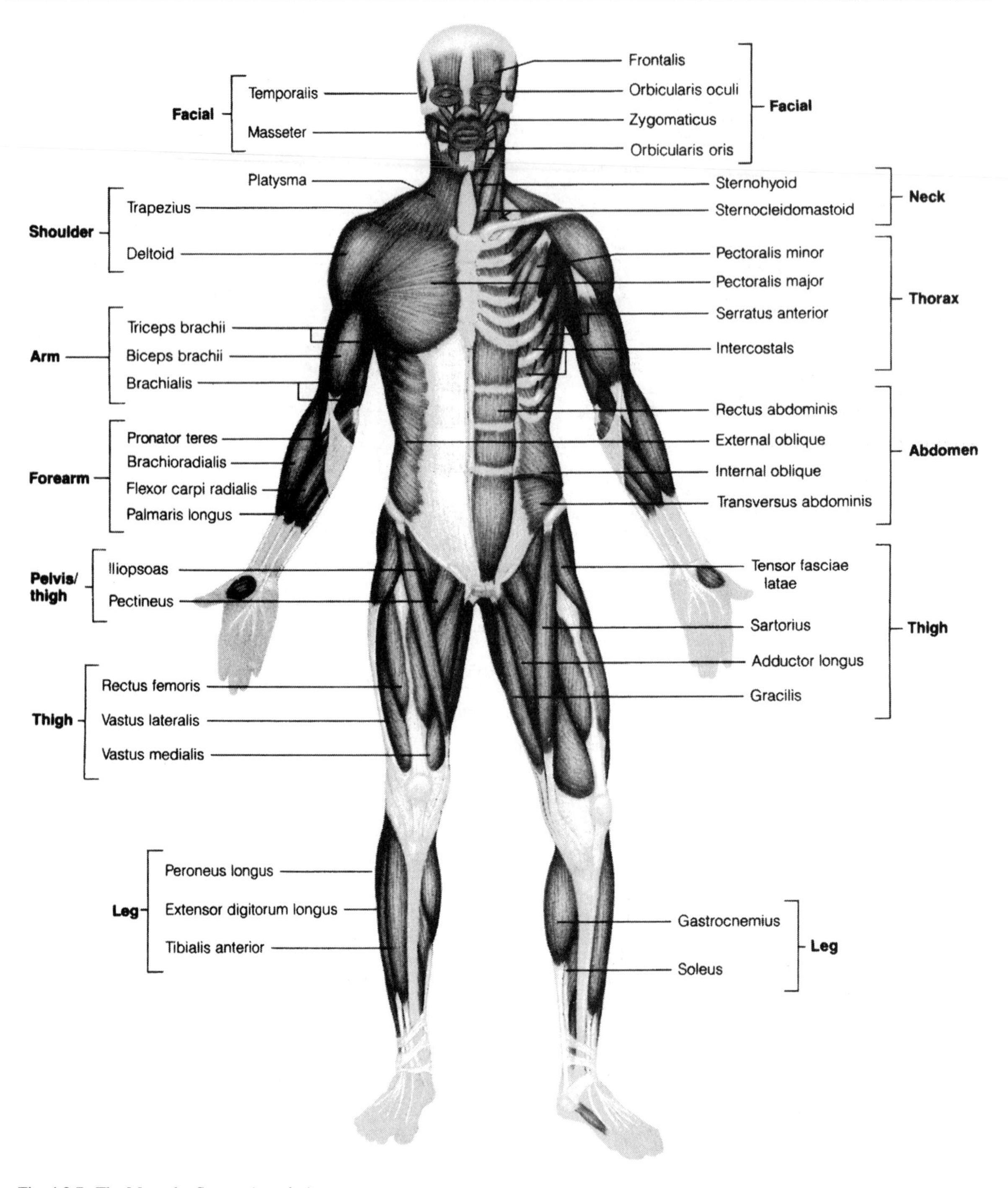

Fig. 4.2.7a The Muscular System (anterior)

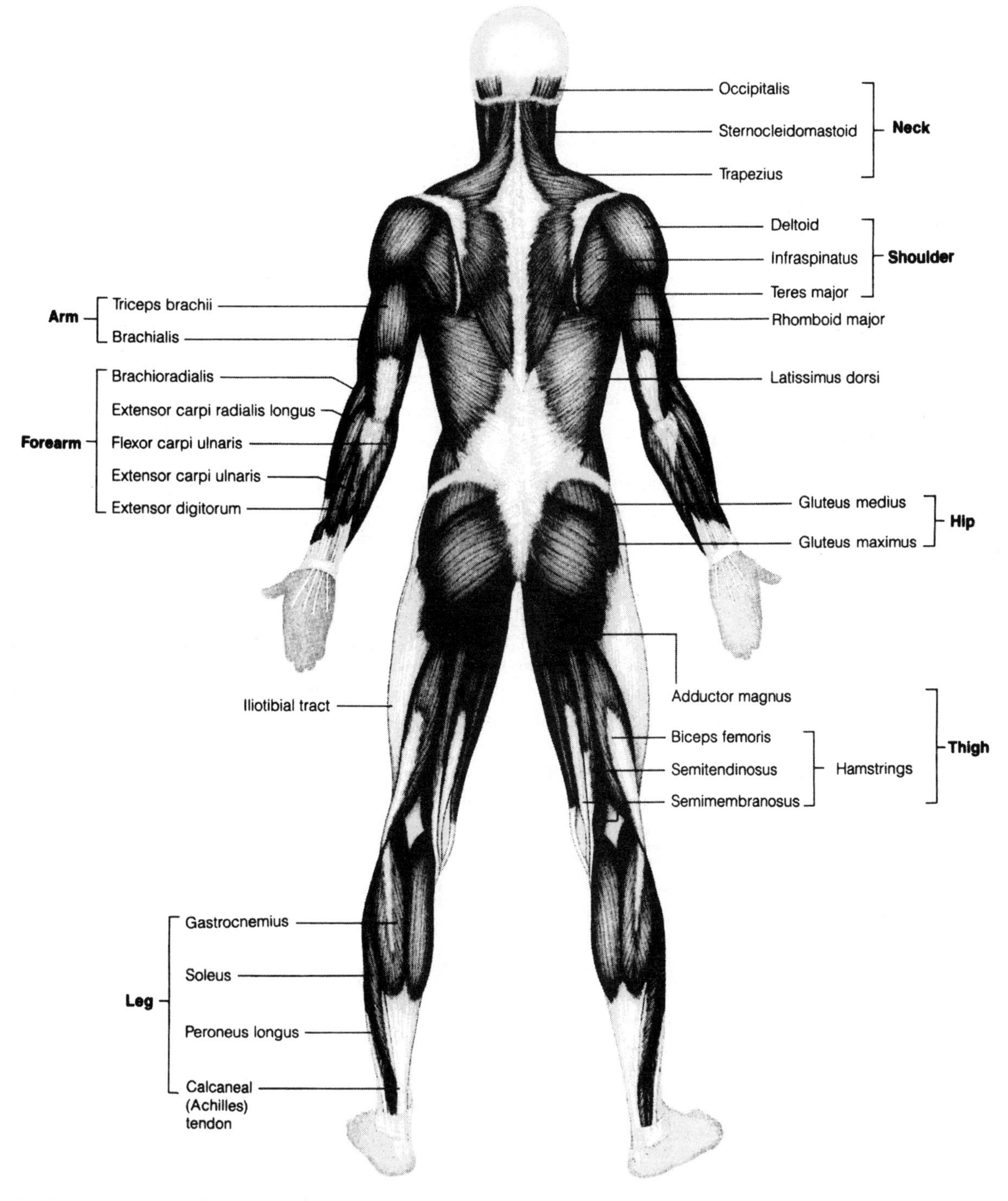

Fig. 4.2.7b The Muscular System (posterior)

The **muscular system** gives us the ability to move and control posture. It also moves things through our bodies (nutrients, vital fluids). This movement requires energy. Energy, in turn, generates heat and metabolic waste.

The body has three types of muscle tissue: **skeletal**, **smooth** and **cardiac**. Skeletal muscles are action muscles; they move bones or skin, and also help maintain body temperature. They are activated by the somatic nervous system, which means that they are under **voluntary** control. When motor neurons (nerve cells) release the neurotransmitter acetycholine (ACh), this causes the depolarization of muscle fiber, which in turn, stimulates the muscles to contract. During contraction, thick filaments composed of myosin, crossbridge with thin filaments containing actin, tropomyosin and troponin. This shortens the length of the **sarcomere**, reducing the distance between Z-lines. The sarcomere must contain enough calcium ions (Ca^{2+}) and ATP for contraction to occur. Skeletal muscle **fibers** (cells) are long, cylindrical and multinucleate with dark bands called **striations**. Because the fibers are intricately wrapped in connective tissue, skeletal muscle is quite strong.

Smooth muscle is found mainly in the walls of internal organs, including veins and arteries where it moves fluids and solids through bodily "plumbing." Smooth muscle is activated and deactivated by the autonomic nervous system, therefore is **involuntary**. Unlike skeletal contractions which are all or none (on or off), smooth muscle contractions are gradual and slow, according to the degree of depolarization. Smooth muscle fibers are small, spindle shaped, and have no striations.

Cardiac muscle pumps blood from the heart throughout the body by means of contraction. Each heartbeat is initiated by the **sinoatrial (SA) node** (pacemaker) found on the posterior wall of the right atrium. The **atrioventricular (AV) node** briefly delays the transmission so that the atria can empty. It then spreads this signal throughout the ventricle walls, stimulating heart contraction. These contractions take place without neural stimulation. Although considered to be involuntary, cardiac muscle can speed up or slow down contractions, depending on arousal level. Arousal level is governed by the autonomic nervous system's vagus nerve. Cardiac fibers are short, fat, branched, striated and have one or two nuclei.

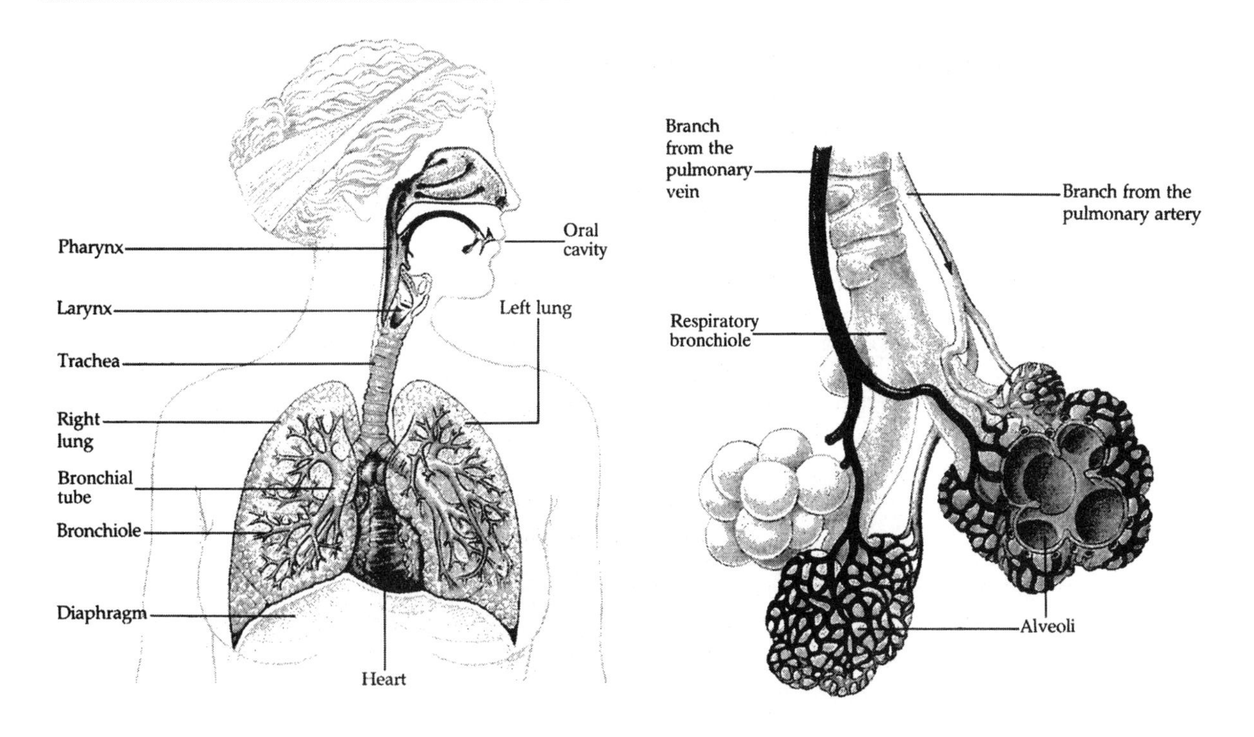

Fig. 4.2.8 The Respiratory System

The **respiratory system** exchanges gases in the atmosphere with gases in the body. It is divided into two zones: the **conducting zone** and the **respiratory zone**. The conducting zone is like a set of pipes. Because no gas exchange occurs there, it's sometimes called dead air space. The conducting zone includes the **nasal cavity** (nose), **oral cavity** (mouth), **pharynx** (throat), **larynx** (voicebox), **trachea** (windpipe) and two bronchial tubes leading to each **lung.**

Each **bronchial tube** branches into smaller and smaller passages until the air reaches the respiratory zone at the **respiratory bronchioles**. The respiratory bronchioles are made of **alveolar ducts**, which bring air to the **alveolar sacs**. Alveolar sacs are made up of small chambers, or alveoli. Picture an alveolar sac as a bunch of grapes (only smaller). In this comparison, each alveole is like an individual grape. This is where gas exchange occurs across a **respiratory membrane** made of squamous epithelial cells. When blue (deoxygenated) blood passes along one side of this membrane, carbon dioxide (CO_2) is released from **hemoglobin** in red blood cells and is replaced with oxygen (O_2), giving blood its distinctive red color. The hindbrain (medulla and pons) carefully monitors how CO_2 affects the blood's acidity levels. When CO_2 levels are high, blood is more acidic, and the breathing increases to compensate.

The **digestive system** can be divided into two main groups of organs: the **gastrointestinal tract** (GI, also known as the alimentary canal) and **accessory digestive organs**. The GI tract is a long, winding muscular tube that connects the **mouth** to the **anus**. It is made up of several organs: **mouth**, **pharynx**, **esophagus**, **stomach**, **small intestine** and **large intestine** (which ends in the anus).

The GI tract digests (breaks down) food, absorbs the nutrients and passes them on to the circulatory system. Most nutrients are absorbed through the small intestine, while remaining water is absorbed through the large intestine. The epiglottis prevents food from entering the lungs.

Accessory digestive organs help the digestion process. These organs include the **tongue**, **teeth**, **salivary glands**, **pancreas**, **liver** and **gall bladder**.

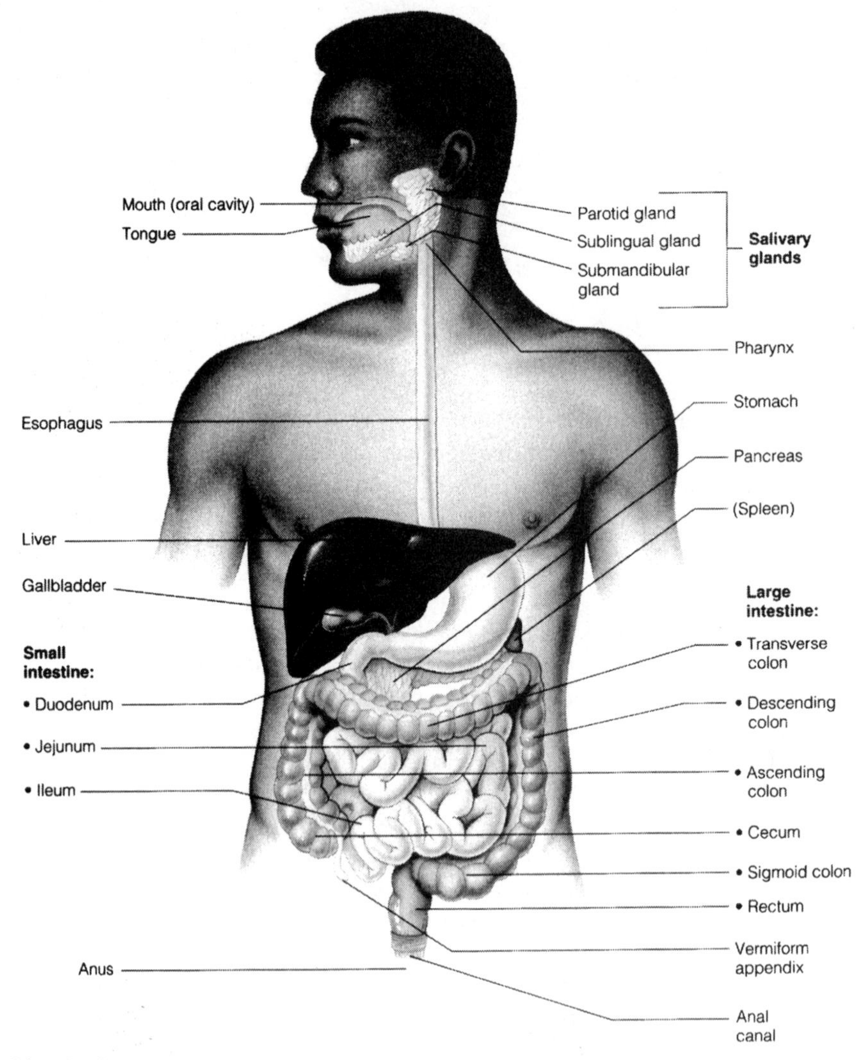

Fig 4.2.9 The Digestive System

The **urinary system** is much like a city's water purification system: it cleans the body of nitrogen-containing wastes, and maintains the composition, pH and volume of the blood. The two **kidneys** act like purification plants, receiving blood from blood vessels, filtering it, and putting it back into circulation. The **nephron** is the functional unit of the kidney. The wastes collected from the blood, pass through the **ureters** to the **bladder** for storage. The bladder is an expandable muscular sac that holds urine, and passes it on to the **urethra**. The urethra is a conduit for secreting urine.

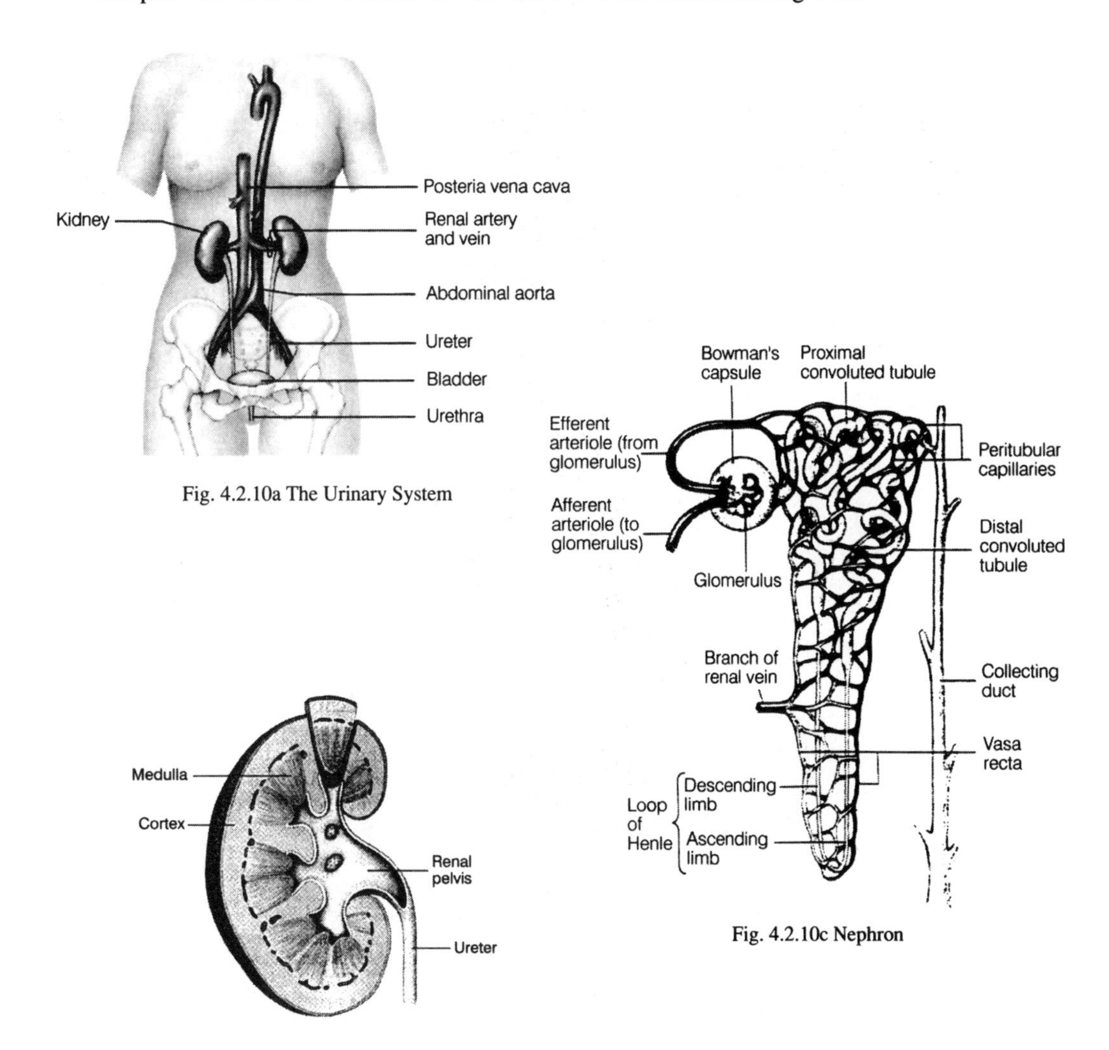

Fig. 4.2.10a The Urinary System

Fig. 4.2.10c Nephron

Fig. 4.2.10b Kidney Cross-Section

We think of the **reproductive system** as a baby production center. Male and female gametes have to be ready at the same time to unite and form a zygote (baby's first cell). The male system is designed to be fertile 100% of the time. It produces and secretes male gametes (sperm) in a fluid called semen. The female system is designed to produce female gametes (eggs), receive sperm for fertilization and house the developing offspring until ready for birth. The female produces an egg that is viable for less than a day during her reproductive cycle (which lasts about a month); however, a female can keep sperm alive for up to five days.

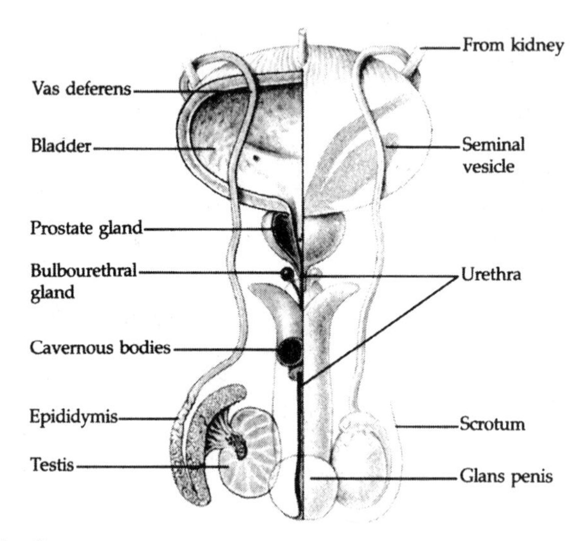

Fig. 4.2.11a Male Reproductive System

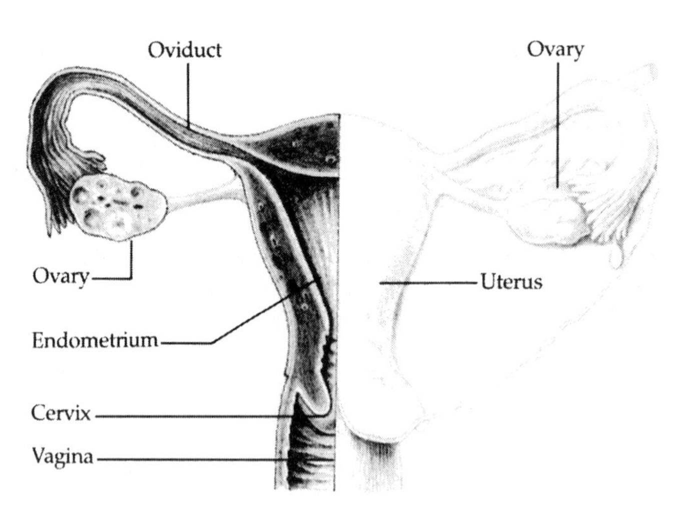

Fig. 4.2.11b Female Reproductive System

The **circulatory system** is made of the **heart, blood vessels** and **blood.** The heart is the pump that moves the blood throughout the body. About the size of a fist, the heart is made up of cardiac muscle. It has four valves by which it controls blood flow.

Blood vessels are the "pipes" through which blood moves. **Arteries** take blood away from the heart to smaller **arterioles**, and then to tiny **capillaries** which nourish cells. Capillaries are so small that individual blood cells must pass through "in single file," allowing gas exchange (O_2– CO_2) to occur. Blood then begins its journey back to the heart, starting in **venules** (smaller veins), and on to **veins**, which complete the circuit.

Blood is made of two components: a liquid medium called **plasma**, and a collection of cells. Plasma consists of water, inorganic salts, proteins, nutrients, hormones, antibodies and wastes. The cellular component of blood is made up of **erythrocytes** (red blood cells), **leukocytes** (white blood cells) and **thrombocytes** (platelets). Red blood cells are numerous (4.5-5 million/mL blood). Mature red blood cells contain no nucleus and are involved in gas exchange.

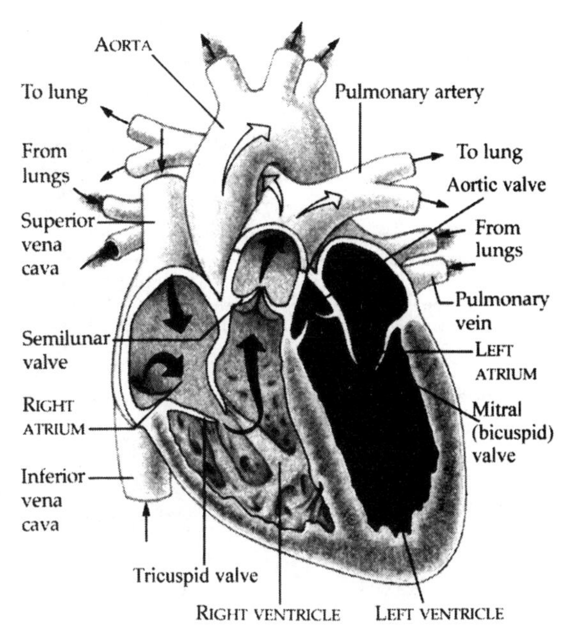

Fig. 4.2.12a The Heart

White blood cells come in many forms and play an integral role in the immune system. White blood cells are larger in diameter than red blood cells (10-20 µm vs. 7 µm), but red cells outnumber white cells (700:1). Platelets are cellular fragments involved in blood clotting and are not normally considered cells.

The **lymphatic system** aids both the circulatory and immune systems. It's a network of vessels, nodes and organs that replenishes fluids in the circulatory system, and houses immune cells (phagocytes and lymphocytes). The lymphatic fluid in the system is called **lymph.**

The **immune system** protects the body from the outside world. It's a collaboration of two lines of defense: **specific** and **nonspecific.** Specific defenses identify an antigen (foreign body) and design antibodies (specific defense proteins) to trap them.

Nonspecific defenses include physical barriers such as skin and mucous membranes. Physical processes also help. Inflammation, for example, increases blood flow so healing can occur. The immune system can also call on chemical defenses: stomach acids, lysozomes and antimicrobial proteins such as interferon, which infected cells release to prevent further infection. White blood cells (phagocytes and natural killer cells) also help the immune system by engulfing and digesting invading bacteria.

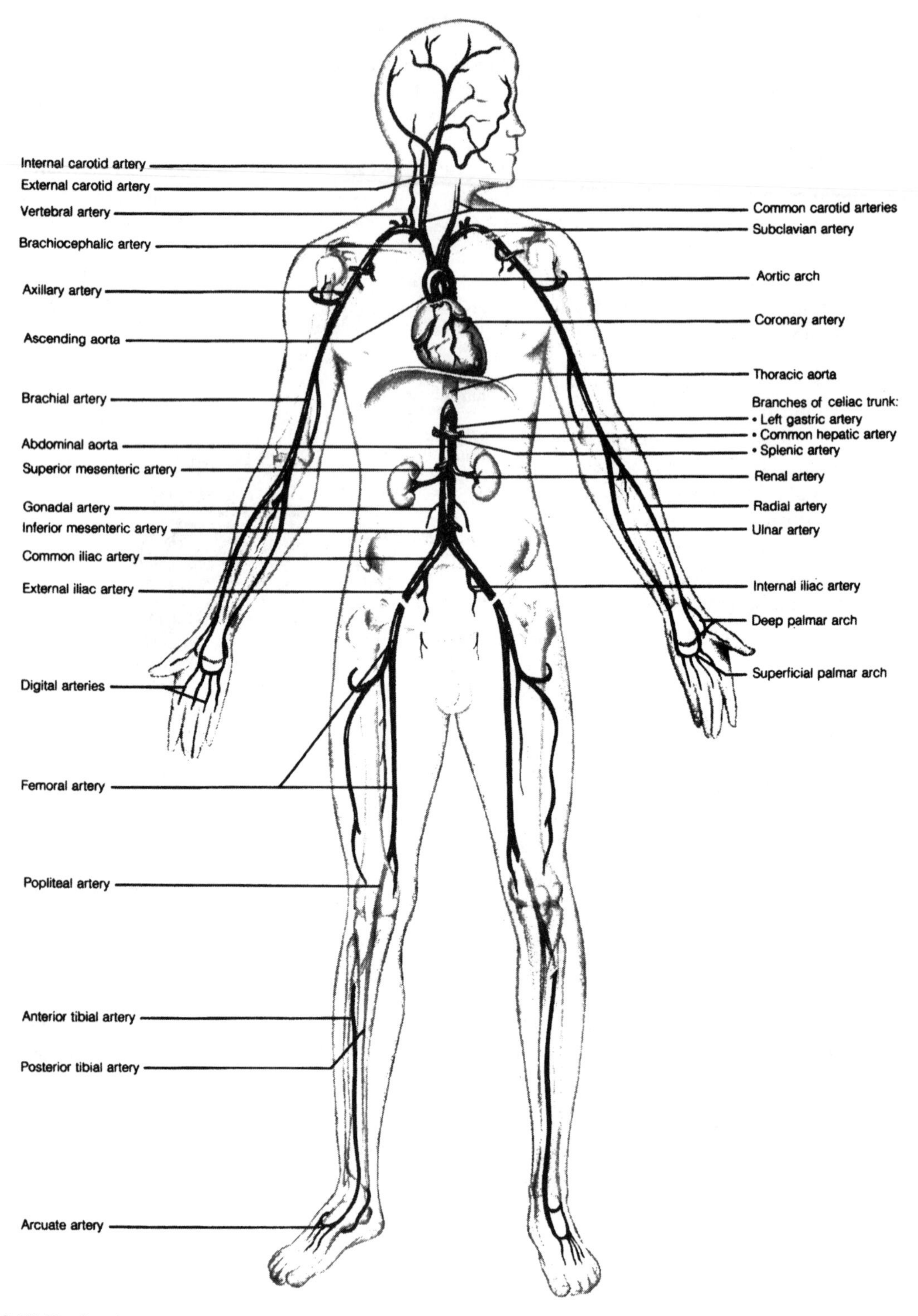

Fig. 4.2.12b The Arteries

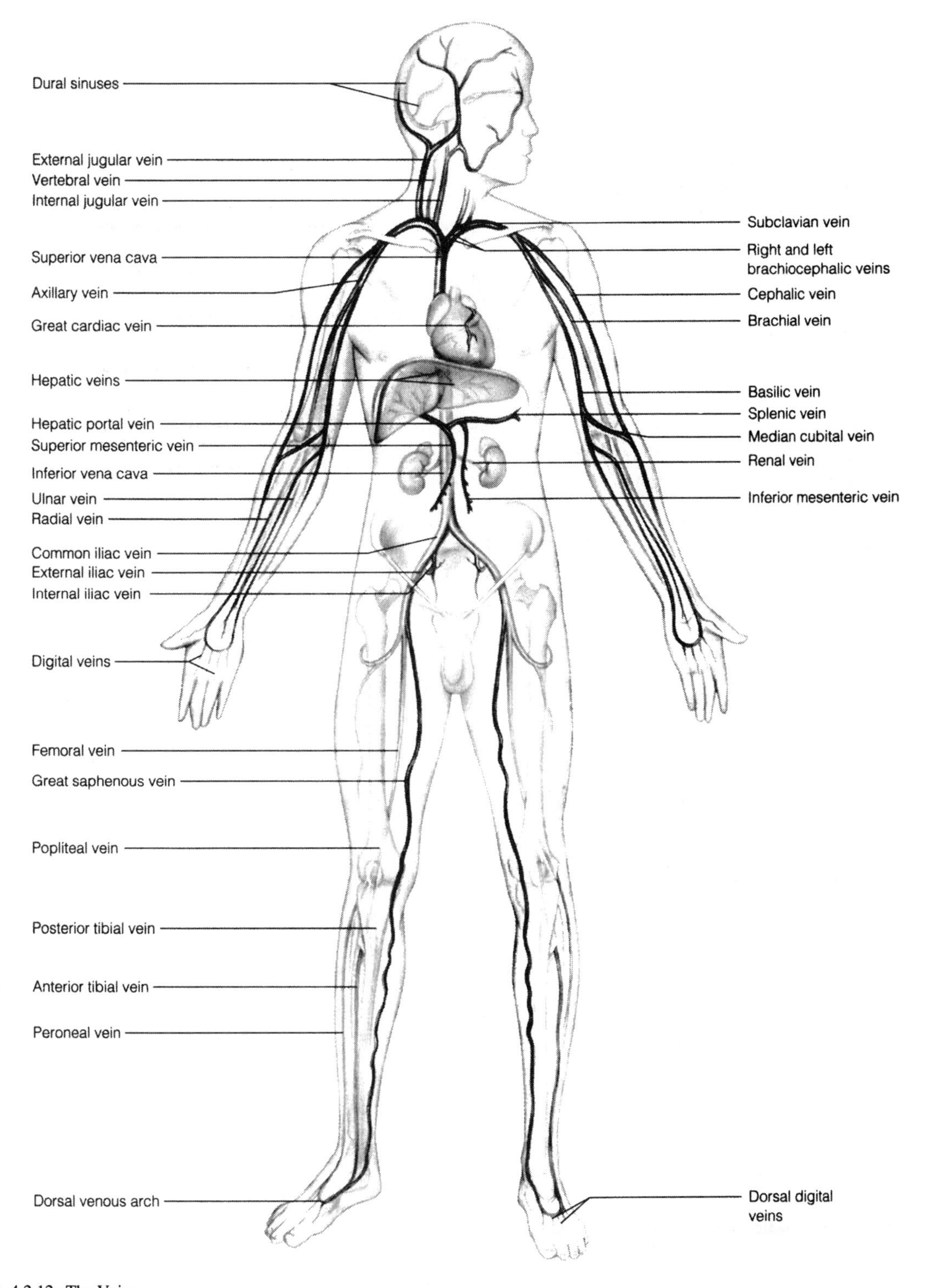

Fig. 4.2.12c The Veins

The nervous and endocrine systems are the body's control centers. The **endocrine system** acts like a post office, sending **hormones** through the bloodstream to control specific organs. It's made up of three types of endocrine tissue: small ductless glands (**pituitary**, **thyroid**, **parathyroid**, **adrenal**, **pineal** and **thymus**), organs with discrete patches of endocrine tissue (**pancreas** and **gonads**), and organs with few hormone-producing cells (small intestine, stomach, kidneys, heart and placenta). The endocrine system secretes hormones into the circulatory and lymphatic systems. Hormones regulate growth and development, reproduction, metabolism, electrolyte nutrient and water balance.

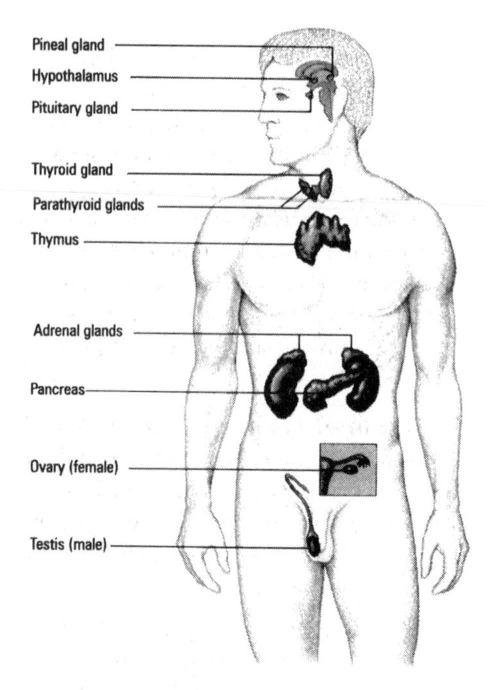

Fig.4.2.13 The Endocrine System

The **nervous system** acts like a highly complex electrical circuit (actually **electrochemical**). The nervous system can be divided into two subsystems: the **central nervous system (CNS)** and the **peripheral nervous system (PNS)**. The CNS consists of the brain and spinal cord. The CNS processes, interprets and stores incoming information. Decisions are made in the CNS and orders are sent through the PNS.

The PNS represents all nervous tissue that is not part of the CNS. The PNS is divided into the **somatic** and **autonomic (ANS)** nervous systems. The **somatic** nervous system represents all **voluntary** controls (via motor neurons) and sensory inputs (via sensory neurons). The **autonomic** system represents involuntary controls, and is further subdivided into the **parasympathetic** and **sympathetic** systems. Sympathetic systems are involved with arousal (pupil dilation, increase in heart and breathing rates, etc.). Parasympathetic systems are involved in relaxation (pupil constriction, decreases in heart and breathing rates, increase of blood flow to digestive tissue, etc.).

The functional unit of the nervous system is the **neuron** (nerve cell). There are three types of neurons: **afferent** (sensory), **efferent** (motor) and **associative** (interneurons). Afferent neurons receive inputs from sensory receptors. Efferent neurons transmit information to muscles and glands. Interneurons within the CNS link afferent and efferent neurons.

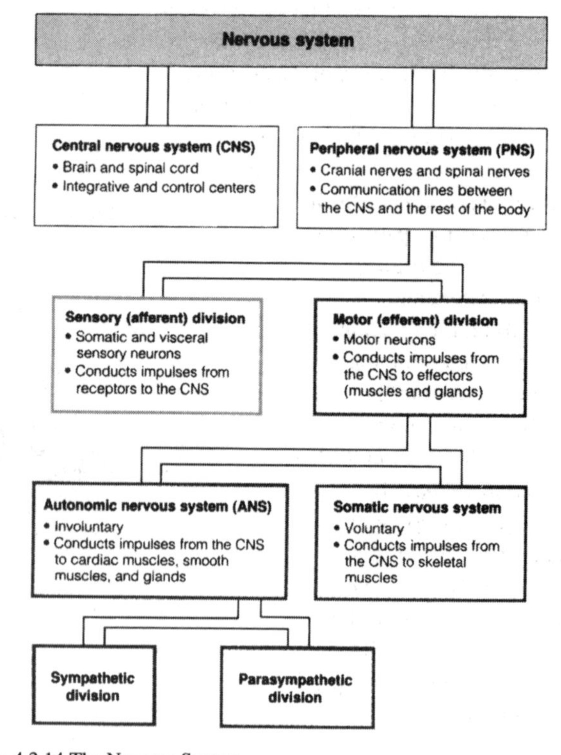

Fig. 4.2.14 The Nervous System

Most neurons have three main parts: **dendrites** (sensory tree), **soma** (cell body) and **axon** (tail), ending in synaptic terminals. Dendrites receive information from chemical messengers called **neurotransmitters.** When stimulated to a sufficient degree, an action potential (neural impulse) races down the axon. This causes the synaptic terminals to release neurotransmitters to the next neuron.

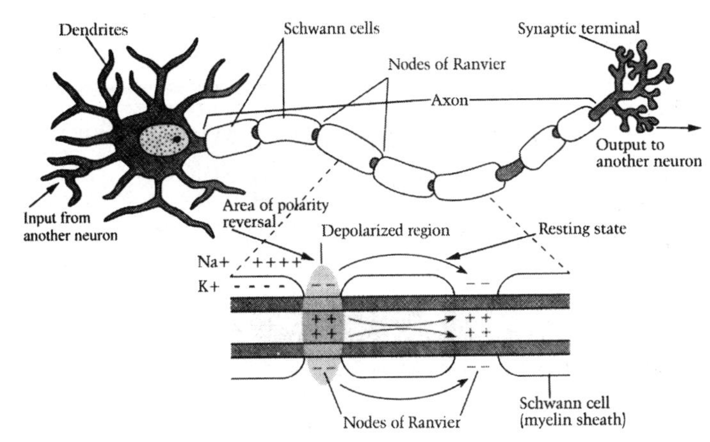

Fig. 4.2.15 The Neuron

The **brain** has been called "the seat of reason." Without a brain you wouldn't be reading this. Composed of about 100 billion neurons, the brain is the center of the CNS. The brain processes information in highly specialized locations.

There are three main parts of the brain: the **hindbrain**, the **midbrain** and the **forebrain**. The hindbrain includes the **medulla, pons, cerebellum** and **reticular activating system**. The hindbrain works to maintain life functions. The midbrain is made up of "neural highways" upon which information is transmitted to and from the forebrain. Made up of the **cerebral cortex** (outer covering), **thalamus, olfactory bulb, hypothalamus, pituitary** gland, **basal ganglia, hippocampus** and **amygdala**, the forebrain governs higher mental functions. Within the forebrain the cerebral cortex is divided into **left** and **right cerebral hemispheres,** joined by a neural bridge called the **corpus callosum**. The cerebral cortex is further divided into four **lobes: frontal, parietal, occipital and temporal.**

Primary motor area
Primary sensory area
Frontal lobe
Parietal lobe
Frontal association area
Somatic association area
Speech (Wernicke's area)
Reading
Speech (Broca's area)
Hearing
Olfaction
Visual association area
Temporal lobe
Smell
Auditory association area
Occipital lobe

Fig. 4.2.16 The Cerebral Cortex

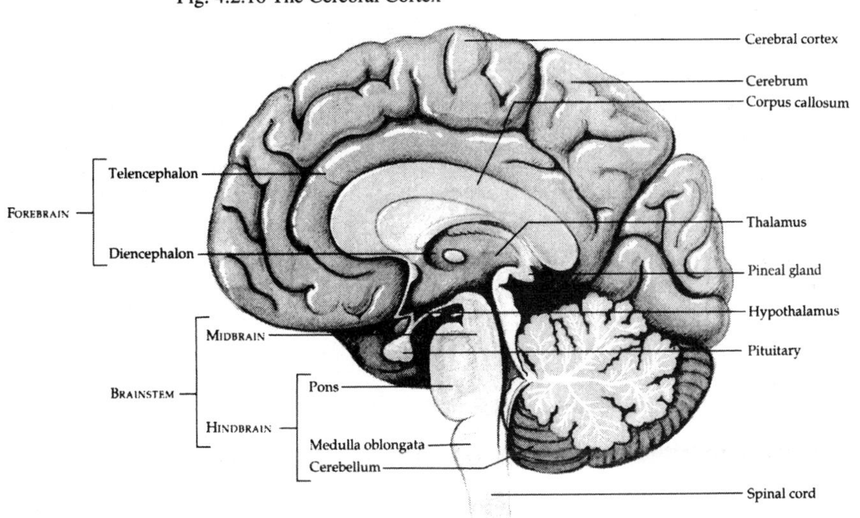

Fig. 4.2.17 The Brain (midsagittal)

4.3 Ecological Cycles

An Aid to Understanding

Biomass	The total amount of living material in a region.
Community	All the organisms that have a potential for interaction in a habitat.
Consumer	Organisms that eat other organisms for food (a type of heterotroph).
Decomposer	An organism that eats dead organisms and organic wastes (detritovore, a type of heterotroph).
Ecosystem	A community of organisms and their environment.
Eutrophic	Having an excess supply of a nutrient in water; occurs naturally or as a result of pollution.
Food web	The relationship of "what eats what" in an ecosystem.
Limiting factor	An agent whose presence (or absence) affects growth within an environment. (Example: phosphates in a lake.)
Omnivore	Animals that eat both plants and animals.
Producer	Autotroph. An organism that makes its own food from energy and chemicals in the environment. Photoautotrophs (plants, algae) are the most common form, getting their energy from sunlight.
Trophic levels	Feeding levels in an ecosystem.

Ecology is the study of relationships among organisms and the environment. Energy and chemicals within the environment are passed from organism to organism through complex feeding patterns. Biologists classify these feeding patterns as **trophic levels**. Trophic levels are broken down into two major levels: producers and consumers.

Producers are ranked at the first trophic level because they supply all other levels directly or indirectly with nourishment (energy and chemicals). Producers are organisms that can trap energy from the environment and store it chemically.

The second trophic level includes consumers, and is subdivided into primary, secondary and tertiary consumers. **Primary** consumers such as herbivores eat only producers, getting their nourishment from the first level. **Secondary consumers** get their nourishment from primary consumers (a carnivore that eats an herbivore). **Tertiary** consumers receive nourishment from secondary consumers (a carnivore that eats another carnivore).

Decomposers are nature's recyclers. They break down dead matter and organic waste created in the trophic levels. They return chemicals to the environment, where they enter the cycle again.

Feeding patterns in an ecosystem are highly complex and form an intricate pattern called a **food web**. Organisms often feed at more than one trophic level. Omnivores, for example, eat plants (producers) and animals (consumers). Some protists are both photoautotrophs and consumers.

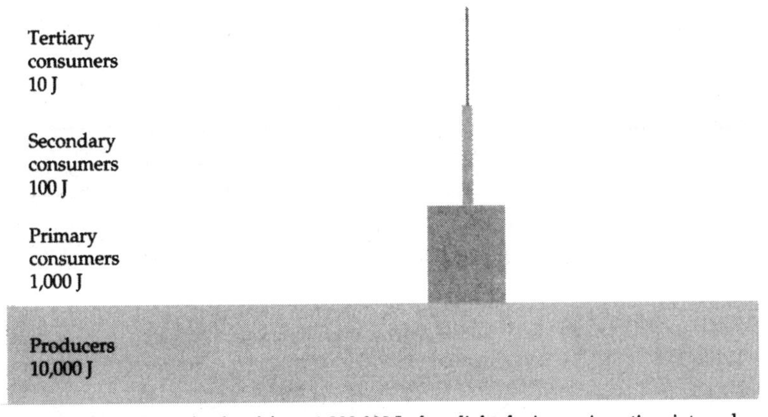

Energy at each trophic level from 1,000,000 J of sunlight during a given time interval

Fig. 4.3.1 Energy Pyramid

All chemicals within organisms cycle throughout ecosystems. **Biogeochemical** cycles trace the path of chemicals as they cycle through organisms and the environments in which they live.

The water cycle, also known as the hydrologic cycle, describes the circulation of water (H_2O) in nature. Over the oceans, **evaporation** is greater than **precipitation** (rain or snow), leading to an excess of water in the air. The wind blows this moisture-laden air over the land where it precipitates. Because precipitation over land exceeds evaporation and **transpiration** (water lost from plants), water flows back to the ocean through surface water runoff (rivers, streams) and groundwater. Water is very abundant on Earth, covering about 75% of the surface. Two percent of all water is trapped: frozen, locked in soil and integrated within organisms.

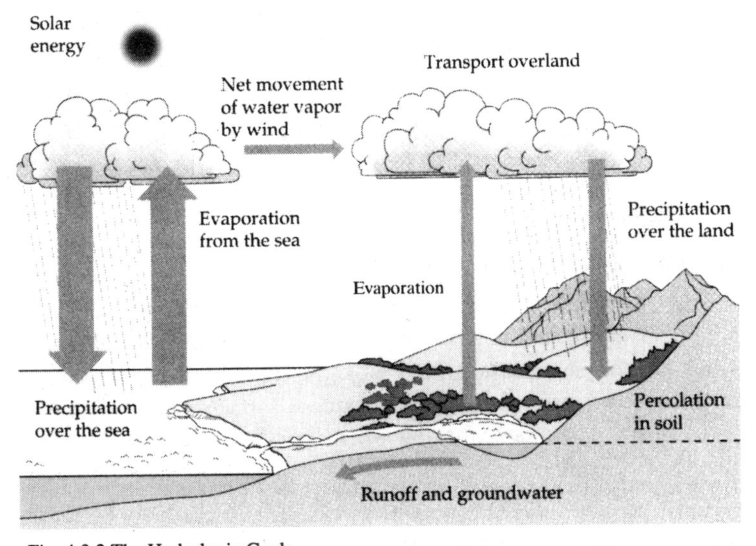

Fig. 4.3.2 The Hydrologic Cycle

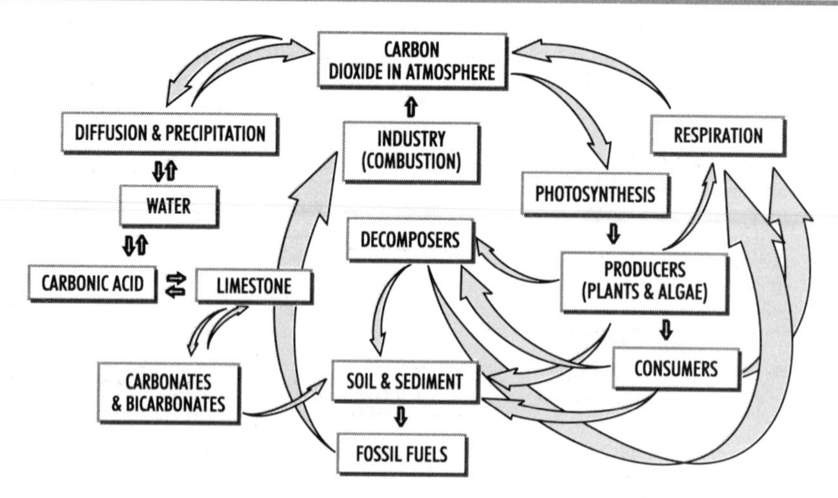

Fig. 4.3.3 The Carbon Cycle (The size of the arrows is not indicative of amount.)

The **carbon cycle** describes the circulation of carbon-containing molecules in nature. The major source of carbon in the atmosphere is in the form of carbon dioxide gas (CO_2). Photosynthetic organisms (plants) take carbon dioxide from the atmosphere and through the process of carbon-fixation incorporate it into living matter (biomass). When animals eat plants, they add this carbon to their biomass. Through the metabolic process of respiration, both plants and animals convert carbon (in organic molecules) back to carbon dioxide. This CO_2 is released back to the atmosphere.

Respiration and photosynthesis occur almost everywhere on Earth. After consuming organic waste and dead organic matter, decomposers in the soil release carbon dioxide back into the atmosphere through respiration. When carbon dioxide dissolves in water, some of it reacts chemically to form carbonic acid (H_2CO_3). Carbonic acid in turn, reacts with limestone ($CaCO_3$) to form carbonates (CO_3^{2-}) and bicarbonates (HCO_3^-):

i) $H_2CO_3 + CaCO_3 \rightleftarrows Ca(HCO_3)_2 \rightleftarrows Ca^{2+}\ 2HCO_3^-$

ii) $2HCO_3^- \rightleftarrows 2H^+ + 2CO_3^{2-}$

Bicarbonate Carbonate

Carbonates and bicarbonates store carbon in water.

In nature, respiration and photosynthesis are closely balanced. Human activities are throwing this balance off. For millions of years, vast amounts of CO_2 remained inert, locked within oil, coal and natural gas. When these fossil fuels are burned for energy, they release CO_2 at a rate too great for photosynthesis to absorb. Some scientists fear that this excess carbon dioxide in the atmosphere could have drastic effects on the global climate.

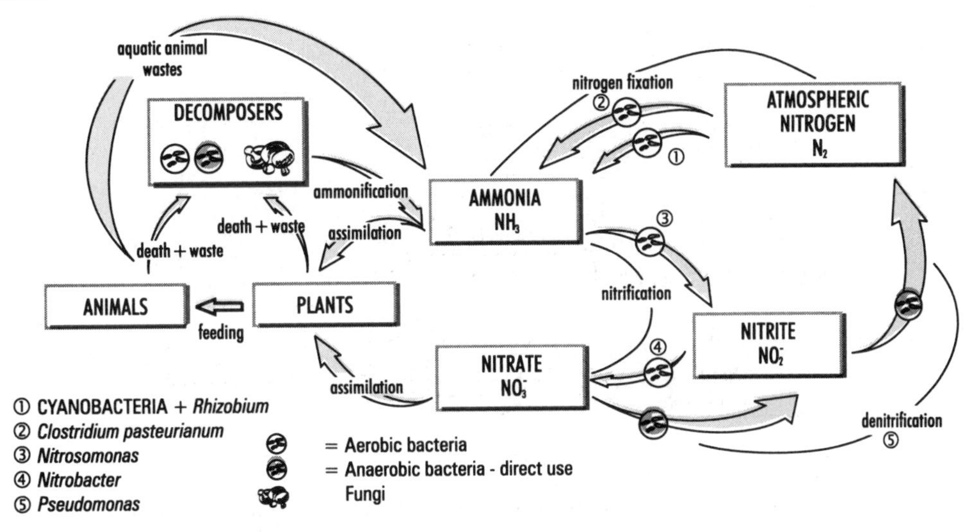

Fig. 4.3.4 The Nitrogen Cycle

The nitrogen cycle describes the circulation of nitrogen-containing molecules in nature. Most of the world's nitrogen is in the form of nitrogen gas (N_2), which makes up 78% of our atmosphere.

Some nitrogen-containing molecules, such as amino acids, are essential to life. Only a few species of bacteria can use nitrogen directly from the atmosphere. Without these bacteria we wouldn't be here. These bacteria use a process known as **nitrogen fixation** to change nitrogen gas to ammonia (NH_3). Nitrogen-fixing is an energy intensive process, but without it, nitrogen-containing molecules could not enter the food web.

Nitrogen-fixing bacteria represent the beginning of metabolic processes that eventually bring nitrogen into living matter (biomass). They are often found in symbiotic relationships with the roots of certain plants (legumes).

Nitrifying bacteria are different from nitrogen-fixing bacteria. They are important because they convert ammonia to nitrites (NO_2^-) and nitrates (NO_3^-) through a process known as **nitrification**. Nitrates are also important for the production of amino acids and proteins in plants.

Fixed nitrogen is often a limiting factor in plant growth. Animals cannot use ammonia or nitrates, so they have to get useable nitrogen products from plants or other animals.

Decomposers (aerobic bacteria, anaerobic bacteria and fungi) break nitrogen products back down to ammonia. This is a process known as **ammonification**.

Denitrification is the process that completes the nitrogen circuit. Denitrifying bacteria break down nitrates trapped in the soil, and return them to the atmosphere as nitrogen gas. Denitrifying bacteria live in anaerobic environments such as mud, bogs and the sea floor.

When you aim for perfection, you discover it's a moving target. George Fisher

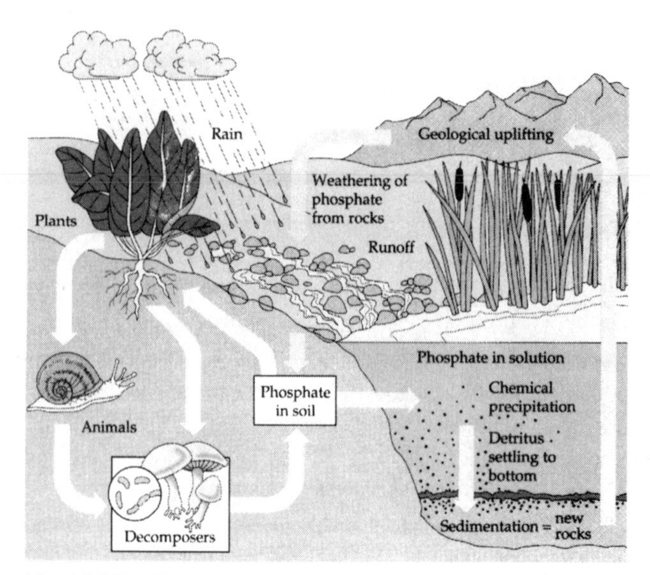

Fig. 4.3.5 The Phosphorous Cycle

The **phosphorous cycle** describes the circulation of phosphorous-containing compounds throughout nature. It is representative of how other minerals cycle through the environment. The phosphorous cycle is of major importance, because phosphorous is an essential component of many biological compounds. It is a necessary component of ATP, phospholipids and nucleic acids (DNA and RNA).

Phosphate (PO_4^{3-}) is the only useable source of phosphorous in plants, and is often a limiting factor in plant growth in soil and freshwater. It originally comes from certain types of weathered rock and, after a gradual process, ends up in soil. Plants absorb the phosphate and use it to synthesize important organic compounds. Animals receive these compounds by eating plants or other animals. Then, animals return phosphate to the soil.

Among other things, phosphorous is necessary for healthy teeth and bones. In fact, first century Roman doctors recommended brushing teeth with urine. Early alchemists used to prepare phosphorous by slowly heating buckets of urine for weeks.

Decomposers (fungi and bacteria) consume dead plant and animal matter, returning phosphate to the soil and water. The phosphate cycle is not entirely balanced. Every year, 17 million tons of phosphorous are lost to the oceans. Only half of this is due to natural erosion; the rest results from human activities (domestic and industrial use, sewage). Calcium dihydrogen phosphate ($Ca(H_2PO_4)_2$) is a common plant fertilizer.

4.4 Cell Structure

Every cell is either prokaryotic or eukaryotic. These two types of cells have very different internal morphologies (structures). **Eukaryotic** cells have a complex internal organization. They contain many membrane-bound compartments called **organelles** that allow biochemical reactions to be isolated. The genetic material, DNA, is stored in a membrane-bound **nucleus**. The presence of a nucleus is one of the major differences between prokaryotes and eukaryotes.

Prokaryotes are typically much smaller than eukaryotes, and have a simpler internal organization. They lack a nucleus or other membrane-bound organelles. Prokaryotes have inner membranes where many important reactions take place. In photosynthetic bacteria, these membranes are sometimes arranged into chlorosomes, sacs containing chlorophyll. Most prokaryotes are surrounded by a rigid cell wall that provides protection and support for the cell.

An Aid to Understanding

Endocytosis	A process of taking material into the cell (*endo*, within; *cyto*, cell).
Exocytosis	Exporting material from the cell (*exo*, out).
Hypertonic	A solution with a higher concentration of solute outside the cell; this causes the cell to dehydrate (lose water).
Hypotonic	A solution with a lower concentration of solute outside the cell; this causes water to flow into the cell and burst unprotected membranes.
Isotonic	A solution with an equal concentration of solute on both sides of a semi-permeable membrane. Saline solutions are isotonic.
Osmosis	Diffusion of water across a semi-permeable membrane. Water moves from the side with a lower solute to the side with a higher solute concentration.
Permeability	The degree to which small and large molecules can pass through a membrane. A semi-permeable membrane, or selectively permeable membrane (such as the plasma membrane of a cell), allows only water and some small solutes through, blocking others.
Phagocytosis	Taking solid materials into the cell (*phage*, eat; *cyto*, cell).
Pinocytosis	Taking fluid materials into the cell (*pino*, to drink; *cyto*, cell).
Solute	A substance dissolved in a solvent.

Life is a sexually transmitted terminal disease. Graffito

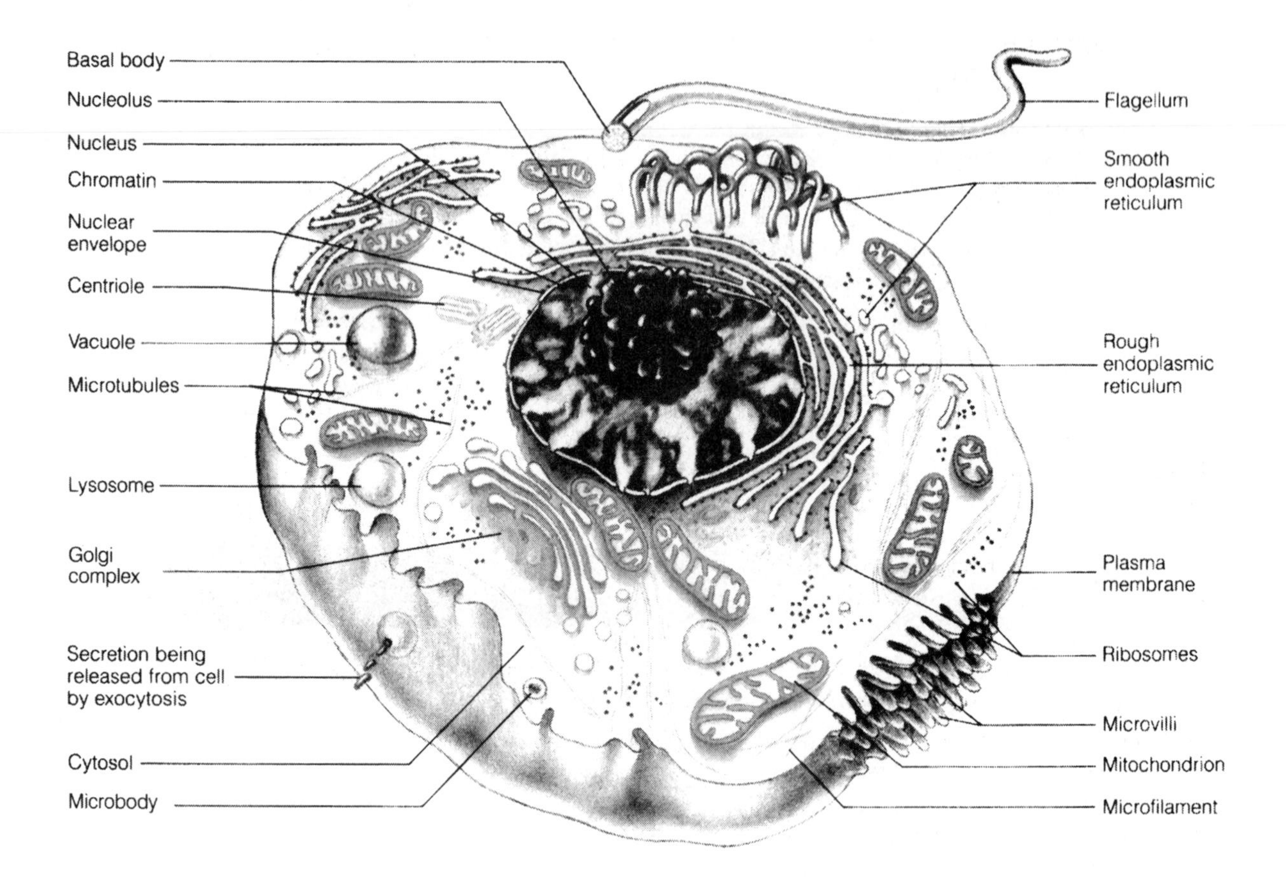

Fig. 4.4.1 Eukaryotic Cell. You'll never see a cell that looks like the one above, because cells are specialized. Each contains different types of cellular stuctures in different proportions.

Endosymbiotic theory. Chloroplasts and **mitochondria** contain DNA and ribosomes and have two membranes. Although they cannot survive outside the eukaryotic cell, many biologists believe that at one time, they lived independently as prokaryotes. Both chloroplasts and mitochondria are structurally similiar to existing prokaryotes. Chloroplasts resemble photosynthetic eubacteria. Mitochondria resemble aerobic heterotrophic bacteria. Through the process of endocytosis, these organelles could have been trapped within primitive eukaryotes that eventually evolved into modern eukaryotic cells.

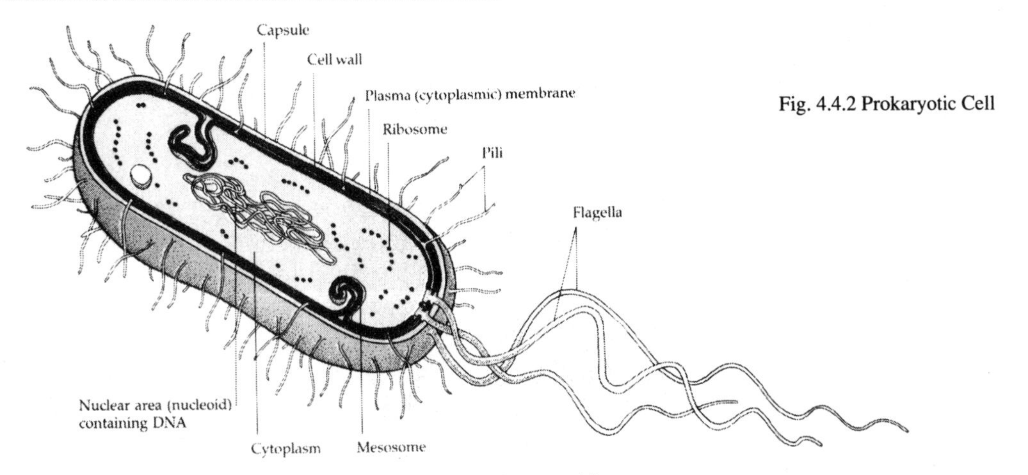

Fig. 4.4.2 Prokaryotic Cell

Cell membrane. Also known as the plasma membrane, this is present in all cells. It consists of proteins and a bilipid layer. It is semi-permeable with active transport sites that allow it to regulate what travels in and out of the cell. Proteins on the membrane's outer surface have various functions, one of which involves cell recognition.

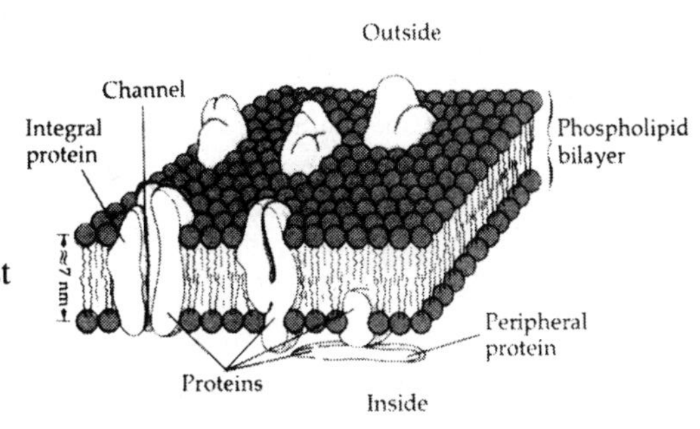

Fig. 4.4.3 Cell Membrane

Cell wall. Present in prokaryote, plant, fungi and some protist cells but absent in animal cells. Made of cellulose in plants and algae, chitin in fungi (some have cellulose), and peptidoglycan in eubacteria. Eubacteria have two types of cell wall. Following Gram staining procedures, Gram-positive cell walls turn purple. These cell walls have a complex structure, and a high peptidoglycan content. Gram-negative cell walls are colored red. These cell walls have a simple structure and low peptidoglycan content. Many prokaryotes also secrete a sticky protective outer layer called the **capsule. Pili** and **fimbria** are sticky strands outside the cell wall that allow the prokaryote to stick to other cells or other structures. The cell wall functions to protect and support the cell. It also prevents a cell from bursting in a hypotonic solution.

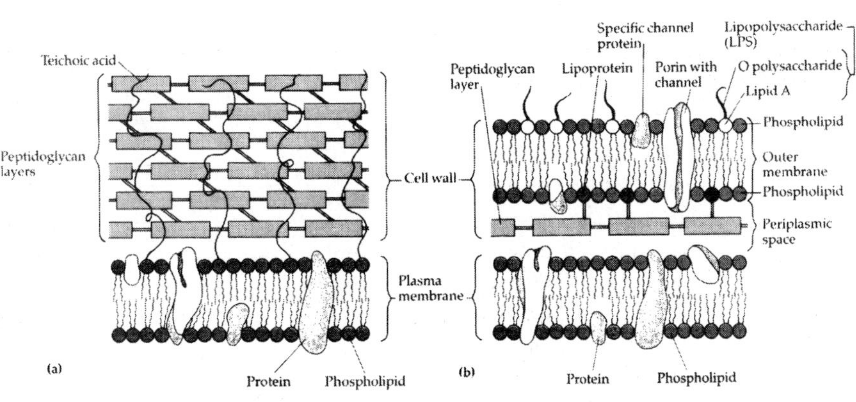

Fig. 4.4.4 Cell Walls (a) Gram-positive (b) Gram-negative

Centrioles. Present in animals but not in higher plants or prokaryotic cells. A centriole is an assembly of nine sets of microtubule triplets ordered in a ring. Centrioles occur in pairs. During cell division, the centrioles are involved in the movement of chromosomes.

Cilia and flagella. Used for motion in fluids or to move fluids about the cell. May be present in prokaryotes, protists and animals but absent in plants, except for the sperm of a few species. In prokaryotes, flagella are made of a protein filament called flagellin, and are anchored to the cell by a **basal body**. The basal body is structurally similar to a centriole allowing the filament to spin like a propeller. In eukaryotes, the core of cilia and flagella is an assemby of nine microtuble doublets, with two microtubules in the center (9+2 arrangement). Eukaryote cilia and flagella beat back and forth, but do not rotate.

Chloroplasts. Present in plant cells, and in some protists (green algae). Because they contain DNA, and have two membranes, they are believed to have originated endosymbiotically. It is the center of the cell's photosynthetic activity. The chloroplast contains chlorophyll in flattened membrane discs called thylakoids. These discs are organized into stacks called grana. The structures convert light energy into ATP (chemical energy).

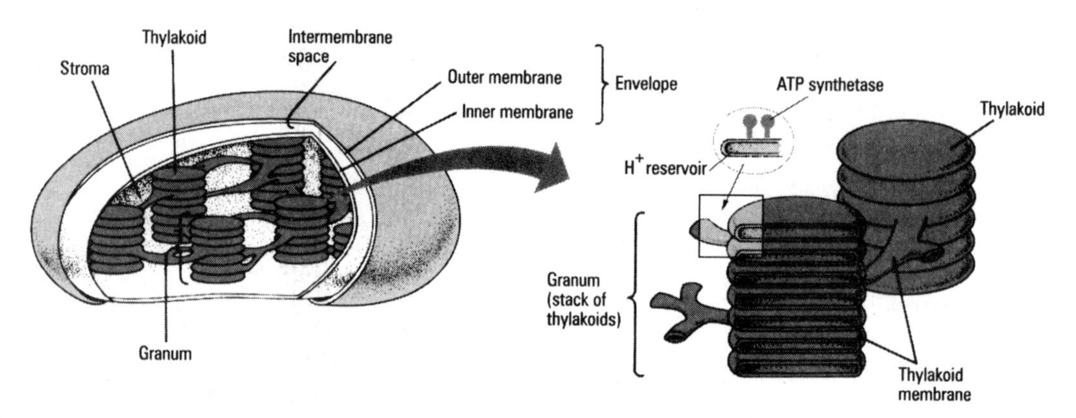

Fig. 4.4.5 Chloroplast

Cytoskeleton. The cytoskeleton is a network of protein fibers found only in the cytoplasm of eukaryotic cells. Three different types of fibers make up the cytoskeleton: microtubules, intermediate filaments and microfilaments. Its function is to support the cell and aid in movement by allowing the cell to change shape. Organelles and other cytoplasmic enzymes may also be bound in place or moved about the cell by the cytoskeleton.

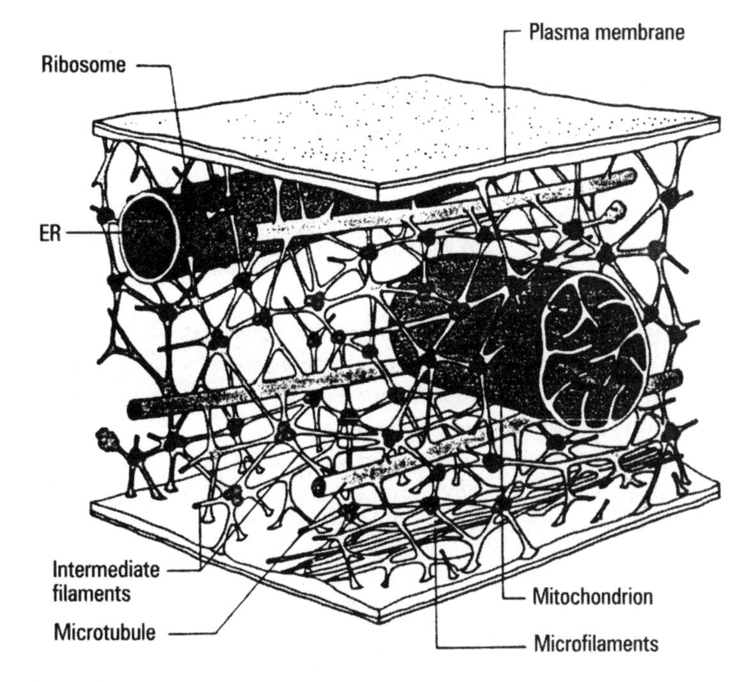

Fig. 4.4.6 Cytoskeleton

Golgi complex. Found in eukaryotic cells. Animal cells typically contain 10 to 20, plants may contain hundreds, protists a few. The Golgi complex is a processing station that modifies substances and then packages them in vesicles which are usually exported from the cell. They are made up of stacks of flattened membranes called dictyosomes. The Golgi complex refers to all the dictyosomes in the cell. It collects, packages and distributes the molecules synthesized in the endoplasmic reticulum of the cell. Molecules are collected at the *cis* end (top right in the diagram). After processing and modification, vesicles pinch off from the *trans* face (left-hand side).

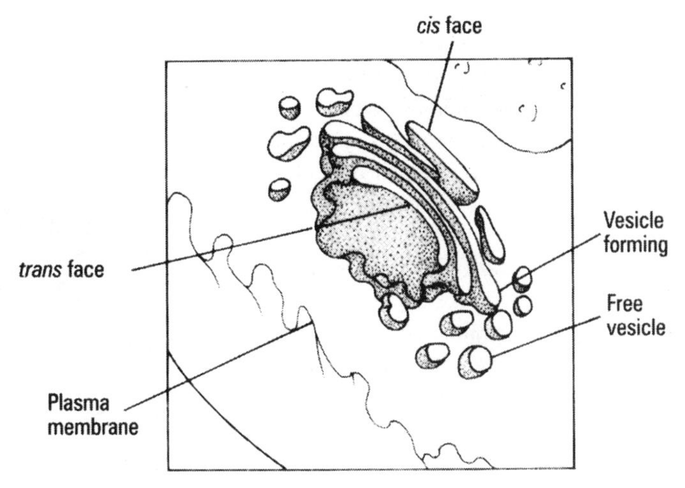

Fig. 4.4.7 Golgi Complex

Lysosomes. Present in animals, plants and a few protists but not found in prokaryotic cells. These are vesicles produced by the Golgi complex and contain hydrolytic digestive enzymes. They break down unneeded macromolecules and damaged cell constituents so that their components can be recycled. They are also used to digest material obtained through endocytosis.

Microbodies. Found in most eukaryotic cells (see Fig. 4.4.1). Microbodies are vesicles that carry specialized enzymes for different metabolic pathways. There are two important types of microbodies: glycoxyomes, which convert fats to carbohydrates, and peroxisomes, which detoxify alcohol and other molecules that can be harmful.

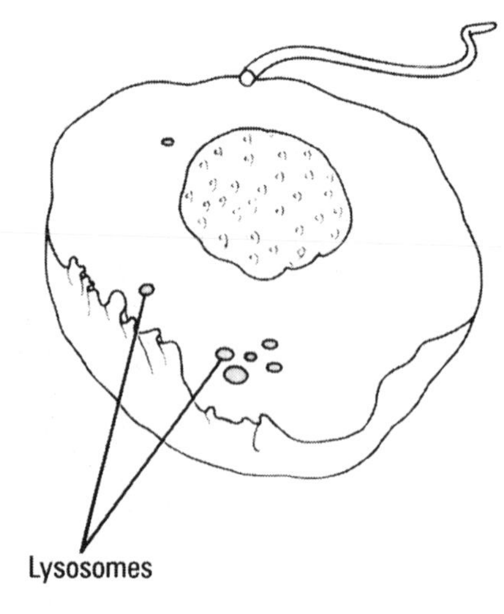

Fig. 4.4.8 Lysosomes

Mesosomes. Found in prokaryotes only (see Fig. 4.4.2). Mesosomes are infoldings of the cell membrane in bacteria. Although their function is not quite clear, many biologists believe that they are involved in cell division.

Mitochondria. Found in eukaryotic cells only. Mitochondria are involved in energy generation and respiration. These "powerhouse furnaces" are believed to have an endosymbiotic origin because they have a double membrane, DNA, ribosomes and are bacterium-sized. The outer membrane of mitochondria is smooth; the inner membrane is elaborately folded to increase internal surface area. The infoldings are called **cristae** and form a compartment known as the **matrix**, where **cellular respiration** occurs. Cellular respiration breaks down pyruvate obtained from glycolysis, into water and carbon dioxide, while changing ADP to ATP. ATP is the cell's main energy source. The number of mitochondria in each cell varies according to the metabolic demand placed on the cell.

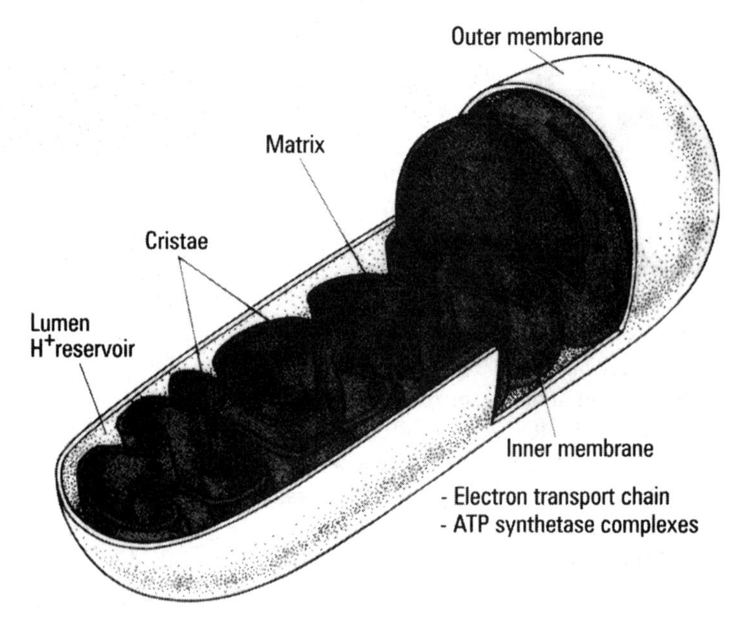

Fig. 4.4.9 Mitochondrion

Serving God is doing good to man, but praying is thought an easier service and therefore more generally chosen.
Benjamin Franklin

Nucleus. Found in eukaryotic cells only. It houses the genetic blueprint of the cell in the form of DNA. Nuclei are surrounded by a double membrane called the nuclear envelope. Material can pass into and out of the nucleus through nuclear pores in the nuclear envelope. Its DNA is found in the form of chromosomes during cell division. During the cell's resting state, DNA is stored as chromatin. Chromatin is a system of beaded threads where DNA is wrapped around clusters of proteins called **histones** (see Fig. 4.5.1). The nucleolus is a region in the nucleus involved in the production of ribosomal RNA (rRNA).

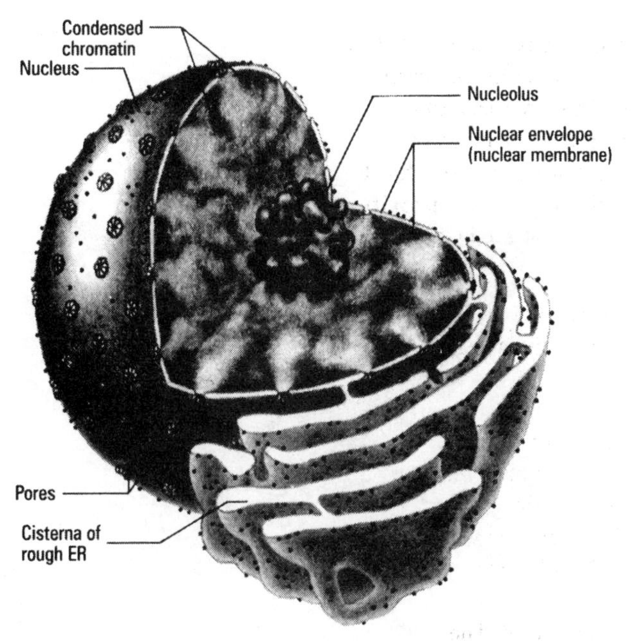

Fig. 4.4.10 The Nucleus

Ribosomes. Found in prokaryotes (70s) and eukaryotes (80s). Ribosomes receive protein-building instructions in the form of messenger RNA (mRNA). Ribosomes are made of large and small subunits of proteins and mRNA. In prokaryotes, they are smaller and structurally different from eukaryotic ribosomes.

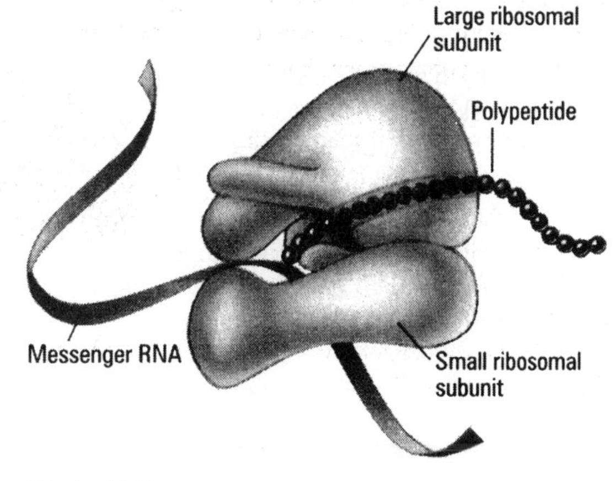

Fig. 4.4.11 Ribosome

Rough endoplasmic reticulum. Found in eukaryotic cells only. A large network of internal membranes that forms compartments known as **cisternae**. Rough endoplasmic reticulum (rough ER) manufactures proteins, and then passes them to the Golgi complex. The bumps on the surface of the membrane are anchored ribosomes.

Smooth endoplasmic reticulum. Found in eukaryotic cells only. Also a large network of internal membranes forming cisternae. It is involved with a variety of metabolic activities. Smooth endoplasmic reticulum contains enzymes that catalyze the synthesis of fats, steroid molecules, lipids and phospholipids. It also detoxifies drugs and poisons.

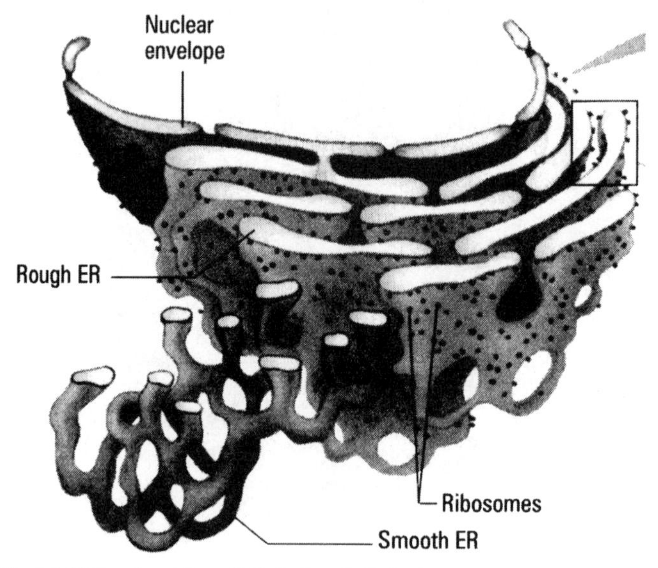

Fig. 4.4.12 Rough and Smooth Endoplasmic Reticulum

Vacuoles. Absent in bacteria, small (or absent) in animals, and present in plants. Vacuoles are membrane-bound and fluid-filled. Acting as a food reservoir, they may also serve as an excretory mechanism for toxic substances. They are small in growing plants and large in mature plants. In animal cells, vacuoles are usually used only for temporary storage.

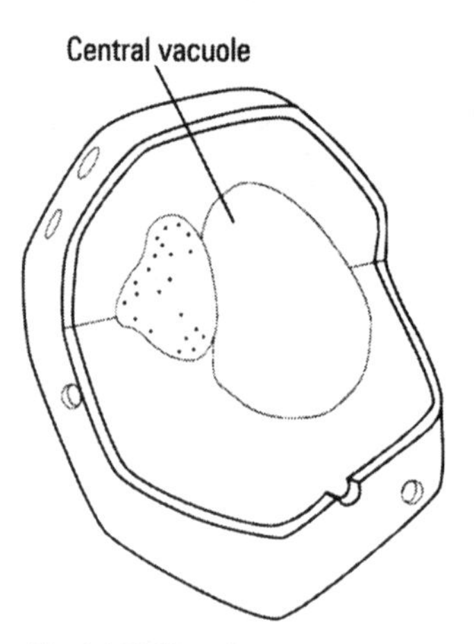

Fig. 4.4.13 Vacuole

4.5 Cell Cycles

Cell cycles describe stages in the life of a cell. The majority of cells grow, prepare to divide and then divide. The cycle begins again with each newly divided cell.

An Aid to Understanding

Autosomal chromosome	Any chromosome that is not a sex chromosome.
Centromere	The constricted region on a chromosome.
Chromatid	A chromosome attached to its duplicate at the centromere.
Chromatin	Non-condensed chromosomes in the nucleus of a eukaryotic cell.
Chromosome	A thread-like structure in the nucleus of eukaryotes that contains the genetic material, DNA and proteins called histones.
Diploid	Having two copies of each type of chromosome.
Haploid	Having half the normal number of chromosomes in the somatic (body) cells of a particular species.
Homologous chromosomes	Chromosomes containing the same type of genetic information. For example, two homologous chromosomes may code for eye color but each may code for a different color.
Kinetochore	This structure is found on a chromosome at or near the centromere. Microtubules join at the kinetochore to separate two chromatids.
Oogenesis	The process of producing ova (eggs).
Ovum	A female gamete (egg).
Polar bodies	Non-fertile cells produced during oogenesis that contain a nucleus and very little cytoplasm.
Sex chromosomes	A set of chromosomes responsible for sexual characteristics (X,Y in humans).
Spermatocytes	Cells that undergo the process of meiosis to become sperm.
Spermatogenesis	The process of producing sperm.
Zygote	A fertilized egg.

Prokaryotes grow rapidly and then reproduce through a process called **binary fission** (splitting down the middle). The **genetic material** in prokaryotes exists as a single circular strand of DNA known as a bacterial chromosome. The bacterial chromosome is attached to the cell membrane. Many bacteria also have plasmids, smaller DNA rings that carry accessory genes. Enzymes in the cell work to generate a second copy of the genetic material. When finished, the cell pinches in half (or close to it). The way it divides ensures that each half gets a copy of the genetic material. Each daughter cell then grows and prepares to divide. This typically takes 30 minutes under ideal conditions.

To ensure genetic variability, bacteria have three mechanisms for gaining new genetic material: **conjugation**, **transformation**, and **transduction**. In **conjugation**, fragments of DNA (such as **plasmids** or pieces of the bacterial chromosome) are transferred by cell to cell contact. In **transformation**, cells pick up free-floating naked DNA found outside the cell. This DNA usually comes from other cells that have **lysed** (split open). This naked DNA binds to **DNA-binding proteins** outside the cell and is imported into the cell.

Cell wall
Plasma membrane
1. DNA divides
DNA (nuclear area)
2. Cell wall and plasma membrane begin to divide
3. Cross wall forms completely around divided DNA
Cells separate

Fig. 4.5.1a Binary Fission

Transduction involves DNA transfer mediated by viruses known as **bacterio-phages**. Viruses walk the fine line between living and nonliving. Essentially, they're protein-coated bundles of genes, and may be crystallized. They reproduce by invading hosts. In transduction, bacteriophages join to the cell, into which they insert their genetic material.

Some bacteria have a protective mechanism that allows them to survive harsh conditions. They form **endospores** which are very resistant to heat, dryness, disinfectants and even radiation. Endospores can remain dormant for years, yet can convert back to a **vegetative** (active) cell in minutes. These bacteria are commonly found in soils, and are Gram-positive.

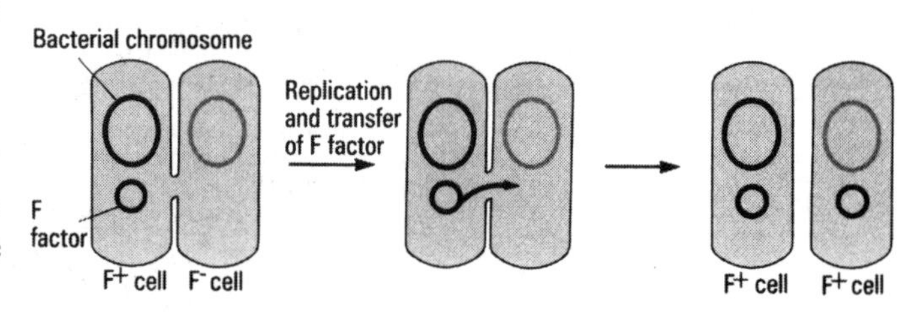

Fig. 4.5.1b Conjugation

Nothing in life is so exhilarating as to be shot at without result. Sir Winston Churchill

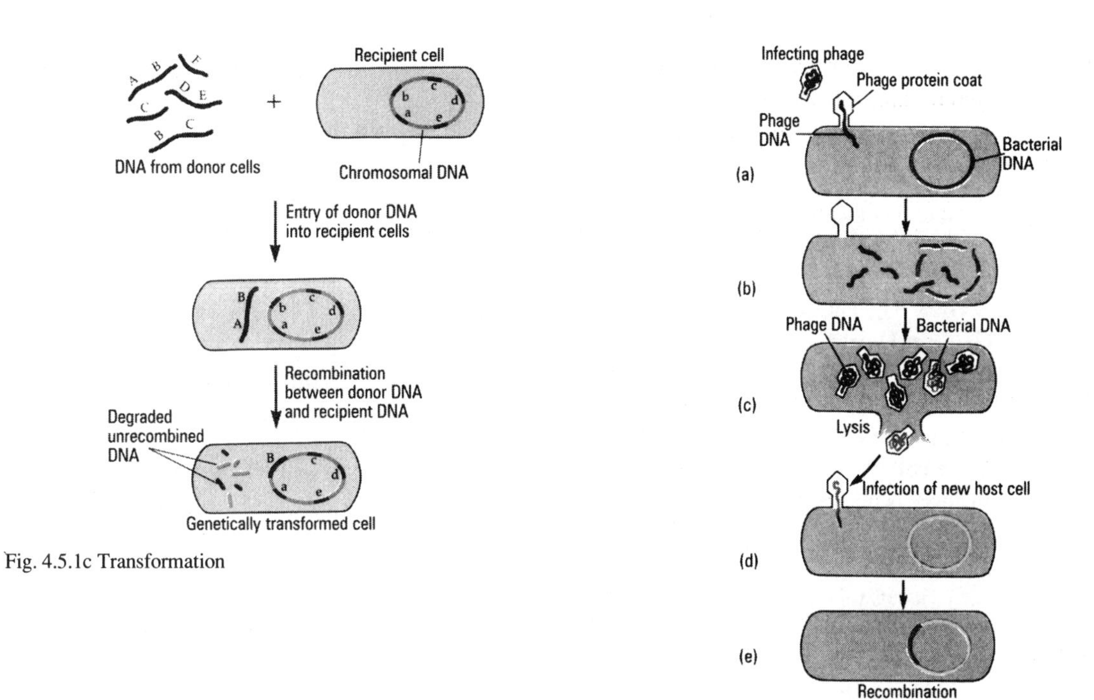

Fig. 4.5.1c Transformation

Fig. 4.5.1d Transduction

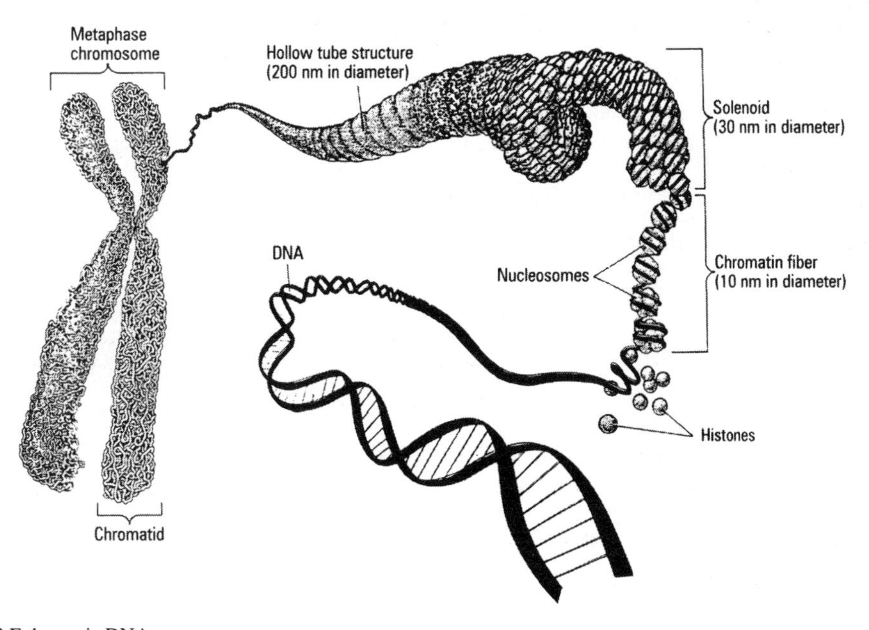

Fig. 4.5.2 Eukaryotic DNA

Every time I have to do homework, I hear my bed call my name. Gordon Coleman

The eukaryotic cell cycle usually involves three major stages: **interphase**, **mitosis** and **cytokinesis**. Together, mitosis and cytokinesis are sometimes referred to as the **M phase**.

A newly formed cell begins its life in **interphase**. During this stage, the cell grows to roughly twice its size, then prepares to divide. It moves through three distinct periods: **G_1, S, G_2**. The cell spends most of its time in the **G_1 (Gap 1)** period. During this primary growth period, the cell grows in size. G_1 varies greatly among cell types, from minutes to years. Once a cell moves to the next period, there is no turning back, the cell must divide. The next period, **S** or **synthesis**, is marked by DNA synthesis. In mammal cells, this is a 6-8 hour period when chromosomes are replicated. When DNA synthesis stops, the cell has twice as much DNA, and the **G_2 (Gap 2)** period begins. Now the cell starts to manufacture all the proteins it will require for division. When completed, this marks the end of interphase, and the beginning of mitosis.

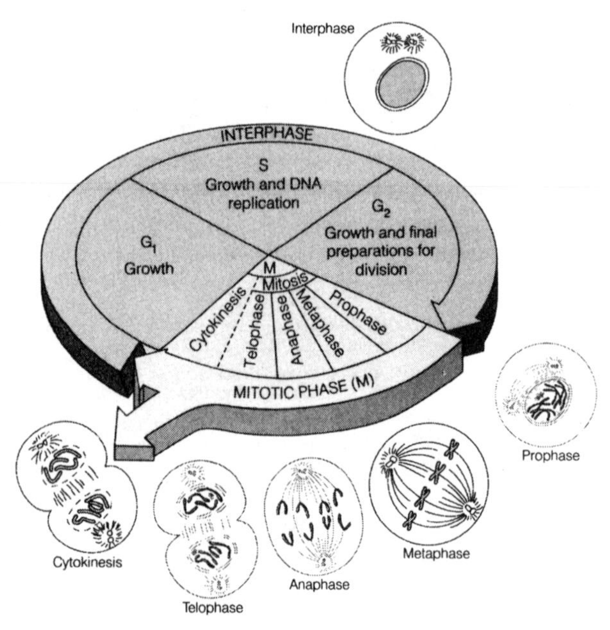

Fig. 4.5.3 Eukaryotic Cell Cycle (mitosis)

The process of **mitosis** is geared toward distributing duplicated genetic material so that two identical daughter cells can form. It is divided into four periods: **prophase**, **metaphase**, **anaphase** and **telophase**. Memorize this sequence by the phrase **P**olish **M**en **A**re **T**all.

During **prophase**, chromosomes condense and become visible (under magnification). Now the **nuclear envelope** and **nucleolus** begin to disappear. A framework of **mitotic spindles** forms; their ends (poles) move to opposite sides of the cell. In animal cells, a pair of centrioles occupies the spindle poles.

Metaphase can be recognized by the alignment of chromosomes along the equator.

At **anaphase**, the centromeres divide and the chromatids are separated. At this point, biologists call the separated chromatids **chromosomes**. These new chromosomes are pulled toward each pole.

Telophase is reached as soon as the chromosomes stop moving. This period resembles prophase, but in reverse. Two nuclei begin to reform around the genetic material as it decondenses. Microfilaments in the cell begin to constrict the equator. Each cell then restarts the cycle in interphase at G_1.

Cytokinesis, which usually begins at anaphase, is marked by the cell pinching in the middle until two smaller daughter cells form.

In eukaryotic organisms, genetic variability is the result of **sexual reproduction**. Not all eukaryotic organisms reproduce sexually, however. For those that do, special reproductive cells called **gametes** form with half the normal number of chromosomes. When these cells unite they form a **zygote**, which marks the beginning of a new organism.

Meiosis is the process that generates these specialized sex cells. This process resembles cycling through mitosis twice. The goal of meiosis is to reduce the genetic material to half its normal amount. It takes place within the organism's sex organs. Meiosis has two major stages: **meiosis I** and **meiosis II**. **Meiosis I** involves the separation of paired homologous chromosomes. **Meiosis II** separates joined chromatids into individual chromosomes. **Meiosis I** is subdivided into **prophase I**, **metaphase I**, **anaphase I** and **telophase I**.

Prophase I. All chromosomes condense and pair with their homologues. Genetic **recombination** (crossing over) also occurs. By the end of this stage the **nucleolus** and **nuclear envelope** disappear.

Metaphase I. Spindle fibers move chromosomes, forming two rows along the equator. This ensures that each daughter cell will receive only one chromosome from each homologous pair.

Anaphase I. Each chromosome moves to either pole.

Telophase I. The cytoplasm begins to divide into two **haploid** cells. Plant cells usually run through a short resting period called **interkinesis**. In interkinesis, the nuclear envelope reappears and the chromosomes decondense. Cells that do not pass through interkinesis skip to metaphase II.

Meiosis II is ready to begin. It also is divided into **prophase II**, **metaphase II**, **anaphase II** and **telophase II**. The process is similar to meiosis I, except only half the genetic information is present.

Prophase II. For cells exiting interkinesis, prophase II involves the recondensing of chromosomes, and the disappearance of the nuclear envelope.

Metaphase II. Spindles form and then move the chromosomes to the equator so that individual chromatids can be separated.

Anaphase II. Centromeres divide and the separated chromatids are now referred to as chromosomes.

Telophase II. The nuclear envelopes begin to reform.

After cytokinesis, four haploid gametes are formed.

I have never let my schooling interfere with my education. Mark Twain

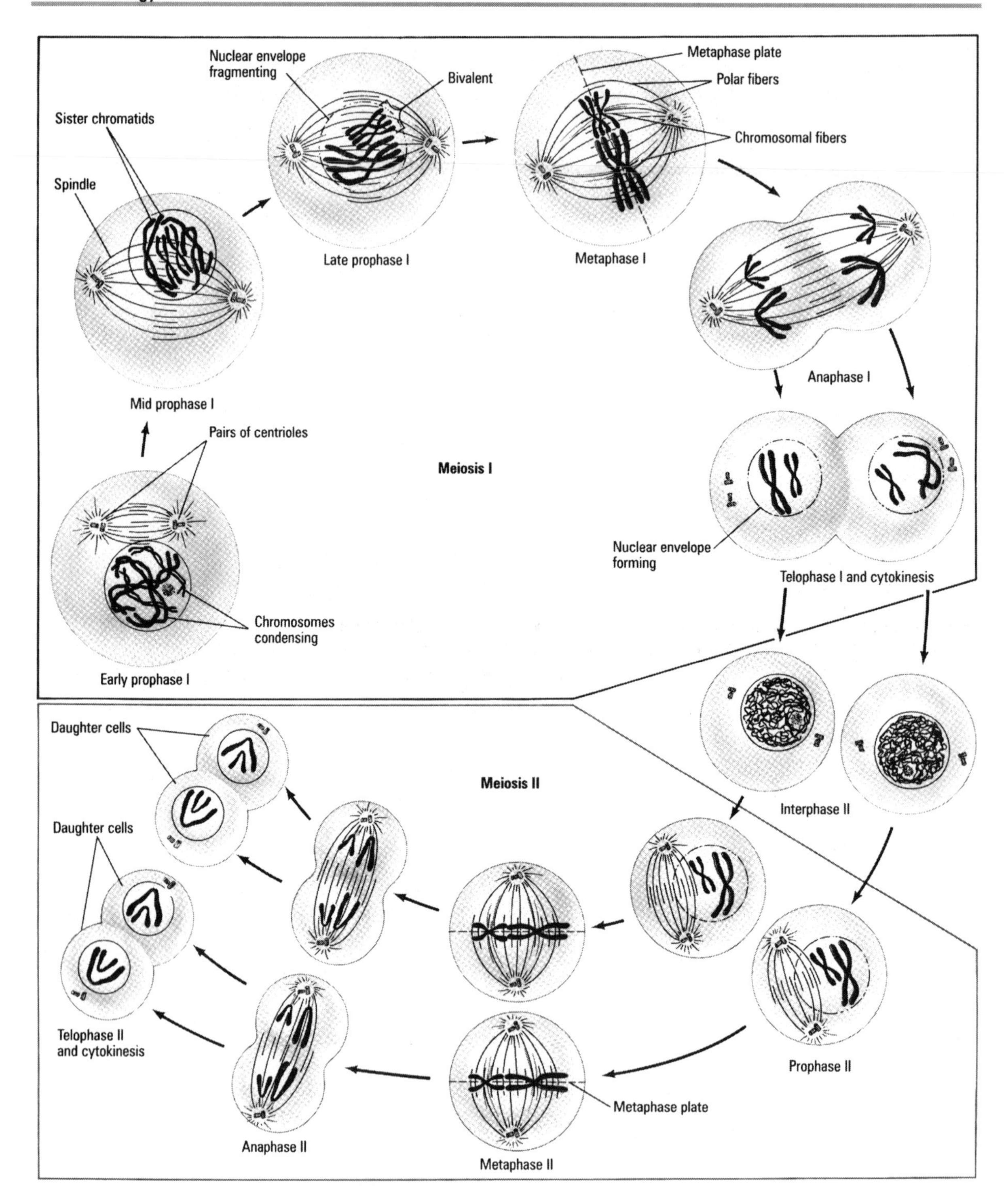

Fig. 4.5.4 Meiosis

Major Differences Between Mitosis and Meiosis in Humans*

Stages	Mitosis	Meiosis I	Meiosis II
Prophase	(2N) 46 chromosomes present	(2N) 46 chromosomes present	(N) 23 chromosomes present
Metaphase	chromosomes line up to form one row	chromosomes line up to form two rows	chromosomes line up to form one row
Anaphase	joined chromatids separated into chromosomes	paired homologous chromosomes separated	joined chromatids separated into chromosomes
Telophase	two (2N) cells each with 46 chromosomes	two (N) cells each with 23 chromosomes	four (N) cells each with 23 chromosomes

* Statements represent end of each stage.

Cell division is not always perfect. Sometimes chromosomes become fragmented or are improperly separated among cells. This is usually fatal, but in some cases, survivable. When this happens in gametes, developmental difficulties often occur.

Results of Abnormal Chromosome Distribution

Syndrome name	Chromosomal abnormality	Sex	Symptoms
Down	Trisomy 21 (extra copy or fragment of chromosome 21)	M or F	Physical: short, stocky build, flattened forehead, protruding tongue, almond-shaped eyes, crease across palm. Cognitive: mental retardation, poor verbal skills, slow motor development.
Turner	XO	F	Physical: short build, webbed neck. Cognitive: above average student, spatial and orientation deficits.
Metafemale or Triple X	XXX	F	Physical: tall tendency. Cognitive: diminished verbal abilities.
Klinefelter	XXY	M	Physical: tall tendency, incomplete development at puberty, overweight with poor muscle development, often sterile. Cognitive: poor verbal skills, tend to be shy.
XYY	XYY	M	Physical: above average height, large teeth, sometimes severe acne. Cognitive: above average criminality.
Fragile X	Abnormal gaps or breaks on X chromosomes	M or F	Physical: mild facial deformities. Cognitive: intellectual deficits, hyperactivity.

4.6 Metabolic Cycles

Metabolism is the sum of all **catabolic** (breaking down) and **anabolic** (building) reactions in an organism. One way to think of catabolism is to associate it with the word, catastrophe. A catastrophe involves, large complex things breaking down into small and simple ones, like a city in an earthquake.

An Aid to Understanding

Anabolism	An energy-requiring process that builds large and complex molecules from small and simple ones.
Carbon fixation	A process which requires a large amount of energy and adds CO_2 to a molecule.
Catabolism	A process that breaks large molecules into smaller ones and releases energy.
Chlorophyll	A colored pigment that absorbs light energy, used by photoautotrophs to fuel sugar production.
Cytochromes	A complex containing an ion bound to a chemical structure known as a porphyrin ring, surrounded by a protein. This complex carries electrons in the electron transport chain.
Decarboxylation	The removal of CO_2 from a molecule.
Deamination	The removal of an amino group (NH_2).
Electron transport chain	A membrane complex containing cytochromes and ATPase. It harvests energy from electrons to chemiosmotically generate ATP.
Enzyme	A special kind of protein catalyst made by cells to speed up specific reactions without being used up itself. (Words ending in "*-ase*" are enzymes.)
Fermentation	A breakdown (catabolism) of organic molecules to liberate energy without oxygen.
Oxidative phosphorylation	A process whereby organic molecules are oxidized, and energy is captured to create ATP in the mitochondrion.
Phosphorylation	Addition of a phosphate group (PO_4^{-3}).
Photophosphorylation	A process that uses light energy to create the high energy phosphate bonds in ATP.
Photosynthesis	A process involving the use of light to add (fix) CO_2 and build sugars (carbohydrates).
Respiration	The aerobic process involving the oxidative breakdown of organic molecules.

Making Sense of Metabolism

Reduction and Oxidation (Redox)

An oxidation or reduction can be performed three different ways. (Refer to the table below.) To remember the effect of electrons, think of LEO and GER. **LEO**: **L**oss of **E**lectrons = **O**xidation. **GER**: **G**ain of **E**lectrons = **R**eduction. It's also useful to understand that hydrogen atoms are electron carriers. Therefore, adding hydrogen to a compound reduces, while removing hydrogen oxidizes. Anytime a compound is reduced, another is also oxidized. In addition, stating that compound (B) was reduced by compound (A) is the same as stating that compound (A) was oxidized by (B).

$$Ae^- + B \rightarrow A + Be^-$$

Ae^-	B	A	Be^-
reducing agent (electron donor)	oxidizing agent (electron acceptor)	oxidized	reduced

How It's Done		The Differences	
Reduction	**Oxidation**	**Oxidized state**	**Reduced state**
Remove oxygen	Add oxygen	NAD^+	NADH
Add hydrogen	Remove hydrogen	FAD	$FADH_2$
Add an electron	Remove an electron	$NADP^+$	NADPH
		Energy poor	Energy rich

Inconsistencies in names can cause confusion. For example, citric acid and citrate. For the most part, they're the same thing. The difference has to do with the pH level of the molecule's environment.

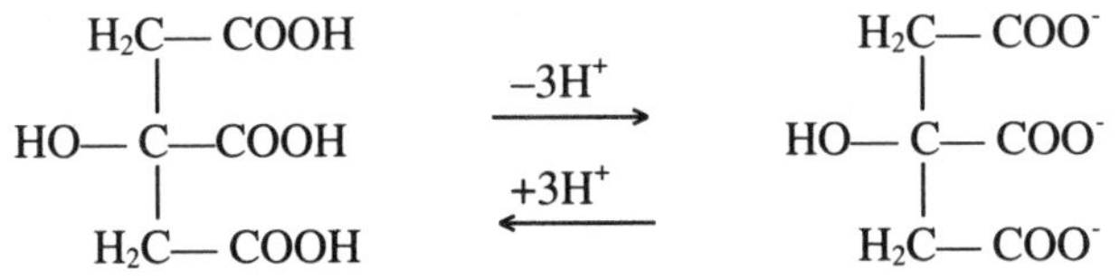

$$\begin{array}{c} H_2C—COOH \\ | \\ HO—C—COOH \\ | \\ H_2C—COOH \end{array} \quad \underset{+3H^+}{\overset{-3H^+}{\rightleftarrows}} \quad \begin{array}{c} H_2C—COO^- \\ | \\ HO—C—COO^- \\ | \\ H_2C—COO^- \end{array}$$

Citric acid in an acidic environment — Citrate in an alkaline environment

Abbreviations and Their Meanings

Symbol	Meaning	Symbol	Meaning
P_i	Inorganic phosphate - *phosphate* (PO_4^{-3})	ATP	Adenosine triphosphate
NAD^+	Nicotinamide adenine dinucleotide	ADP	Adenosine diphosphate
FAD	Flavin adenine dinucleotide	AMP	Adenosine monophosphate
$NADP^+$	Nicotinamide-adenine dinucleotide phosphate		

Respiration

Respiration is a catabolic process that breaks down organic molecules (food) for energy. The most familiar form of respiration involves the oxidative breakdown of glucose molecules to form carbon dioxide (CO_2), water (H_20) and energy. This energy generates heat and ATP. In eukaryotes, respiration involves three integrated pathways: glycolysis, the citric acid cycle and the electron transport chain.

$$C_6H_{12}O_6 + 6\,O_2 \rightarrow 6\,CO_2 + 6\,H_2O$$

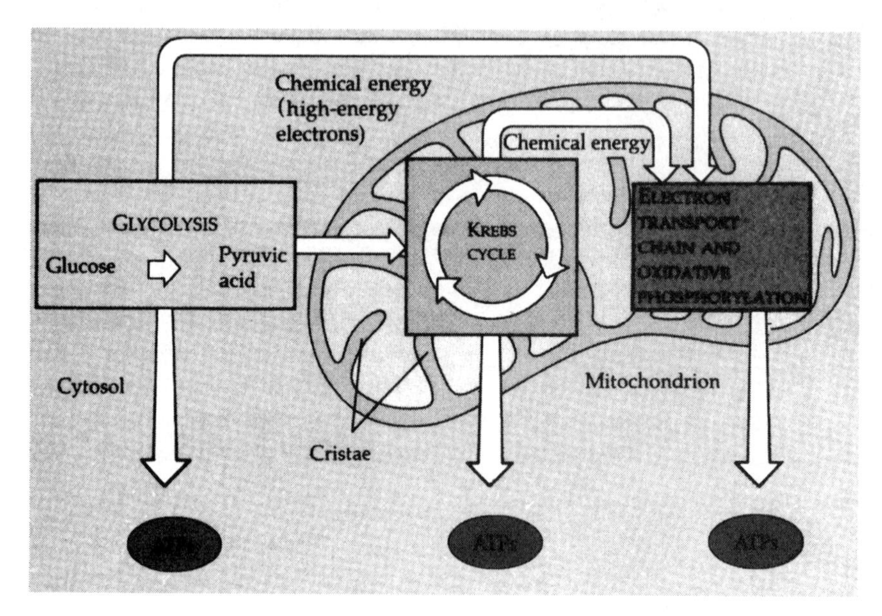

Fig. 4.6.1 Respiration

Glycolysis

Glycolysis is the most common path for glucose catabolism. Glycolysis occurs within the cytoplasm **anaerobically** (without oxygen), and yields two pyruvate (3C) molecules for each glucose (6C) molecule. **Oxidizing** one glucose molecule to two pyruvate molecules yields a net of 2 ATP, 2 NADH and 2 protons (H^+) into the medium. 2 NADH electron carriers move electrons to the electron transport chain where they eventually produce 6 ATP (3 ATP each). Although glycolysis is anaerobic, NADH cannot be oxidized back to NAD^+ by the electron transport chain when oxygen is not present. Under these conditions, NADH builds up while the amount of NAD^+ required for glycolysis is depleted. One way to solve this problem is to pass the two electrons tied up in NADH to pyruvate. This produces NAD^+ but also lactic acid or lactate. You've probably felt the effects of this phenomenon during strenuous exercise. The presence of lactic acid causes your muscles to ache. Some organisms produce ethanol instead of lactic acid when faced with this problem, a process many of us envy, since ethanol is what makes us drunk.

1) The beginning of glycolysis is marked by the transfer of a phosphate group from ATP by hexokinase to glucose. Other carbohydrates such as galactose and sucrose must first be converted into glucose 6-phosphate before entering glycolysis at step 2.

2) The connectivity of glucose 6-phosphate is rearranged by phosphoglucoisomerase, yielding the **isomer** fructose 6-phosphate, required by the next enzyme.

3) Phosphofructokinase uses an additional ATP molecule to **phosphorylate** (add phosphate) fructose 6-phosphate, making fructose 1,6-diphosphate. (Remember, you have to invest money to make money!)

4a) Aldolase splits the molecule in half (sugar-splitting) into two products, **PGAL** and **DHAP**, which are isomers.

4b) An enzyme called triosephosphate isomerase interconverts DHAP to PGAL and back. The next enzyme uses only PGAL, causing isomerase to replenish PGAL by using DHAP.

5) The enzyme glyceraldehyde-3-phosphate dehydrogenase holds PGAL in its active site, and PGAL becomes **oxidized** by NAD^+, yielding NADH + H^+. **Dehydrogenases** oxidize molecules by removing hydrogen from the molecule. Inorganic phosphate (P_i) from the medium is then added to create a high-energy bond, denoted by a squiggle (~).

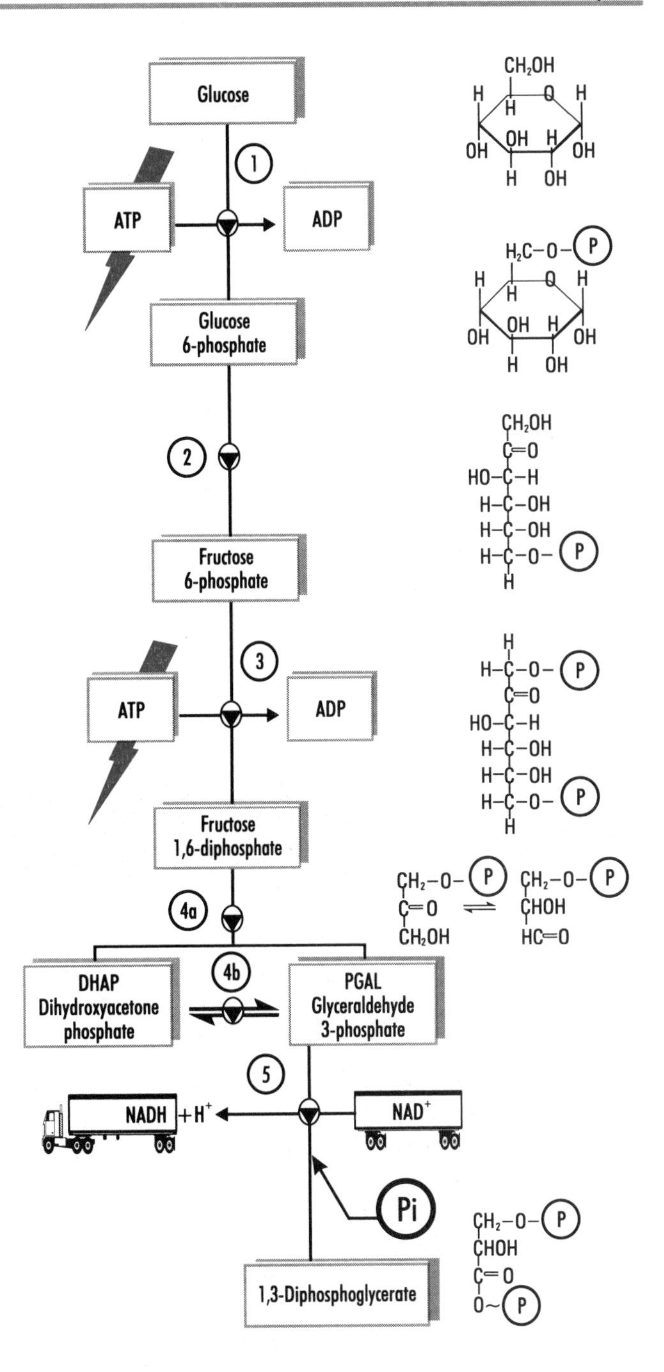

Fig. 4.6.2 Glycolysis

6) Phosphoglycerate kinase allows ADP to pick up the high-energy bond, forming ATP. Remember, this happens twice for every glucose molecule because the sugar was split in two during step 4.

7) The remaining phosphate group is relocated by phosphoglyceromutase.

8) Enolase generates a high-energy bond by removing an equivalent of water (H_2O).

9) The last reaction of glycolysis uses pyruvate kinase to form ATP and pyruvate.

(*) Note that four ATPs are actually formed, since the original glucose molecule was split in two, and each half eventually passes through steps 5-9. The total ATP yielded in glycolysis is therefore, 2 ATP.

(4 formed – 2 used = 2 ATP).

(*) Because energy can still be utilized from pyruvate in anaerobic conditions (absence of oxygen), it can be further oxidized to either ethanol or lactic acid. Lactic acid is what makes you feel sore after exercising too much. When oxygen is available (aerobic conditions), pyruvate is passed to the citric acid cycle.

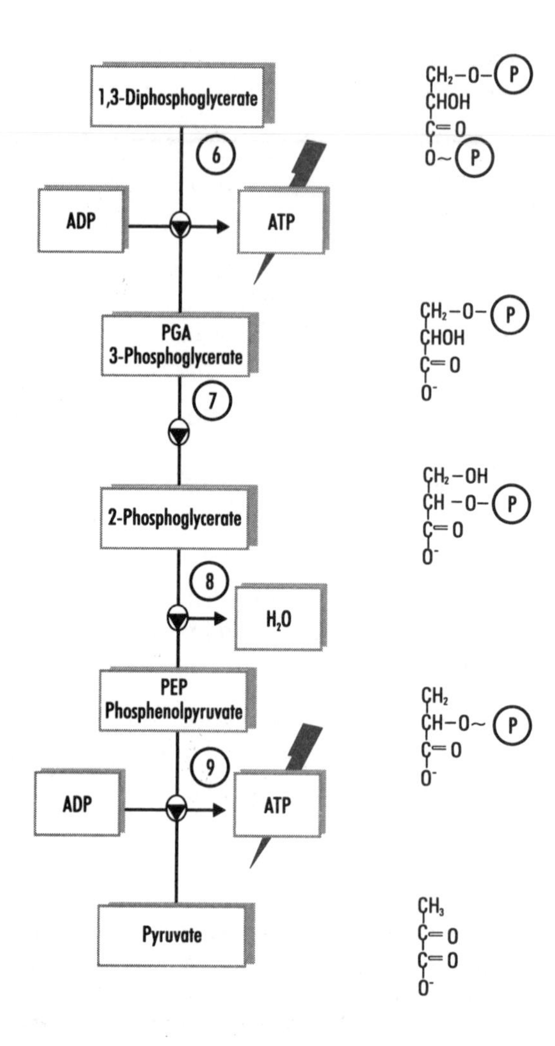

Fig. 4.6.2 Glycolysis (cont.)

Animal Metabolism

The figure below shows some pathways involved in processing food. Food is made of precursors such as nucleotides, fatty acids, sugars and amino acids. When in abundance, they are stored in the form shown in the rectangles above the circles. When food is not abundant, the reserves convert back to their precursors in order to restore essential levels.

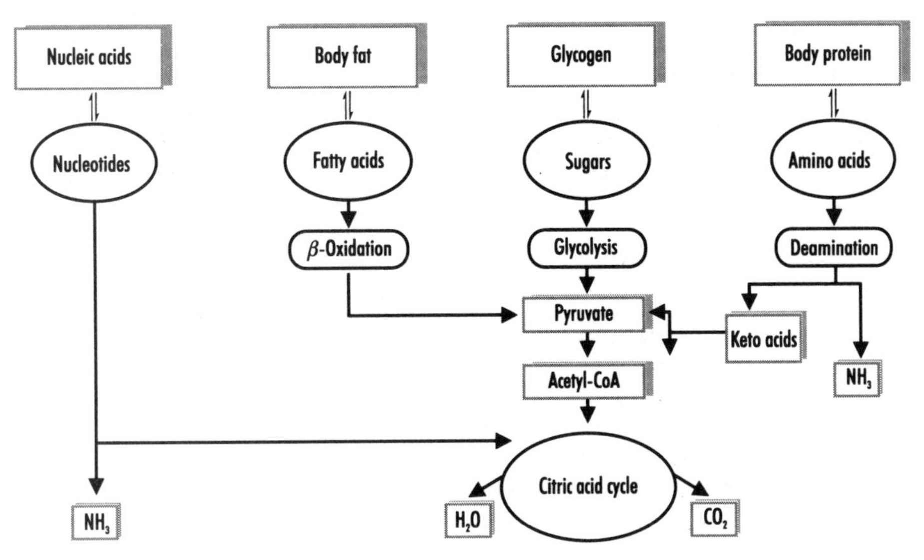

Fig. 4.6.3 Animal Metabolism

The table below summarizes the energy yield utilized from one glucose molecule as it passes through various pathways.

A Simplified Summary of ATP Production From Respiration

Pathway	ATP formed	ATP used	Reduced carriers	Oxidized carriers	ATP formed per oxidized carrier
Glycolysis	4	2	2 NADH	–	–
Preparation of acetyl-CoA from pyruvate	–	–	2 NADH	–	–
Citric acid cycle	2	–	6 NADH 2 $FADH_2$	–	–
Electron transport chain	34	–	–	10 NAD^+ 2 FAD	3 2
Balance 40 ATP formed – 2 ATP used = 38 ATP yield*					

* There are costs to bring NADH from the cytoplasm to the electron transport chain. Including these costs, the net ATP yield is 36 ATP.

A camel can drink 45 liters in 5 minutes.

The Pentose Pathway

Also known as the hexose monophosphate shunt, this pathway operates simultaneously with glycolysis. It's an alternative pathway for the oxidation of glucose; it breaks down 5-carbon sugars called pentoses. Its importance arises from many of its intermediates, which are precursors of other synthetic pathways used to form nucleic acids and amino acids.

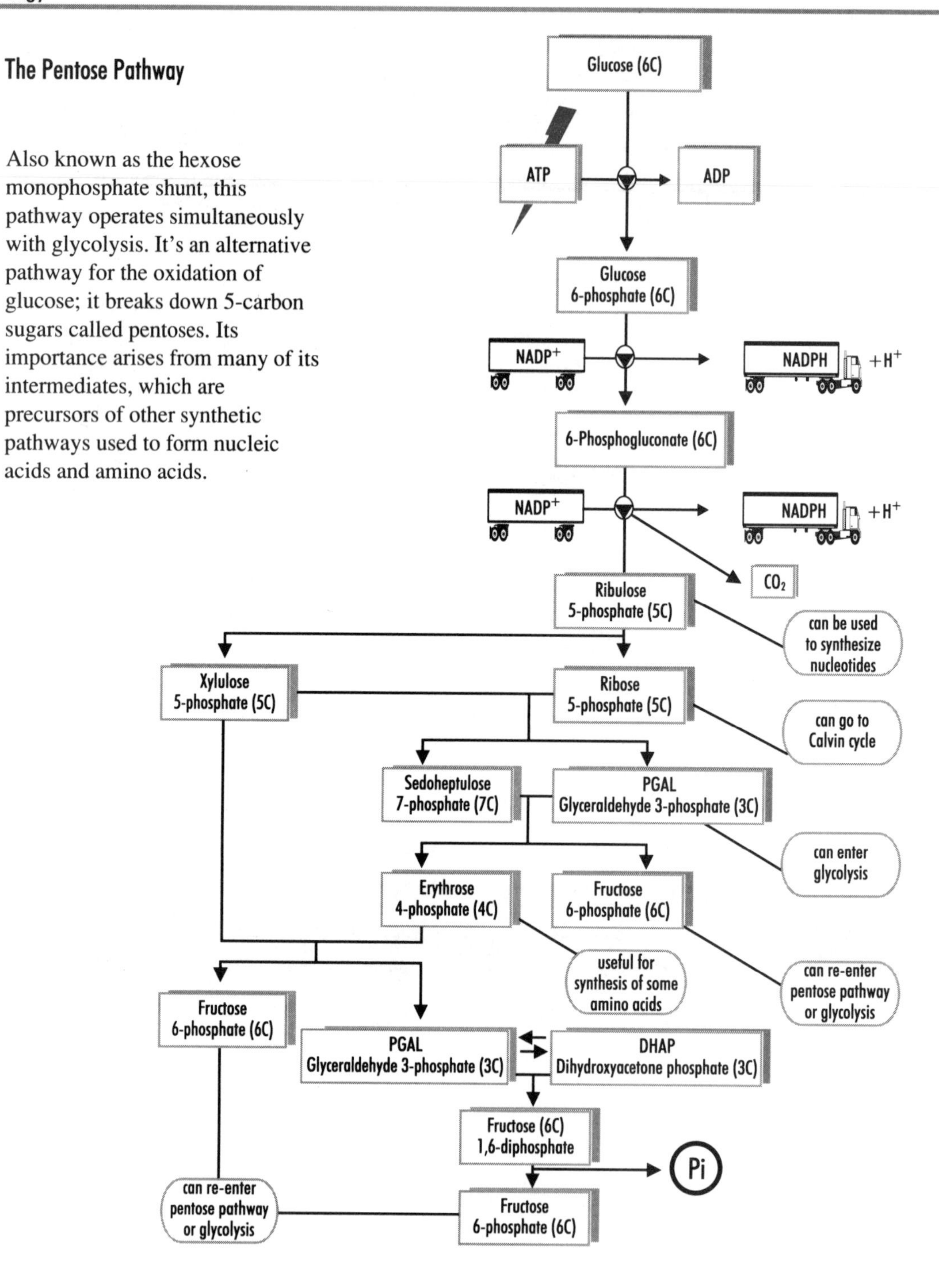

Fig. 4.6.4 Pentose Pathway

The Citric Acid Cycle

This cycle, also known as the **Krebs cycle** or **tricarboxylic acid cycle (TCA)**, completes the breakdown of pyruvate formed from glucose through glycolysis. The cycle liberates carbon dioxide (CO_2) and a wealth of hydrogen atoms which are picked up by the electron transport carriers NAD^+ and FAD in the mitochondria of eukaryotes.

1) The reaction between acetyl-CoA and oxaloacetate is catalyzed by the enzyme citrate synthetase. CoA leaves spontaneously after the molecules join.

2) Aconitase repositions a hydroxyl (OH) group to a more active location. To do this, a water molecule is removed (**dehydration**), then relocated to a different carbon atom (**hydration**), forming an isomer called isocitrate.

3) Isocitrate is **oxidized** (loses electrons) and **decarboxylated** (loses carbon dioxide). This reaction occurs on the surface of an enzyme known as isocitrate dehydrogenase. NAD^+ first oxidizes isocitrate changing it to oxalosuccinic acid. After this, CO_2 is given off spontaneously and alpha-ketoglutarate is formed.

4) Alpha-ketoglutarate dehydrogenase then catalyzes the following reaction: CO_2 leaves, CoA is joined to the decarboxylated molecule, then NAD^+ is reduced to form $NADH + H^+$.

5) The energy stored in the ester bond between succinyl and CoA is comparable to the phosphate bond of ATP. Succinyl-CoA synthetase uses this potential energy to drive the **phosphorylation** (addition of a phosphate group) of GDP to GTP. This then converts ADP to ATP (except in plants where ATP is formed directly).

6) FAD readily oxidizes succinate to fumarate. NAD^+ is not the **coenzyme** in this step because it is not an oxidizer of sufficient strength.

7) Fumarase hydrates fumarate, making malate.

8) In the final step, newly formed malate is oxidized by NAD^+. The reaction governed by malate dehydrogenase yields oxaloacetate, which is now free to start the cycle again.

The human brain has as many cells as there are stars in our galaxy. David Dewar

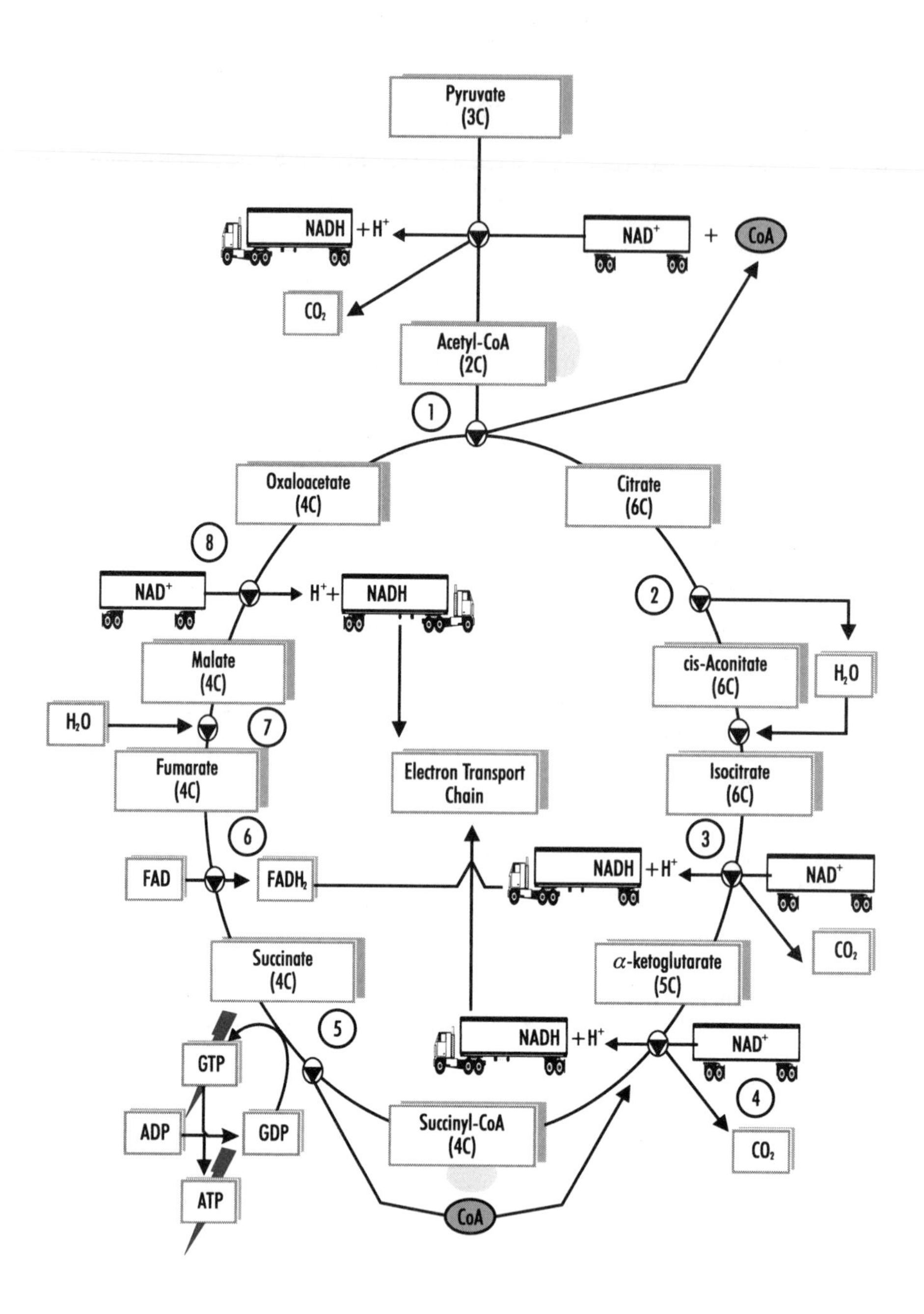

Fig. 4.6.5 Citric Acid Cycle

The Electron Transport Chain

The electron transport chain is a group of associated electron carriers that generates most of the ATP in our cells. It functions by carrying electrons from donor to acceptor (oxygen), and generates energy in the form of ATP. Much of the process occurs in the mitochondrion's inner membrane via **membrane associated electron carriers**. Here, there are three classes of carriers: flavoproteins, cytochromes and ubiquinones. **Flavoproteins** contain flavin, and are derived from riboflavin (vitamin B_2). **Cytochromes** are proteins containing a porphyrin ring complex that holds iron (heme). **Ubiquinones** (coenzyme Q) are nonprotein carriers.

Since the electron transport chain is located at the mitochondrion's **inner membrane**, NADH has to travel there; $FADH_2$ is already attached to the inner membrane. NADH transfers its electrons to FMN. This transfer of electrons to the electron transport chain involves one hydrogen atom from NADH, two electrons, and a proton from the medium. If you're wondering how NADH donates one pair of electrons, remember that one electron was part of the hydrogen atom and the other electron came from NAD to yield NAD^+. Then NADH becomes NAD^+, and FMN is reduced to $FMNH_2$. $FMNH_2$ passes 2 of its electrons to coenzyme Q, and then successively to each cytochrome while the 2 liberated protons are passed into the lumen.

The electrons eventually join with oxygen (final electron acceptor) and certain protons in the medium to form water. Liberated protons are pumped to the outside of the inner membrane, resulting in an H^+ gradient across the membrane. The high concentration of protons in between the two membranes causes them to flow through special protein channels known as **ATP synthetase** enzyme complexes. This travel is known as **chemiosmosis**. It gives enough energy to attach a phosphate group to ADP, making ATP. Three ATP are generated for each NADH + H^+ and two ATP for each $FADH_2$ that are oxidized to NAD^+ and FAD respectively.

The electron transport chain regenerates NAD^+ and FAD. These coenzymes give off protons and high-energy electrons when being regenerated from their reduced form. The energy of the electrons is used to pump the protons out of the matrix and into the lumen. The concentration of protons in the lumen is so high that diffusion pushes the protons through special channels and takes energy from the protons to generate ATP. The used protons and electrons combine with oxygen to form water.

A shark can detect a drop of blood from 200 meters away.

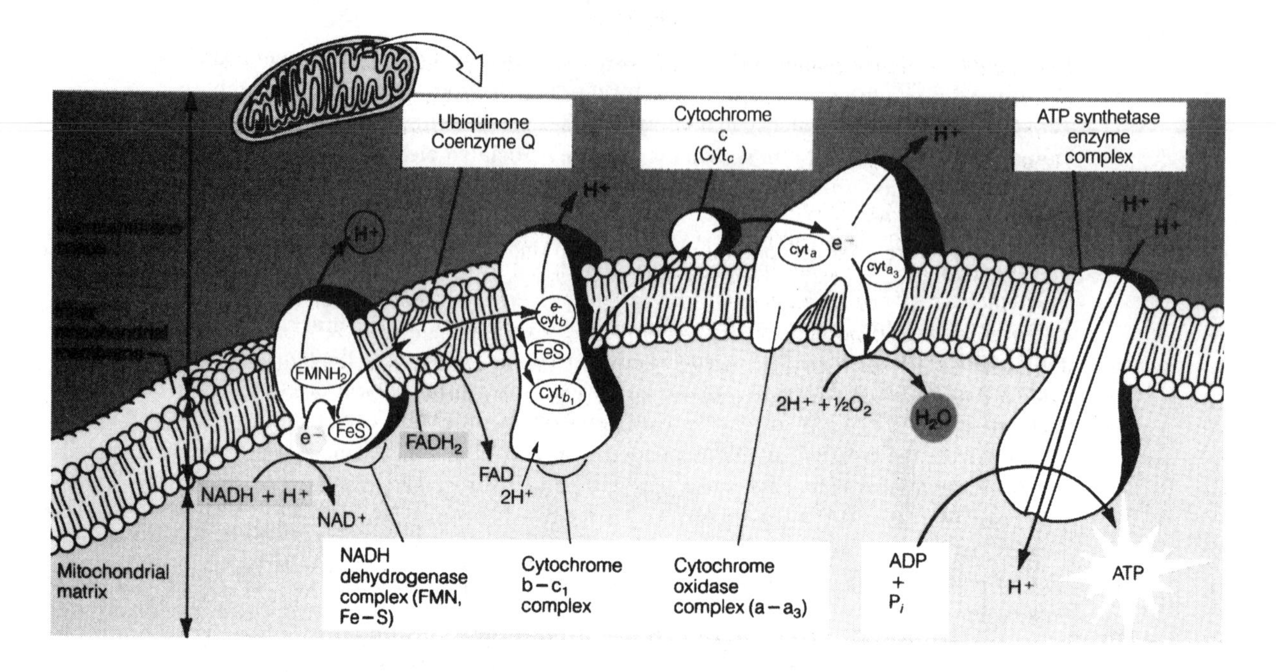

Fig. 4.6.6 Oxidative Phosphorylation

Photosynthesis

Photosynthesis converts light energy into a stored chemical energy form. The raw materials used are carbon dioxide (CO_2) and water (H_2O); the products are carbohydrates (sugars) and oxygen (O_2).

$$6\ H_2O + 6\ CO_2 \rightarrow C_6H_{12}O_6 + 6\ O_2$$

There are two major interdependent reactions in photosynthesis: **light-dependent reactions** and **dark reactions**. Light reactions capture light energy and store it as ATP. Dark reactions use ATP to make carbohydrates (sugars).

In Canada, the Queen is forbidden to enter the House of Commons.

Light Reactions

In more detail, light-dependent reactions occur at the **thylakoid membrane** in chloroplasts where both photosystems (I and II) are present. Each system specializes in absorbing a specific light wavelength. Photosystem I best absorbs light at 700 nm; photosystem II at 680 nm.

Light energy is gathered via chlorophyll **antenna pigments** in the form of **excitation energy**. This energy migrates from molecule to molecule within the photosystem until it reaches the **light trap,** which is a special chlorophyll molecule in the photosystem's **reaction center**. This prompts an electron transfer to an electron acceptor in the transport chain from water. Water is split into electrons (e^-), oxygen and protons (H^+). Oxygen diffuses out of the cell. Protons and electrons are used to drive the production of ATP, and to reduce $NADP^+$ to NADPH.

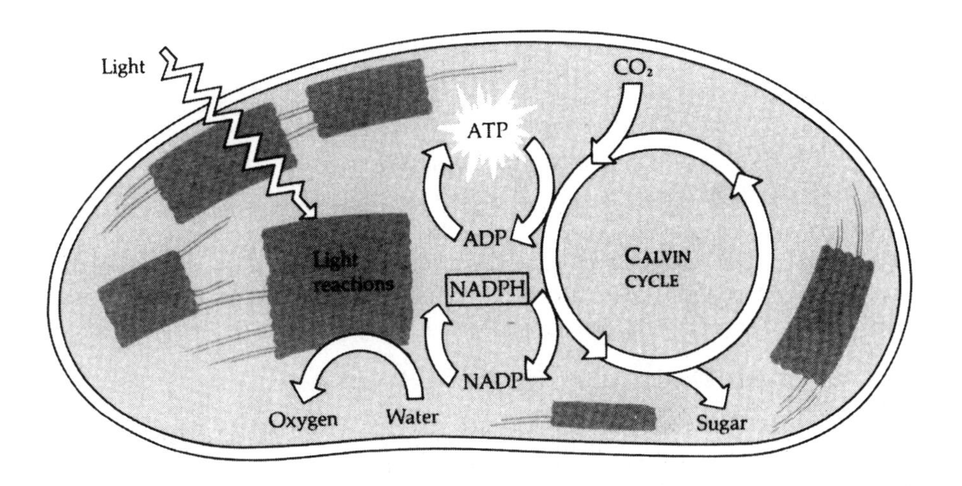

Fig. 4.6.7 Summary of Photosynthesis in C_3 Plants

Fools never open their mouths without subtracting from the sum of human knowledge. Thomas B. Reed

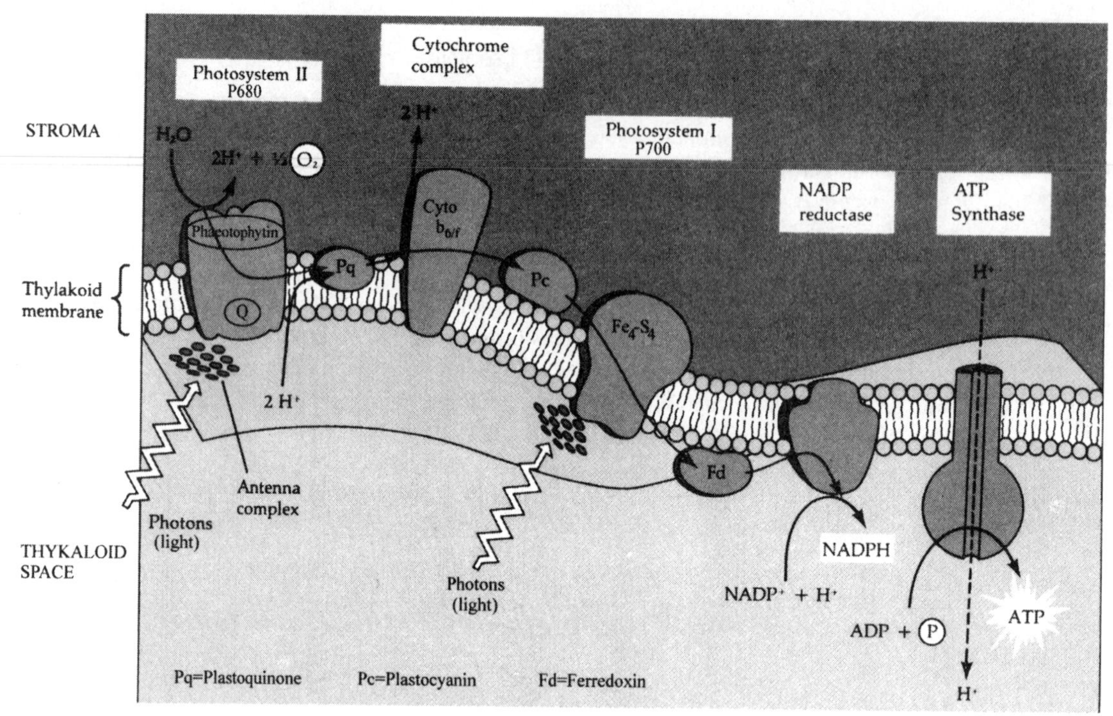

Fig. 4.6.8 Photophosphorylation

The Dark Reactions

Dark reactions such as the Calvin cycle are a series of reactions that produce six carbon sugars (such as glucose) by a process called **carbon fixation**. It occurs in the **stroma** without requiring direct light. This **anabolic** construction of sugars is shown in more detail in the diagram illustrating the Calvin (or C_3) cycle (see Fig. 4.6.9).

C_3 photosynthesis. The dark reactions form a cycle with three major steps: the reformation of ribulose 1,5-diphosphate (5C) (steps 1 and 2), carbon fixation, and a reverse of glycolysis. To re-form ribulose 1,5-diphosphate, DHAP must first be converted to ribulose 5-phosphate via several complicated steps. The ribulose 5-phosphate is prepared to receive carbon dioxide after it has been phosphorylated by ATP to form ribulose 1,5-diphosphate.

During carbon fixation (step 3), carbon dioxide is incorporated into the molecule and the newly formed (6C) sugar is split into two PGA (3C) molecules using water.

Carbon fixation results in two extra PGAL being formed. The cycle starts with 10 DHAP; after one turn of the cycle, 12 PGAL are formed. Ten PGAL are converted back to DHAP to replenish the cycle, leaving 2 extra PGAL. These two extra PGAL result from the fixation of carbon, and are used for the production of glucose and other hexose (6C) sugars.

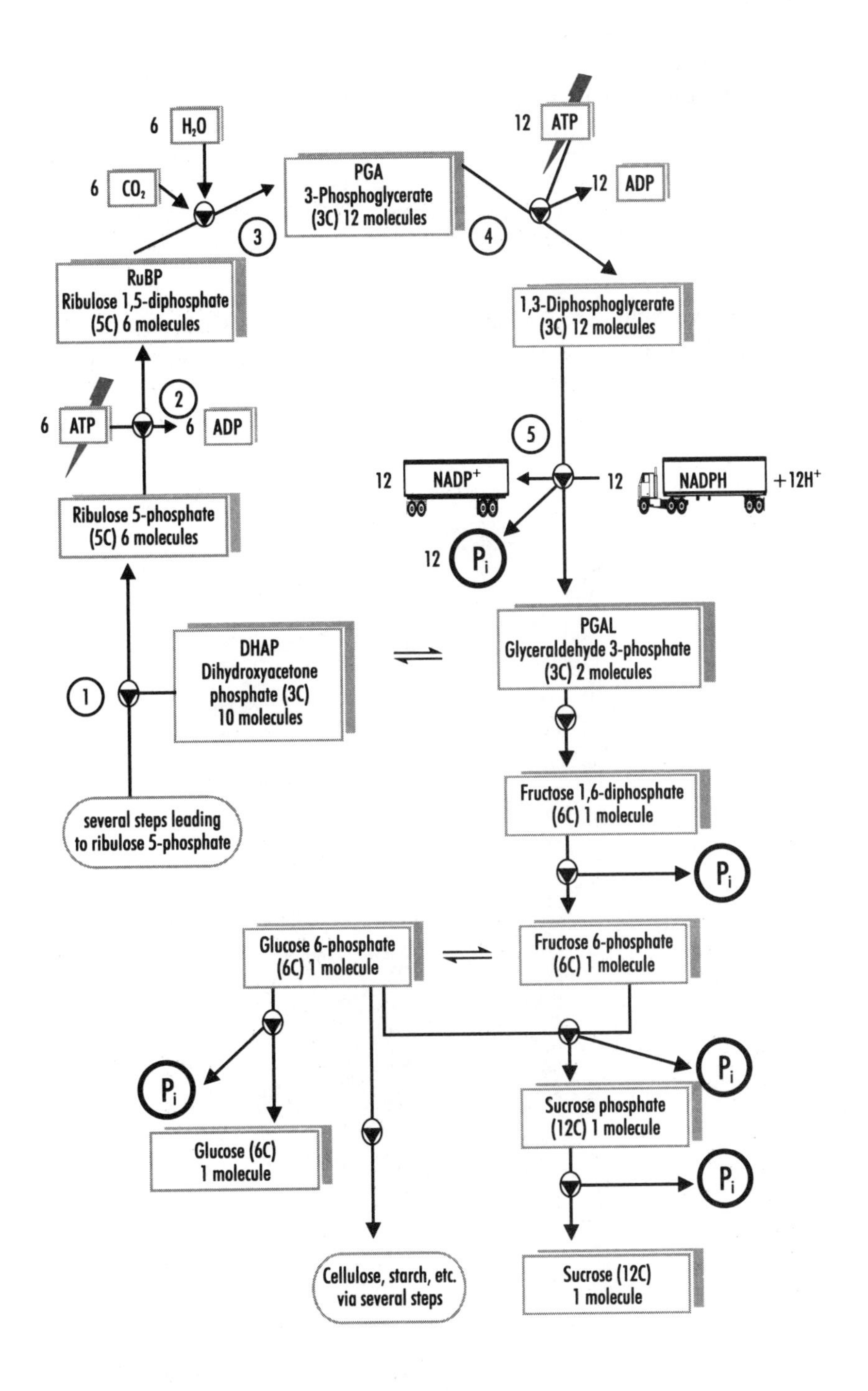

Fig. 4.6.9 Calvin Cycle

The reverse of glycolysis (steps 4 and 5) then begins with PGA becoming rephosphorylated by ATP. The resulting product is reduced by NADPH + H^+ and loses a phosphate group to form PGAL.

Two molecules of PGAL (3C) are joined to form one (6C) fructose 1,6-phosphate. After inorganic phosphate (P_i) leaves, fructose 6-phosphate (F6P) is the result. F6P is also interconvertible with glucose 6-phosphate (G6P). When both G6P and F6P are joined, and each lose their phosphate groups, the resulting disaccharide is sucrose, or table sugar.

C_4 photosynthesis. In warmer climates, plants such as corn, sugar cane and sorghum have evolved to make the process more efficient. Warm temperatures cause plants to lose carbon dioxide without forming ATP or NADPH, a process called **photorespiration**. C_4 plants overcome this by fixing carbon dioxide until it is needed by the Calvin cycle.

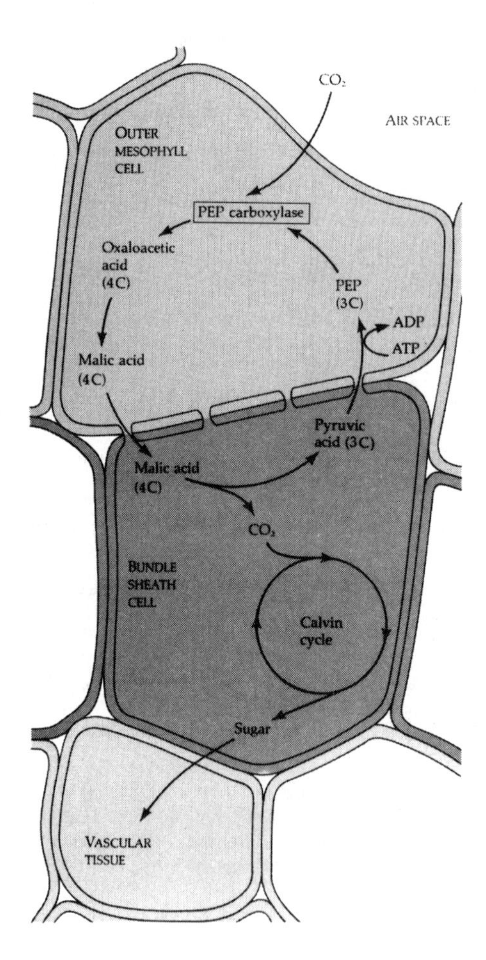

Fig. 4.6.10 C_4 Pathway

4.7 Protein Synthesis

Genetic information, stored as **deoxyribonucleic acid** (DNA), acts like an organism's architectural blueprint. DNA tells RNA what proteins to build for the cell. Three types of RNA are used to build proteins from 20 different amino acids. Think of DNA as the architect, RNA as the workers, amino acids as the bricks, and proteins as structures in a building. The protein-building process occurs in both eukaryotes and prokaryotes (with some variations). The process has two stages: **transcription and translation**.

An Aid to Understanding	
Downstream	Toward the 3′ end of a nucleic acid strand.
Exons	A gene's coding segment. Exons are joined together to form mature mRNA after introns have been cut out.
Frameshift mutation	A shift in the reading frame resulting from the deletion or insertion of a nucleotide. This causes the coding segments of the genetic code to change.
Intron	A gene's non-coding segment. When a gene is transcribed to mRNA the introns are cut out. Mature mRNA is exported from the nucleus of eukaryotes. Introns are generally only seen in eukaryotes.
Missense	A point mutation resulting in the coding of a different amino acid.
Mutation	Any change in the genetic material.
Non-sense mutation	A specific change in genetic material that results in premature termination (stop command) of protein synthesis.
Nucleoside	A nucleotide missing a phosphate group.
Nucleotide	The monomer subunit from which nucleic acids such as DNA are made. One nucleotide consists of a phosphate group joined to a sugar joined to a base. In DNA this sugar is deoxyribose; in RNA, ribose.
Operator	A region of DNA where a repressor binds. The repressor prevents RNA polymerase from binding to the promoter.
Operon	A cluster of related genes controlled by an operator. Eukaryotes do not have operons.
Point mutation	A mutation involving a base substitution in DNA.
Polypeptide	A peptide is the bond of two joined amino acids. A polypeptide has many of these bonds; it's another way of saying many joined amino acids.
Promoter	A segment of DNA where RNA polymerase joins to begin transcription.
Transcription	The process of synthesizing mRNA from DNA.
Translation	The process of translating the genetic language of mRNA to amino acid language, producing proteins.
Upstream	Toward the 5′ end of a nucleic acid strand.

Some folks are wise, and some are otherwise. Tobias Smollett

The information in DNA and RNA is written like a book containing letters and words. The "alphabet" used by DNA is made up of four letters: **A**, **C**, **G** and **T**. Each letter stands for the bases **Adenine**, **Cytosine**, **Guanine** and **Thymine**. In RNA, however, the letter **U** (**Uracil**) is used instead of **T** (**Thymine**). The "words" that make up DNA and RNA are three letter **codons**. Unlike English, each word is not separated by a space. For example, the message, "AUGGGC" is from RNA because only RNA has the letter "U." The message contains two words, because each word can only be three letters long. Each word is like a command in computer code: start, stop, place a certain amino acid here, etc.

DNA to mRNA (Transcription)

RNA polymerase binds to a **promoter site** on DNA and begins to move along the blueprint strand. RNA polymerase moves along DNA in a 3′ to 5′ fashion while placing and joining complementary RNA bases in a fashion very similar to DNA synthesis. Once a **termination signal** is reached, RNA polymerase interprets this as the end of the gene, even if it's not. Then, polymerase separates from the DNA strand and the newly formed single strand of mRNA is released. RNA polymerase is an enzyme that reads DNA and generates a copy of RNA, according to the following rules.

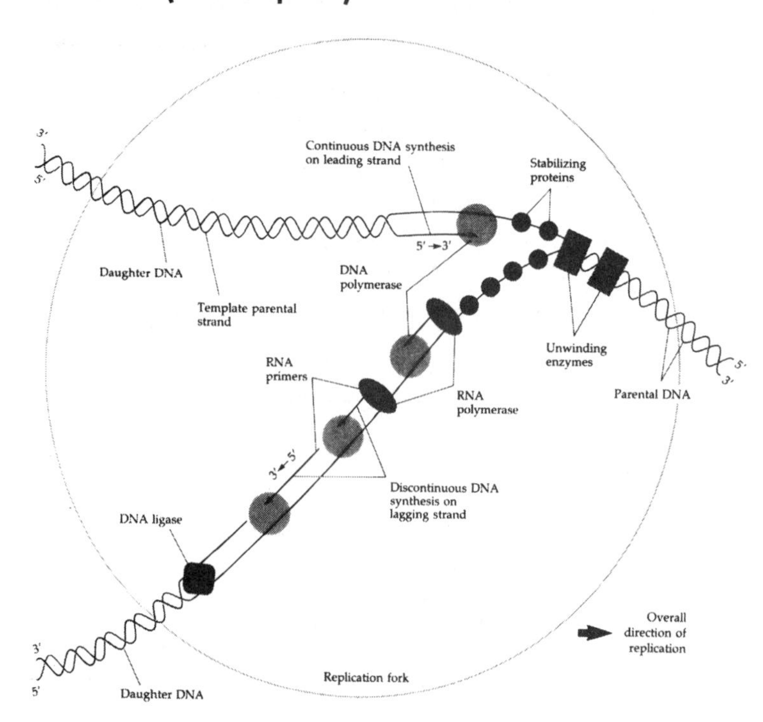

Fig. 4.7.1 Replicative Fork

Complementary Base Pairing		
DNA – DNA*	– Transcription →	RNA*
A = T		A
C ≡ G		C
G ≡ C		G
T = A		U

RNA and Its Many Forms

Messenger RNA (mRNA). This single-stranded intermediate in protein synthesis brings the information from DNA and binds to a ribosome. Once bound, it is then translated into a protein.

There are some differences in the properties of mRNA in prokaryotes and eukaryotes. Newly synthesized mRNA in prokaryotes is translated in the cytoplasm as it is transcribed. It usually lasts about five minutes, does not undergo modification and commonly codes for many proteins.

In eukaryotes, newly synthesized mRNA is considered immature, and is called heterogeneous nuclear RNA (hnRNA). This precursor of mature mRNA is synthesized in the nucleus. It undergoes different types of modification called **capping**, **methylation** and **polyadenylation**. After modification, the strand's lifetime is longer, ranging from minutes to hours. Then, non-coding regions called **introns** must be removed before mature mRNA is exported to the cytoplasm where it is translated.

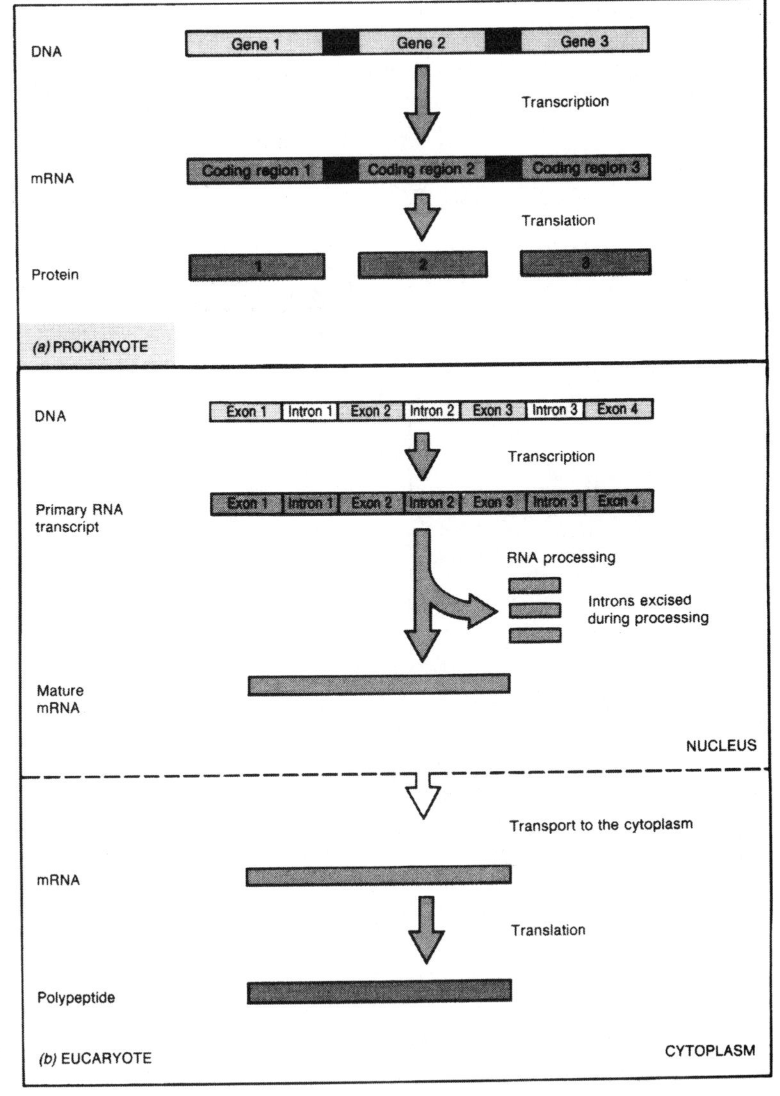

Fig. 4.7.2 DNA Transfer in Prokaryotic vs. Eukaryotic Cells

Transfer RNA (tRNA). tRNA can be recognized by its distinctive clover leaf shape. There is one tRNA for every kind of amino acid. It has two important regions: the **anticodon**, and the **aminoacyl attachment site** at the 3′ end. The anticodon is where tRNA joins the codon on an mRNA strand. This occurs by complementary base pairing. The aminoacyl site is where the amino acid is attached to tRNA. Precursors of tRNA are called pre-tRNA, and undergo modification of many of their bases to create the required three-dimensional shape.

Ribosomal RNA (rRNA). This type of RNA forms a complex with certain proteins to make **ribosomes**. It functions to provide a site where mRNA can bind to the ribosome to allow translation. Pre-rRNA is the precursor to rRNA, and is modified through methylation and specific cleavage. After modification, many different mature products are formed. In eukaryotes, these products are (5S, 5.8S, 18S, 28S); in prokaryotes, (5S, 16S, 23S, spacer tRNA).

The Genetic Code

First base	U	C	A	G	Third base
U	UUU Phe	UCU Ser	UAU Tyr	UGU Cys	U
U	UUC Phe	UCC Ser	UAC Tyr	UGC Cys	C
U	UUA Leu	UCA Ser	UAA Stop	UGA Stop	A
U	UUG Leu	UCG Ser	UAG Stop	UGG Trp	G
C	CUU Leu	CCU Pro	CAU His	CGU Arg	U
C	CUC Leu	CCC Pro	CAC His	CGC Arg	C
C	CUA Leu	CCA Pro	CAA Gln	CGA Arg	A
C	CUG Leu	CCG Pro	CAG Gln	CGG Arg	G
A	AUU Ile	ACU Thr	AAU Asn	AGU Ser	U
A	AUC Ile	ACC Thr	AAC Asn	AGC Ser	C
A	AUA Ile	ACA Thr	AAA Lys	AGA Arg	A
A	AUG Met or start	ACG Thr	AAG Lys	AGG Arg	G
G	GUU Val	GCU Ala	GAU Asp	GGU Gly	U
G	GUC Val	GCC Ala	GAC Asp	GGC Gly	C
G	GUA Val	GCA Ala	GAA Glu	GGA Gly	A
G	GUG Val	GCG Ala	GAG Glu	GGG Gly	G

(Column headers U, C, A, G = Second base)

Fig. 4.7.3 The Genetic Code

Amino Acids

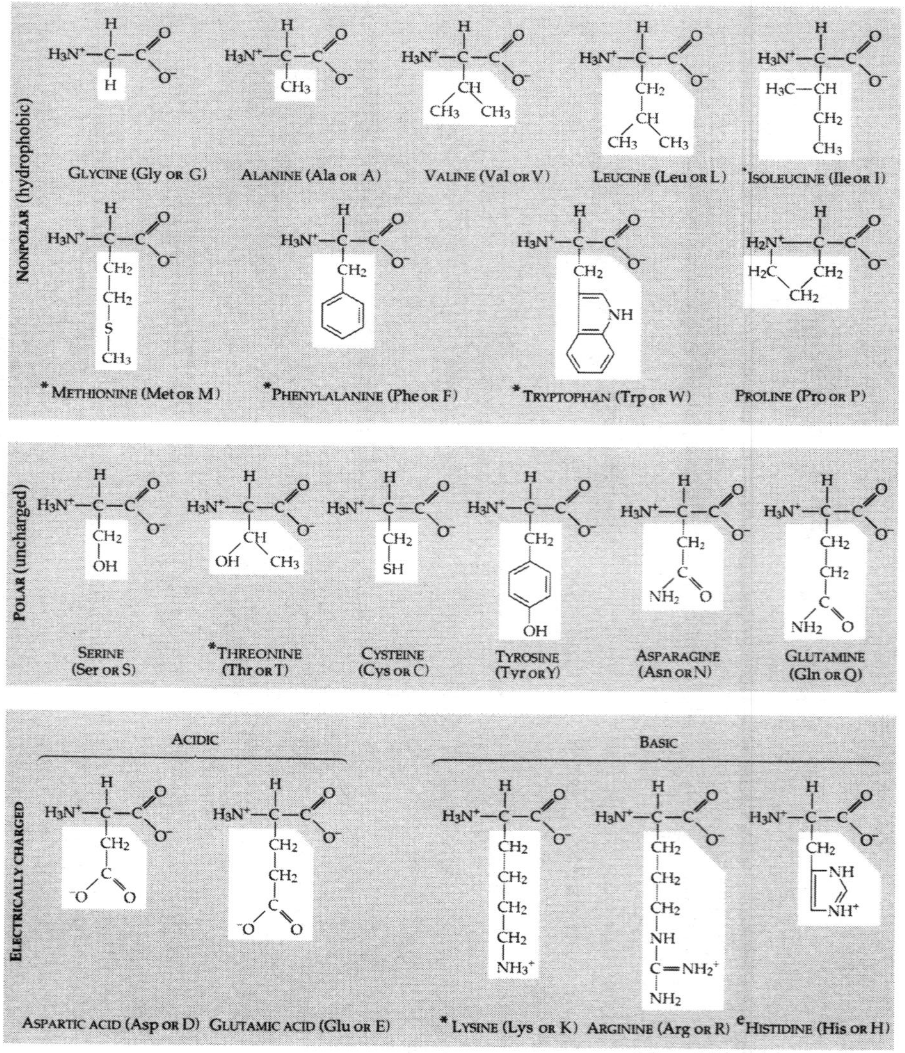

Nonpolar (hydrophobic)

Glycine (Gly or G)

Alanine (Ala or A)

Valine (Val or V)

Leucine (Leu or L)

*Isoleucine (Ile or I)

*Methionine (Met or M)

*Phenylalanine (Phe or F)

*Tryptophan (Trp or W)

Proline (Pro or P)

Polar (uncharged)

Serine (Ser or S)

*Threonine (Thr or T)

Cysteine (Cys or C)

Tyrosine (Tyr or Y)

Asparagine (Asn or N)

Glutamine (Gln or Q)

Electrically charged

Acidic

Aspartic acid (Asp or D)

Glutamic acid (Glu or E)

Basic

*Lysine (Lys or K)

Arginine (Arg or R)

ᵉHistidine (His or H)

Fig. 4.7.4 Amino Acids

* Essential for adults

e Essential for children

mRNA to a Protein (Translation)

Translation encompasses three main events: **initiation**, **elongation** and **termination**. This process resembles sewing. Use the following loose analogy to help you remember. Initiation is like threading a needle and tying a knot. Elongation is the actual sewing. Termination is tying the final knot and clipping the thread.

Initiation

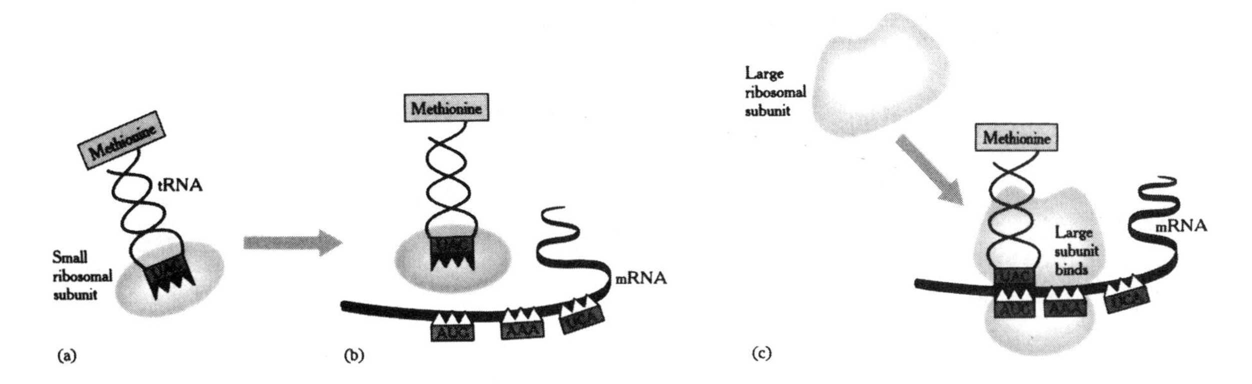

Fig. 4.7.5 Initiation

The beginning of initiation is marked by mRNA binding to the small ribosome subunit at the **ribosome binding site** (a, b). The anticodon of tRNA carrying methionine then binds to the mRNA at the start codon, usually AUG (b). Then the large subunit binds to the small subunit (c). Methionine is not always the first amino acid incorporated. This marks the end of initiation. Now elongation can begin.

Elongation

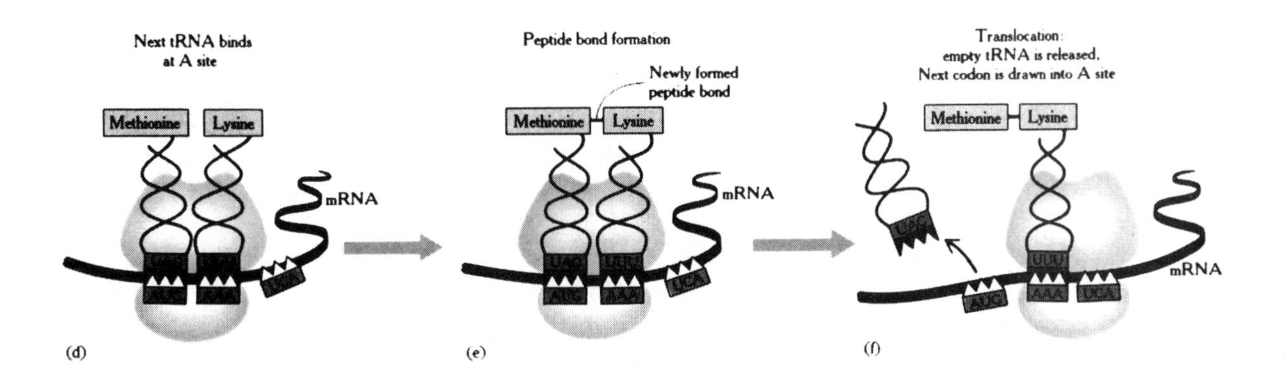

Fig. 4.7.6 Elongation

Elongation is cyclic and occurs in three steps. The next tRNA is accepted into the A-site (d). A **peptide bond** between the two amino acids in each site forms and tRNA in the P-site releases its amino acid (e). This results in the **polypeptide** attaching to tRNA in the A-site. A peptide bond results when the amino (NH_2) group reacts with a carboxylic acid (COOH) group. Now tRNA is released from the P-site. Then the ribosome's movement (**translocation**) along mRNA to the next codon results in the tRNA holding the peptide to occupy the P-site (f). Elongation repeats the first step until termination.

Termination

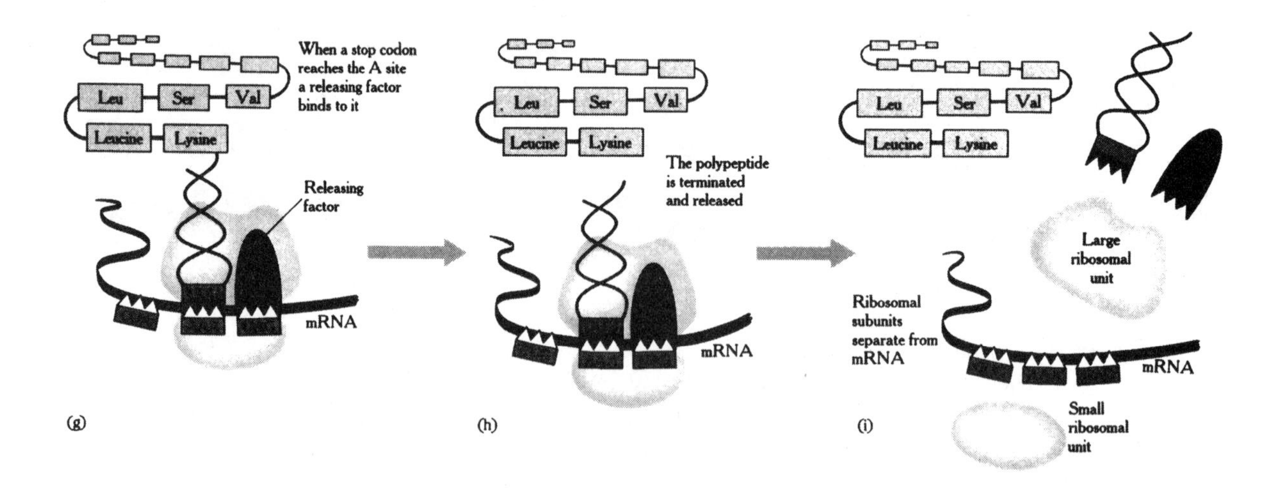

Fig. 4.7.7 Termination

When the next codon in the A-site is a **termination codon** (UAA, UAG, UGA), a protein called a **releasing factor** binds to the A-site (g). This causes the polypeptide or protein attached to the tRNA in the P-site to be released upon the addition of water (h). After the protein is released, the complex splits apart into its single components (i).

In prokaryotes, approximately 15 amino acids are joined per second. In eukaryotes, translation halts if the **leader peptide** (first few amino acids) is recognized by a **signal recognition particle** (**SRP**). The entire complex is then directed to the rough endoplasmic reticulum (rER) where it binds to a docking protein. Translation resumes and the polypeptide emerges inside the rER. Proteins targeted to the rough endoplasmic reticulum are products usually exported by that particular cell.

4.8 Stains & Tests

Common Tests for Biological Molecules

Test	Molecules	Comments
Benedict's	reducing sugars	Monosaccharides stain green to red to orange, depending on concentration. Disaccharides stain green.
Biuret	proteins	A small number of peptides stain pink. Many peptides stain purple. Blue indicates negative results.
Fehling's	reducing sugars	Forms red precipitates with monosaccharides and some disaccharides.
Iodine	starch	Cellulose stains blue-grey. Glycogen stains red-brown. Amylose stains blue-black.
Millon's	proteins	Brick-red precipitate indicates positive results.
Sudan III	lipids	Red layer indicates positive results.
Sudan black	lipids	Darkly stained upper layer indicates positive results against lightly stained aqueous layer.

Common Stains Used in the Laboratory

Stain	Comments
Aniline blue	Usually used to stain fungi and sieve plates.
Congo red	Nuclei stain blue, amyloid stains brick red.
Eosin Y	Stains cellulose and cytoplasm.
Feulgen	Useful for staining DNA purple.
Gentian violet	Stains nuclei.
Gram's	Gram-positive bacteria stain purple. Gram-negative bacteria stain red.
Iodine	Stains short starch chains red-brown. Stains long chains blue.
Leishman's	Stains nuclei of white blood cells.
Methylene blue	Stains bacteria, nuclei.
Periodic acid-Schiff's (PAS)	Stains cell walls and starch grains red.
Toluidine blue O	Companion cells stain pink. Xylem cell elements stain green-turquoise. Phloem stains blue.

It is later than you think. Chinese proverb

4.9 Important Molecules

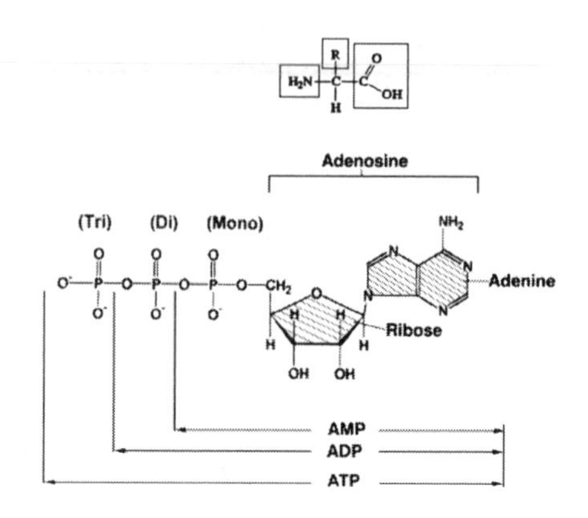

Amino acids. The building blocks of proteins. Each different amino acid has a different chemical group (R) which gives it unique properties. (Refer to section 4.7 for a complete list of amino acids.)

ATP. Adenosine triphosphate stores energy for use during cellular metabolism in high energy phosphate bonds. These bonds release their energy when hydrolyzed.

Carbohydrates (sugars). The simplest type of sugar is monosaccharide. Monosaccharides are building blocks for more complex disaccharides and polysaccharides.

Common monosaccharides

D-glucose D-galactose D-fructose D-deoxyribose D-ribose

Common Disaccharides

Disaccharide	Subunit	Linkage type (*)	Sources
Sucrose	D-glucose, D-fructose	$\alpha(1,2)\beta$	all photosynthetic plants, especially sugar cane and sugarbeets
Maltose	D-glucose	$\alpha(1,4)$	hydrolyzed starch
Cellobiose	D-glucose	$\beta(1,4)$	hydrolyzed cellulose
Lactose	D-glucose, D-galactose	$\beta(1,4)$	milk of most mammals

Common Polysaccharides

Polysaccharide	Subunit	Linkage type (*)	Sources
Starch	amylose, amylopectin (D-glucose)	$\alpha(1, 4)$	roots, tubers, seeds of plants
Amylose	D-glucose (linear coil)	$\alpha(1, 4)$	starch
Amylopectin	D-glucose (branched)	$\alpha(1,4)$ and $\alpha(1,6)$	starch
Glycogen	D-glucose (highly branched)	$\alpha(1,4)$ and $\alpha(1,6)$	energy storage in animal cells
Cellulose	D-glucose (bundled chains)	$\beta(1,4)$	cell walls

* All linked by glycosidic bonds.

Carbon dioxide. A gas in air used by photoautotrophic organisms to build carbon-containing molecules such as carbohydrates. It's produced as a by-product of respiration.

Chlorophyll. A group of related molecules (a,b,c,d) made of a porphyrin ring (containing a Mg ion) and a hydrocarbon tail. Found in photosynthetic organisms, these molecules are pigments that absorb light energy, and pass it on to the light reactions of photosynthesis.

Depressants. Molecules that retard the activity of the central nervous system (CNS). Types of depressants include alcohol, tranquilizers and barbiturates.

Ethanol Valium (diazepam) 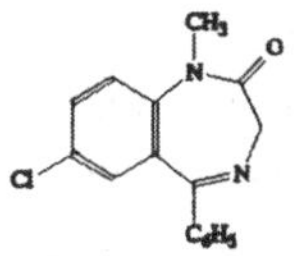 Phenobarbital

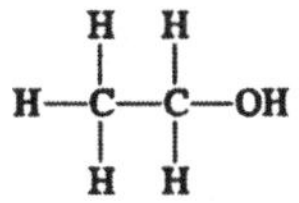

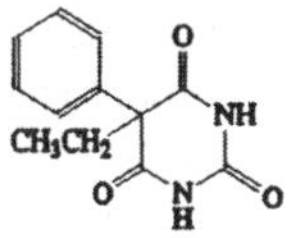

Fatty acids. Molecules with a carboxylic acid head bound to a long unbranched nonpolar hydrocarbon chain. Saturated fatty acids contain the maximum number of hydrogens on the hydrocarbon chain and have no double bonds between carbons. Unsaturated hydrocarbons lack hydrogen atoms, creating double bonds between carbons.

Common Fatty Acids

Saturated		Unsaturated		
Number of carbons	**Common name**	**Number of carbons**	**Number of double bonds**	**Common name**
12	Laurate	16	1	Palmitoleate
14	Myristate	18	1	Oleate
16	Palmitate	18	2	Linoleate
18	Stearate	18	3	Linolenate
20	Arachidate	20	4	Arachidonate
22	Behenate			
24	Lignocerate			

Functional groups. Chemical substituents attached to molecules that give them special properties. (For a list of functional groups refer to Section 5.3.)

Glycerol. A three-carbon alcohol with a hydroxyl group (OH) attached to each carbon. Forms a backbone to many lipids through ester bonds.

Lipids. A diverse group of compounds that are insoluble in water.

Common Lipids

Name	Notes on structure	Examples
Carotenes	Diterpenes linked in a tail-to-tail arrangement.	α, β, γ carotenes
Fats	Ester bonding between glycerol (OH) and a fatty acid R(COOH), solid at room temperature.	triglycerides
Oils	Ester bonding between glycerol (OH) and fatty acid R(COOH), liquid at room temperature.	triglycerides
Phospholipids	Ester bonding between glycerol, phosphoric acid and 2 fatty acids.	cell membranes
Prostaglandins	A C_{20} carboxylic acid, 5 member ring, one or more double bonds, oxygen containing functional groups.	found in all animal tissue; governs heart rate, blood pressure, clotting, fertility, allergic response
Sphigolipids	Sphingosine which may be bonded to sugars, fatty acids or phosphoric acid.	sphingomyelin, cerebroside
Steroids	Based on the perhydrocyclopentanophenthrene ring system (6,6,6,5 carbon ring structure).	biological regulators such as estradiol, testosterone, and cholesterol
Waxes	Ester bonding between long chain fatty acids and alcohols.	beeswax, protective coatings on animals and plants

Neurotransmitters. The nervous system's chemical messengers. A few examples:

Dopamine (DA) Norepinephrine (NE) Serotonin (5-HT) Glutamate

Nucleic acids. DNA and RNA are long polymers made up of nucleotides. Nucleotides are made of a five-carbon sugar (pentose) called deoxyribose (DNA) or ribose (RNA). They also contain a heterocyclic nitrogen base and a phosphate group. The structure of a single strand of DNA resembles the molecule shown on the right.

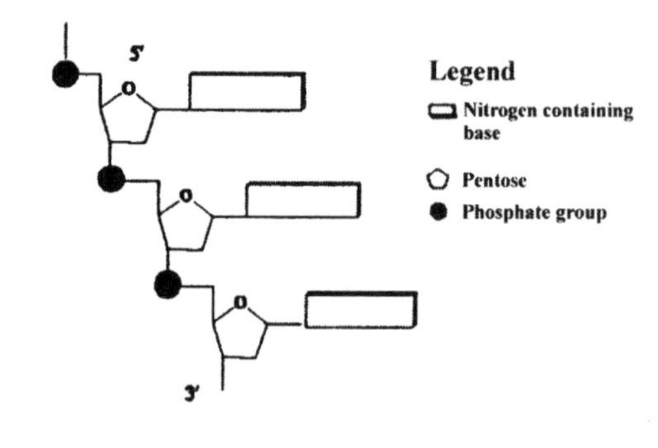

Nitrogen-containing bases of nucleic acids:

Adenine (A) Guanine (G) — Purines

Cytosine (C) Thymine (T)* Uracil (U)* — Pyrimidines

* Note that thymine is found only in DNA, uracil only in RNA.

Minerals. Inorganic substances required for an organism's proper functioning. Technically, they're not molecules, but are listed due to their biological importance.

Common Minerals

Name	Common functions	RDA
Calcium	Bones, teeth, clotting, muscles and nerves	1000-1500 mg
Chlorine	Gastric juice, normal digestion	700-5100 mg
Chromium[t]	Works with insulin and glucose metabolism	0.03-0.2 mg
Copper[t]	Brain, blood cell function (hemoglobin formation), cytochromes	2 mg
Fluorine[t]	Bones, teeth	1-4 mg
Iodine[t]	Component of thyroid gland hormones	90-150 μg
Iron[t]	Found in the cytochromes, hemoglobin and myoglobin	10-18 mg
Magnesium	Body metabolism	200-300 mg
Manganese[t]	Bone, tendons, enzymes	1.5-5 mg
Molybdenum[t]	Enzyme: xantine oxidase	0.06-0.5 mg
Phosphorous	Bones, teeth and metabolism	1000 mg
Potassium	Body fluid balance, nerves	775-5600 mg
Selenium[t]	Works with vitamin E and in fat metabolism	0.03-0.2 mg
Sodium	Body fluid balance, nerves	450-3300 mg
Sulfur	Component of many body molecules such as proteins	Unknown
Zinc[t]	Element of enzymes	10-15 mg

[t] Indicates trace minerals.

Opiates and related molecules. A group of molecules that inhibit pain perception.

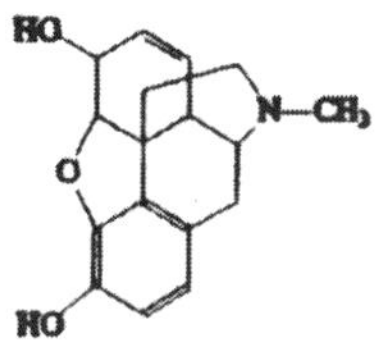

Morphine 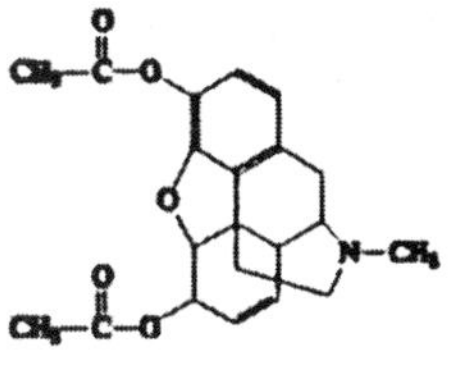 | **Heroin (diacetylmorphine)** 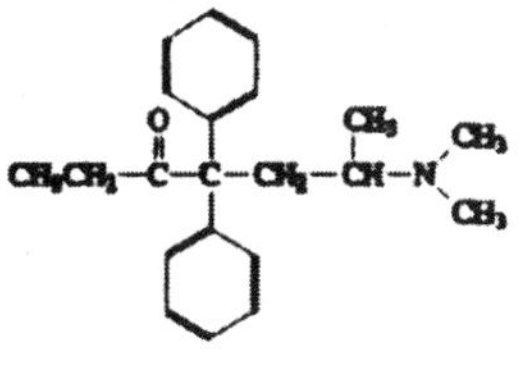 | **Methadone**

There is no sin except stupidity. Oscar Wilde

Proteins. A long chain of peptide-bonded amino acids that serve many functions in organisms. (See Section 4.7 for more information.)

Protein Properties

Protein type	Function	Protein type	Function
Albumin	Osmotic pressure balance	Hormone	Regulation of body functions
Surface antigen	Cell-to-cell recognition	Immunoglobin	Bodily defense
Channel	Transport across membranes	Ion-binding	Storage of ions
Enzyme	Aid in metabolism	Contractile	Contraction of muscle
Structural (fiber)	Structure and support	Repressor	Regulate gene activity
Transport (globin)	Bind molecules in blood	Toxin	Bodily defense/offense

Levels of protein structure. Proteins fold into shapes which may be analyzed at four levels: **primary structure**, the amino acid sequence; **secondary structure**, local amino acid interactions (i.e. α-helix, β-pleated sheets); **tertiary structure**, secondary structure foldings; **quaternary structure**, the protein subunit.

Psychedelics. Molecules that alter consciousness, produce hallucinations and disrupt perception.

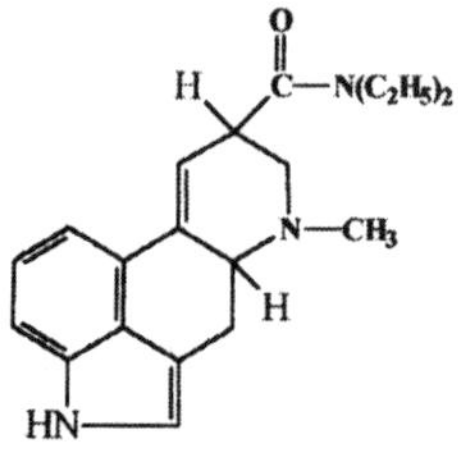

LSD-25
Lysergic acid diethylamide

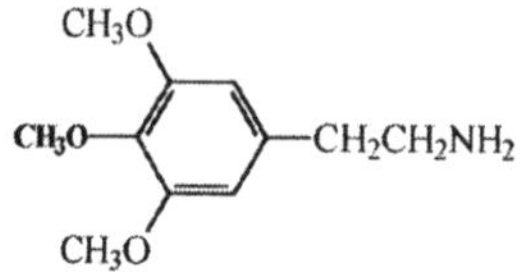

Mescaline

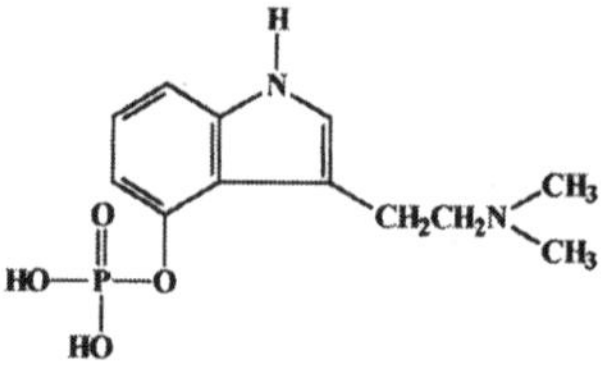

Psilocybin

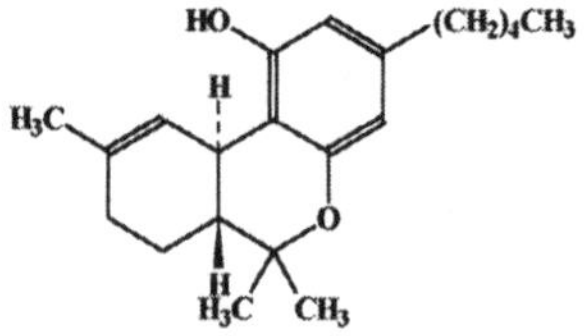

THC
Tetrahydrocannabinol

Stimulants. A diverse collection of molecules that speed up activity in the central nervous system (CNS).

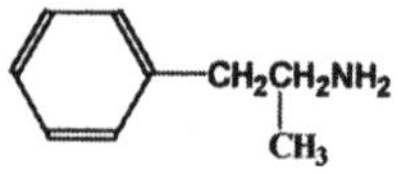

Amphetamine

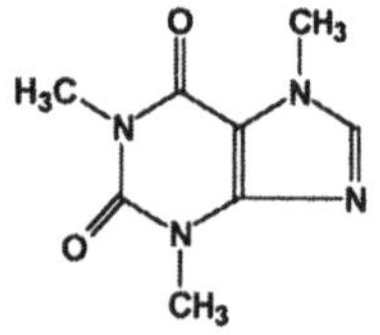

Caffeine

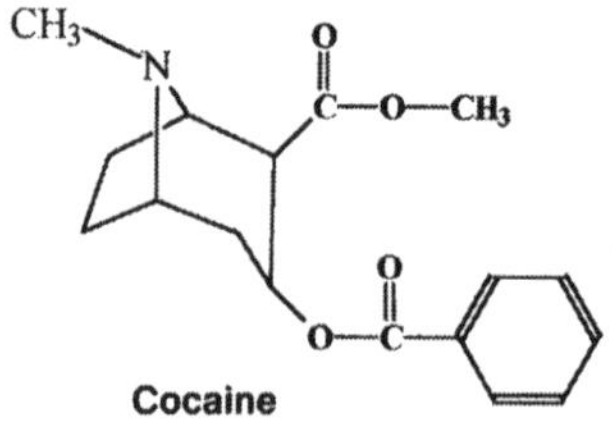

Cocaine

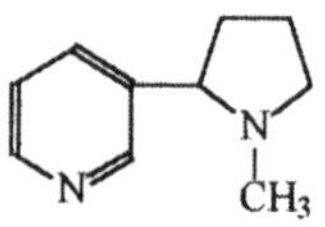

Nicotine

Water (H_2O). Without question, one of the most biologically important molecules on the planet. (For a list of the properties of water, refer to the chemical index, Section 5.12.)

Vitamins. Organic molecules required by the body in small amounts.

Common Vitamins				
Vitamin	**Solubility**	**Common Functions**	**Deficiency Symptoms**	**RDA**
A (retinol)	Fat	Vision, immune system, skin	Night blindness	5000 IU
B_1 (thiamine)	Water	Nerves, muscle, heart, growth	Beriberi	1-1.5 mg
B_2 (riboflavin)	Water	Precursor of FAD	Slow growth	1.2-1.7 mg
B_5 (pantothenic acid)	Water	Precursor of CoA	Muscle spasms, mental depression	10 mg
B_6 (pyridoxine)	Water	Assimilation of protein and fat	Anemia	1.6-2.0 mg
B_{12} (cobalamin)	Water	Red blood cells, nervous system	Pernicious anemia	2 μg
C (ascorbic acid)	Water	Growth and repair of tissues	Scurvy	60 mg
D (calciferol)	Fat	Bones, teeth (Ca^{2+} absorption)	Rickets	200-400 IU
E (tocopherol)	Fat	Vasodilation, anticoagulation	Unknown	8-10 IU
H (biotin)	Water	Metabolism of fat and protein	Dermatitis	100-300 μg
K (menadione)	Fat	Blood clotting (prothrombin)	Slow clotting	65-80 μg
Folic acid	Water	Red blood cells, cell metabolism	Anemia	180-200 μg
Niacin	Water	Precursor of NAD^+ and $NADP^+$	Pellagra	13-19 mg

I'd give my right arm to be ambidextrous. Michael Dewar

4.10 Formulas Used in Biology

Key to Formula Symbols

Symbol	Meaning
a_c	Centifugal acceleration
Avr	Alveolar ventilation rate
c	Concentration
[ ]	Concentration in moles/L
CFU	Colony-forming units
d	Rate of diffusion
D	Diffusion constant
D_f	Dilution factor
e	Extinction coefficient
E	Internal energy
ΔE	Change in internal energy
EDV	End diastolic volume
ESV	End systolic volume
E°	Reduction potential
E_x	Equilibrium potential for an ion
f	Frequency of rotation
fc	Frictional coefficient
F	Faraday's constant (9.65×10^4 C/mol)
ΔG	Change in free energy
$\Delta G^{\circ'}$	Change in free energy where (°) denotes standard pressure, temperature and concentration; (') refers to standard H^+ concentration
I	Resulting intensity
I_c	Initial concentration of competitive inhibitor
I_n	Initial concentration of noncompetitive inhibitor
I_o	Original intensity
I_u	Initial concentration of uncompetitive inhibitor
K_{eq}	Equilibrium constant
K_{ion}	Equilibrium potential for a certain ion
K_m	Michaelis's constant
λ	Wavelength
Mag	Magnification of drawing
N_A	Avogadro's number 6.022×10^{23} molecule/mole
NA	Numerical aperture
p	Frequency of first allele
P	Pressure
P_r	Peripheral resistance
q	Frequency of second allele
r	Radius
R	Gas constant (8.31 J/Kmol)
s	Distance travelled from axis
s_c	Sedimentation constant
[S]	Initial substrate concentration moles/L
ΔS	Change in entropy
t	Time
T	Temperature
t_{gen}	Generation time
v	Initial reaction velocity
V	Volume
V_m	Membrane potential
V_{max}	Maximum reaction velocity
V_x	Velocity at which particle x travels from center
ω	Angular velocity
χ^2	Chi-square
x^2rms	Root mean square displacement
z	Charge of ion

Stoop and you'll be stepped on; stand tall and you'll be shot at. Carlos A. Urbizo

Active Transport

$$\Delta G_{inward} = RT \ln \frac{[inside]}{[outside]}$$

$$\Delta G_{outward} = RT \ln \frac{[outside]}{[inside]}$$

$$\Delta G_{inward} = RT \ln \frac{[inside]}{[outside]} + z FV_m$$

$$\Delta G_{outward} = RT \ln \frac{[outside]}{[inside]} - z FV_m$$

$$\Delta G° = -RT \ln K_{eq}$$

Bioenergetics

$$\Delta E = Energy\ In - Energy\ Out$$

$$\Delta E = \Delta G + T\Delta S$$

$$\Delta E = \Delta G - \Delta PV + \Delta TS$$

$$\Delta G = \Delta E - T\,\Delta S$$

$$\ln \frac{[products]}{[reactants]} = \ln K_{eq}$$

$$\Delta G = RT \ln \frac{[products]}{[reactants]} - RT \ln K_{eq}$$

$$\Delta G°' = -RT \ln K'_{eq}$$

$$\Delta G' = \Delta G°' + RT \ln \frac{[C]^c\,[D]^d}{[A]^a\,[B]^b}$$

Cardiac

Stroke volume $(S_v) = EDV - ESV$

Cardiac output $(C_o) = Heart\ rate \times S_v$

Blood pressure $= C_o\ P_r$

Blood flow $= \dfrac{\Delta Pressure}{Resistance}$

Chromatography

$$Rf = \frac{distance\ traveled\ by\ pigment}{distance\ traveled\ by\ solvent}$$

Data Interpretation

$$\chi^2 = \sum \frac{(Observed - Expected)^2}{Expected}$$

Diffusion

$$t = \frac{x^2 rms}{2D}$$

Enzyme Kinetics

$$v = \frac{V_{max}[S]}{K_m + [S]}$$

$$\frac{1}{v} = \frac{K_m}{V_{max}}\frac{1}{[S]} + \frac{1}{V_{max}}$$

$$v = \frac{V_{max}[S]}{[S] + K_m(1 + [I_c]/K_I)}$$

$$\frac{1}{v} = \frac{1}{V_{max}} + \frac{K_m}{V_{max}}\frac{1}{[S]}\left(1 + \frac{[I_c]}{K_I}\right)$$

$$\frac{1}{v} = \left(\frac{1}{V_{max}} + \frac{K_m}{V_{max}}\frac{1}{[S]}\right)\left(1 + \frac{[I_n]}{K_I}\right)$$

$$\frac{1}{v} = \frac{1}{V_{max}}\left(1 + \frac{[I_u]}{K_I}\right) + \frac{K_m}{V_{max}}\frac{1}{[S]}$$

More university graduates become criminals every year than policemen. Phillip Goodheart

Centrifugation

$\omega = 2\pi f$

$s_c = \frac{v_x}{\omega^2 s}$

$D = \frac{RT}{N_A f_c}$

$f_c = \frac{RT}{N_A D}$

$a_c = \omega^2 r$

Equilibrum Potential

$K_{ion} = \frac{RT}{zF} \ln \frac{[ion]_{Out}}{[ion]_{In}}$

$K_{ion} = 2.303 \frac{RT}{zF} \log \frac{[ion]_{Out}}{[ion]_{In}}$

Microbiology

CFU per mL $= no.\ of\ colonies\ formed \times D_f$

Growth rate constant $(\mu) = \frac{\ln 2}{t_{gen}}$

Growth rate constant $(\mu) = \frac{0.693}{t_{gen}}$

Growth rate constant $(k) = \frac{1}{t_{gen}}$

Dilution rate $= \frac{Flow\ rate}{Volume}$

Photospectrometer

Absorbance (Abs) $= \log \frac{I_o}{I}$

% Transmittance $= \frac{I}{I_o} \times 100$

% Transmittance $= \frac{1}{10^{(\text{Abs-2})}}$

Absorbance $= \log \frac{100}{\%\ Transmittance}$

Absorbance $= e\ c\ path\ length$

Microscope

Mag $= \frac{length\ of\ drawing}{actual\ length}$

Numerical aperture (NA) $= \frac{1}{2} \sin\theta$

Resolving power $(RP) = \frac{\lambda}{2(Numerical\ aperture)}$

Population Genetics

$p^2 + 2pq + q^2 = 1$

Other Equations

Gas flow $= \frac{pressure\ gradient}{resistance}$

$Avr = \frac{frequency\ of\ breaths}{minutes}(TV - dead\ space)$

$E_{total} = (heat + work + E_{storage})$

Concentration of any ion $= \frac{conc.\ in\ (mg/L)\ of\ ions}{atomic\ mass\ of\ ion} z$

Redox Potentials

$\Delta G° = -nF\Delta E°$

A tapeworm can grow to thirty meters in length in your digestive system, now that's weight control!

CHEMISTRY

5

5.1 Periodic Table

Legend	
Noble gas	Halogen
Metal	Rare earth
Trans. metal	Non-metal
Alkali metal	Alkali earth

Key: Average atomic mass — 1.0079; Oxidation states — -1, +1; Element name — Hydrogen; Element symbol — H; Atomic number — 1; Physical state @SATP — gas

	1A	IIA	IIIA	IVA	VA	VIA	VIIA	VIIIA	VIIIA	VIIIA	IB	IIB	IIIB	IVB	VB	VIB	VIIB	VIIIB
1	1.0079 -1, +1 Hydrogen H 1 gas																	4.0026 Helium He 2 gas
2	6.941 +1 Lithium Li 3 solid	9.012 +2 Beryllium Be 4 solid											10.81 +3 Boron B 5 solid	12.011 +2,+4,-4 Carbon C 6 solid	14.007 +1,+2,+3,+4,+5,-1,-2,-3 Nitrogen N 7 gas	15.999 -2 Oxygen O 8 gas	18.998 -1 Fluorine F 9 gas	20.18 Neon Ne 10 gas
3	22.990 +1 Sodium (Natrium) Na 11 solid	24.305 +2 Magnesium Mg 12 solid											26.982 +3 Aluminum Al 13 solid	28.086 +2,+4,-4 Silicon Si 14 solid	30.974 +3,+5,-3 Phosphorous P 15 solid	32.06 +4,+6,-2 Sulfur S 16 solid	35.453 +1,+5,+7,-1 Chlorine Cl 17 gas	39.948 Argon Ar 18 gas
4	39.098 +1 Potassium (Kalium) K 19 solid	40.078 +2 Calcium Ca 20 solid	44.956 +3 Scandium Sc 21 solid	47.89 +3, +4 Titanium Ti 22 solid	50.942 +3,+4,+5 Vanadium V 23 solid	51.996 +2,+3,+6 Chromium Cr 24 solid	54.938 +2,+3,+4,+6,+7 Manganese Mn 25 solid	55.847 +2,+3 Iron (Ferrum) Fe 26 solid	58.933 +2,+3 Cobalt Co 27 solid	58.693 +2,+3 Nickel Ni 28 solid	63.546 +1,+2 Copper (Cuprum) Cu 29 solid	65.38 +2 Zinc Zn 30 solid	69.72 +3 Gallium Ga 31 liquid	72.61 +2,+4 Germanium Ge 32 solid	74.922 -3,+3,+5 Arsenic As 33 solid	78.96 -2,+4,+6 Selenium Se 34 solid	79.904 -1,+1,+5 Bromine Br 35 liquid	83.80 Krypton Kr 36 gas
5	85.468 +1 Rubidium Rb 37 solid	87.62 +2 Strontium Sr 38 solid	88.91 +3 Yttrium Y 39 solid	91.22 +4 Zirconium Zr 40 solid	92.906 +3,+5 Niobium Nb 41 solid	95.94 +2,+3,+4,+5,+6 Molybdenum Mo 42 solid	98 +4,+6,+7 Technetium Tc 43 solid	101.07 +2,+3,+4,+6,+8 Ruthenium Ru 44 solid	102.906 +2,+3,+4 Rhodium Rh 45 solid	106.42 +2,+4 Palladium Pd 46 solid	107.868 +1 Silver (Argentum) Ag 47 solid	112.41 +2 Cadmium Cd 48 solid	114.82 +3 Indium In 49 solid	118.7 +2,+4 Tin (Stannum) Sn 50 solid	121.75 -3,+3,+5 Antimony (Stibium) Sb 51 solid	127.60 -2,+4,+6 Tellurium Te 52 solid	126.905 -1,+1,+5,+7 Iodine I 53 solid	131.3 Xenon Xe 54 gas
6	132.905 +1 Cesium Cs 55 solid	137.3 +2 Barium Ba 56 solid	138.906 +3 Lanthanum La 57 solid	178.49 +4 Hafnium Hf 72 solid	180.95 +5 Tantalum Ta 73 solid	183.85 +2,+3,+4,+5,+6 Tungsten (Wolfram) W 74 solid	186.207 -1,+2,+4,+6,+7 Rhenium Re 75 solid	190.20 +2,+3,+4,+6,+8 Osmium Os 76 solid	192.22 +2,+3,+4,+6 Iridium Ir 77 solid	195.08 +2,+4 Platinum Pt 78 solid	196.97 +1,+3 Gold (Aurum) Au 79 solid	200.59 +1,+2 Mercury (Hydrar-gyrum) Hg 80 liquid	204.38 +1,+3 Thallium Tl 81 solid	207.20 +2,+4 Lead (Plumbum) Pb 82 solid	208.980 +3,+5 Bismuth Bi 83 solid	209* +2,+4 Polonium Po 84 solid	210* -1,+1,+3,+5,+7 Astatine At 85 solid	222* Radon Rn 86 gas
7	223* +1 Francium Fr 87 solid	226.025 +2 Radium Ra 88 solid	227.028 +3 Actinium Ac 89 solid	261* +4 Unnil-quadium Unq 104	262* Unnil-pentium Unp 105	263* Unnil-hexium Unh 106	262* Element 107 Uns 107	(not observed) Uno 108	Element 109 Une 109	Ununnium Unn 110								

Lanthanide Series														
Lanthanide Series	140.115 +3,+4 Cerium Ce 58 solid	140.91 +3,+4 Praseo-dymium Pr 59 solid	144.24 +3 Neodymium Nd 60 solid	145* +3 Promethium Pm 61 solid	150.36 +2,+3 Samarium Sm 62 solid	151.96 +2,+3 Europium Eu 63 solid	157.25 +3 Gadolinium Gd 64 solid	158.925 +3,+4 Terbium Tb 65 solid	162.50 +3 Dysprosium Dy 66 solid	164.930 +3 Holmium Ho 67 solid	167.26 +3 Erbium Er 68 solid	168.934 +2,+3 Thulium Tm 69 solid	173.04 +2,+3 Ytterbium Yb 70 solid	174.967 +3 Lutetium Lu 71 solid
Actinide Series	232.038 +4 Thorium Th 90 solid	213.04 +4,+5 Protac-tinium Pa 91 solid	238.029 +2,+3,+4,+5,+6 Uranium U 92 solid	237.05 +3,+4,+5,+6 Neptunium Np 93 solid	244* +3,+4,+5,+6 Plutonium Pu 94 solid	243* +2,+3,+4,+5,+6 Americium Am 95 solid	247* +3 Curium Cm 96 solid	247* +3,+4 Berkelium Bk 97	251* +3 Californium Cf 98	252* +3 Einsteinium Es 99	257* +3 Fermium Fm 100	258* +2,+3 Mende-levium Md 101	259* +2,+3 Nobelium No 102	260* +3 Lawrencium Lr 103

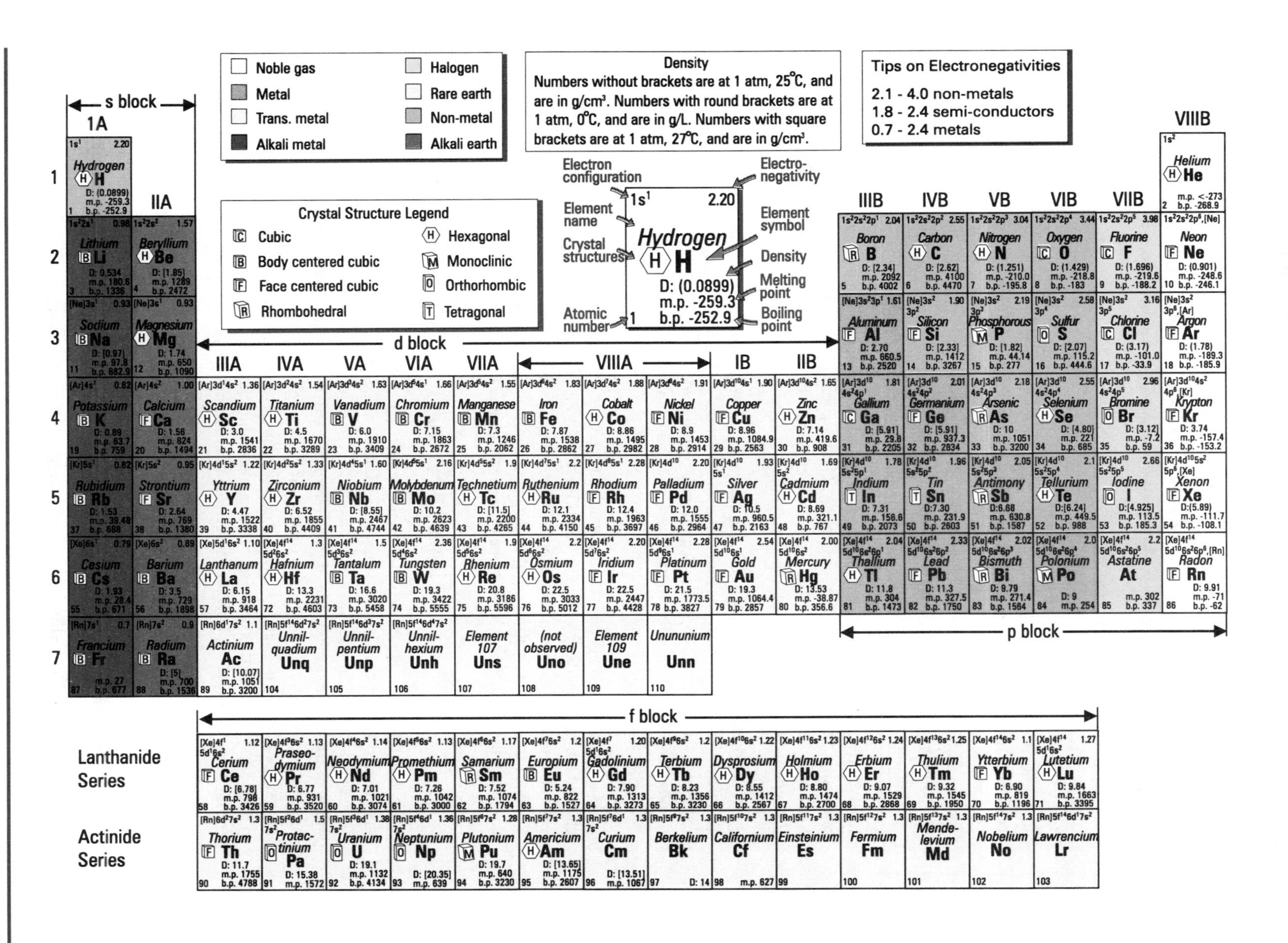

Noble gas
Metal
Trans. metal
Alkali metal
Halogen
Rare earth
Non-metal
Alkali earth
Density
Numbers without brackets are at 1 atm, 25°C, and are in g/cm³. Numbers with round brackets are at 1 atm, 0°C, and are in g/L. Numbers with square brackets are at 1 atm, 27°C, and are in g/cm³.
Tips on Electronegativities
2.1 - 4.0 non-metals
1.8 - 2.4 semi-conductors
0.7 - 2.4 metals
Crystal Structure Legend
Cubic
Body centered cubic
Face centered cubic
Rhombohedral
Hexagonal
Monoclinic
Orthorhombic
Tetragonal
Electron configuration
Element name
Crystal structures
Atomic number
Electro-negativity
Element symbol
Density
Melting point
Boiling point
s block
d block
p block
f block
1A
IIA
IIIA
IVA
VA
VIA
VIIA
VIIIA
IB
IIB
IIIB
IVB
VB
VIB
VIIB
VIIIB
Lanthanide Series
Actinide Series

5.2 Inorganic Nomenclature

Understanding matter is the fundamental goal of chemistry. The nomenclature used in chemistry is a way of labeling differences among types of matter. Nomenclature is a fancy word for naming. There are millions of different chemicals and each one has a different name. However, many molecules have more than one name. To avoid confusion, chemists developed an international system for naming chemicals. **IUPAC** (International Union of Pure and Applied Chemistry) sets the international standard for nomenclature. Still, **trivial names** like water (dihydrogen oxide) and table salt (sodium chloride) are common.

There are two tricks to science. First, try to sound intelligent, even if you don't know what you're talking about. Second, if you don't know what you're doing, do it neatly and safely. With that in mind, we proceed – but with a word of caution: the chemistry lab is full of dangerous materials. Names are important in identifying hazardous materials. Know what you are handling and how to protect yourself. *If you don't know what something is, don't touch it!*

Both this and the following section are a crash course in naming. To name a chemical, you have to know its molecular composition. A molecular formula represents the basic unit of a chemical. It may be **monatomic** (single atom), **diatomic** (two bonded atoms), or **polyatomic** (cluster of bonded atoms). The molecular formula is the simplest way to show molecular composition.

Many inorganic compounds are made up of a positively charged ion and a negatively charged ion held together by an ionic bond. These ionic charges represent the number of electrons one atom gave to or took from another atom. This number is also known as the **valence number** or **oxidation number**. In a neutral molecule, the oxidation numbers of each atom must balance each other (the total overall charge must be zero).

Elemental Molecules

Elemental molecules are made of the same element. They're usually identified by their elemental names (Cl_2 = chlorine). For some elements, more than one molecular form exists (see Section 5.5 on allotropes). To avoid confusion when naming these different forms (allotropes), one of two systems is used: **systematic** (based on the number of atoms), or **trivial** (based on historical naming). For example, to differentiate between O_2 and O_3 the **systematic names** are based on the addition of a prefix: dioxygen and trioxygen (see the common multiples listed in Section 2.2). The **trivial names**, oxygen and ozone, come from old naming practices.

Never attribute to malice what can be adequately explained by stupidity. Nick Diamos

Common Elemental Molecules

Monatomic (noble gases)		Diatomic		Polyatomic	
Ar	Argon	Br_2	Bromine (dibromine)	As_4	Arsenic (tetraarsenic)
He	Helium	Cl_2	Chlorine (dichlorine)	O_3	Ozone (trioxygen)
Kr	Krypton	F_2	Fluorine (difluorine)	P_4	White phosphorous (tetraphosphorous)
Ne	Neon	H_2	Hydrogen (dihydrogen)	S_8	Sulfur (octasulfur)
Rn	Radon	I_2	Iodine (diiodine)	Sb_4	Antimony (tetraantimony)
Xe	Xenon	N_2	Nitrogen (dinitrogen)	Se_8	Selenium (octaselenium)
		O_2	Oxygen (dioxygen)		

Binary Compounds

Binary compounds are made of two different elements: an electropositive element and an electronegative element. When naming binary compounds the electropositive element is always named first; the electronegative element follows, its name modified with the suffix -ide. When the molecule contains a nonmetal (C, N, S, etc.), prefixes are added to each element to indicate the correct stoichiometry, as in the following:

CO = carbon monoxide NO_2 = nitrogen dioxide SO_3 = sulfur trioxide

CO_2 = carbon dioxide N_2O_3 = dinitrogen trioxide

(See Section 2.2 for prefixes.)

Although carbon-containing compounds are organic by definition, chemists sometimes name them as if they were inorganic.

Use the tables on the next two pages to name binary compounds containing metals from a molecular formula. The names in these tables have already been modified. Write their names in the order mentioned and you should have no problems, but beware that not all the combinations result in binary compounds. Notice that some atoms have more than one ionic charge. This is why some electropositive ions have a Roman numeral in brackets. This is **Stock notation**, and is used to indicate the magnitude of the charge. Stock notation is only used when multiple oxidation states are possible. For example, the silver ion (Ag^+) has only one ionic charge. For this reason, "silver" not "Silver(I)" is used. Iron, on the other hand, has two possible ionic charges: Fe^{2+} and Fe^{3+}. "Iron(II)" and "Iron(III)" differentiate the two.

Another, older system of notation is sometimes used too. IUPAC doesn't recommend it but we include it here because you may see it. Positive ions with the lower charge are modified by the suffix -ous. The positive ion with the higher charge is modified by the suffix -ic. To add further complication, Latin names are sometimes used (ferrous, not ironous; ferric, not ironic).

If you have to convert a name to a molecular formula, use the charges to determine the proportions of each element. Iron(III) oxide tells you that Fe^{3+} and O^{2-} are in this compound. Balancing the ionic charges, $(2 \times 3) + (3 \times -2) = 0$, gives Fe2O3.

Common Positively Charged Ions (Cations)			
Ag^{+}	Silver	Hg_2^{2+}	Mercury(I) or mercurous
Al^{3+}	Aluminum	Hg^{2+}	Mercury(II) or mercuric
As^{3+}	Arsenic(III) or arsenious	K^{+}	Potassium
Au^{+}	Gold(I) or aurous	Li^{+}	Lithium
Au^{3+}	Gold(III) or auric	Mg^{2+}	Magnesium
Ba^{2+}	Barium	Mn^{2+}	Manganese(II) or manganous
Bi^{3+}	Bismuth	Mn^{3+}	Manganese(III) or manganic
Ca^{2+}	Calcium	NH_4^{+}	Ammonium
Cd^{2+}	Cadmium	Na^{+}	Sodium
Ce^{3+}	Cerium(III) or cerous	Ni^{2+}	Nickel(II)
Ce^{4+}	Cerium(IV) or ceric	Pb^{2+}	Lead(II) or plumbous
Co^{2+}	Cobalt(II) or cobaltous	Pb^{4+}	Lead(IV) or plumbic
Co^{3+}	Cobalt(III) or cobaltic	Rb^{+}	Rubidium
Cr^{2+}	Chromium(II) or chromous	Sb^{3+}	Antimony(III) or antimonous
Cr^{3+}	Chromium(II) or chromic	Sn^{2+}	Tin(II) or stannous
Cr^{6+}	Chromium(VI)	Sn^{4+}	Tin(IV) or stannic
Cs^{+}	Cesium	Sr^{2+}	Strontium
Cu^{+}	Copper(I) or cuprous	Ti^{3+}	Titanium(III) or titanous
Cu^{2+}	Copper(II) or cupric	Ti^{4+}	Titanium(IV) or titanic
Fe^{2+}	Iron(II) or ferrous	Tl^{+}	Thallium(I) or thallous
Fe^{3+}	Iron(III) or ferric	Tl^{3+}	Thallium(III) or thallic
Ga^{3+}	Gallium	V^{3+}	Vanadium
H^{+}	Hydrogen	Zn^{2+}	Zinc
H_3O^{+}	Hydronium		

Polyatomic Compounds

Polyatomic compounds are made of more than two different elements. Again, using the tables, find the positive ion in the above table, then find the negative ion in the table on the next page. Generally, the negatively charged anion is made up of more than one type of element, known as a **polyatomic anion**. Special nomenclature is used to indicate the presence and proportion of oxygen in the polyatomic anion (prefix hypo- with suffix -ite, or -ate; prefix per- with suffix -ate). The table below summarizes this nomenclature. Although the carbon-containing compounds listed in this section are not strictly inorganic, they may be treated as such for naming.

Names of Polyatomic Anions Containing Oxygen								
Hypo_ite	BrO^{-}		ClO^{-}	IO^{-}		$N_2O_2^{2-}$	$S_2O_4^{2-}$	$PH_2O_2^{-}$
__ite	BrO_2^{-}		ClO_2^{-}	IO_2^{-}		NO_2^{-}	SO_3^{2-}	PHO_3^{2-}
__ate	BrO_3^{-}	CO_3^{2-}	ClO_3^{-}	IO_3^{-}	MnO_4^{2-}	NO_3^{-}	SO_4^{2-}	PO_4^{3-}
Per__ate	BrO_4^{-}		ClO_4^{-}	IO_4^{-}	MnO_4^{-}		$S_2O_8^{2-}$	

Common Negatively Charged Ions (Anions)

As^{3-}	Arsenide	IO_2^-	Iodite
AsO_3^{3-}	Arsenite	IO_3^-	Iodate
AsO_4^{3-}	Arsenate	IO_4^-	Periodate
BO_3^{3-}	Borate	MnO_4^{2-}	Manganate
Br^-	Bromide	MnO_4^-	Permanganate
BrO^-	Hypobromite	N^{3-}	Nitride
BrO_2^-	Bromite	N_3^-	Azide
BrO_3^-	Bromate	NH^{2-}	Imide
BrO_4^-	Perbromate	NH_2^-	Amide
C^{4-}	Carbide	$N_2H_3^-$	Hydrazide
C_2^{2-}	Acetylide	$NHOH^-$	Hydroxylamide
$C_2H_3O_2^-$	Acetate	$N_2O_2^{2-}$	Hyponitrite
$C_7H_6O_2^{2-}$	Benzoate	NO_2^-	Nitrite
CN^-	Cyanide	NO_3^-	Nitrate
CNO^-	Cyanate	O^{2-}	Oxide
CO_3^{2-}	Carbonate	O_2^-	Hyperoxide
$C_2O_4^{4-}$	Oxalate	O_2^{2-}	Peroxide
Cl^-	Chloride	O_3^-	Ozonide
ClO^-	Hypochlorite	OH^-	Hydroxide
ClO_2^-	Chlorite	P^{3-}	Phosphide
ClO_3^-	Chlorate	$PH_2O_2^-$	Hypophosphite
ClO_4^-	Perchlorate	PHO_3^{2-}	Phosphite
CrO_4^{2-}	Chromate	PO_4^{3-}	Phosphate
$Cr_2O_7^{2-}$	Dichromate	$P_2H_2O_5^{2-}$	Pyrophosphite or diphosphite
D^-	Deuteride	$P_2O_7^{4-}$	Pyrophosphate
F^-	Fluoride	S^{2-}	Sulfide
$Fe(CN)_6^{3-}$	Hexacyanoferrate(III) or ferricyanide	S_2^{2-}	Disulfide
$Fe(CN)_6^{4-}$	Hexacyanoferrate(II) or ferrocyanide	SCN^-	Thiocyanate
H^-	Hydride	$S_2O_4^{2-}$	Dithionite
HCO_3^-	Hydrogen carbonate or bicarbonate	SO_3^{2-}	Sulfite
$HC_2O_4^-$	Hydrogen oxalate or bioxalate	$S_2O_3^{2-}$	Thiosulfate
$HC_8H_4O_4^-$	Hydrogen phthalate or biphthalate	SO_4^{2-}	Sulfate
HF_2^-	Hydrogen difluoride	$S_2O_5^{2-}$	Pyrosulfite or disulfite
HPO_4^{2-}	Monohydrogen phosphate	$S_2O_8^{2-}$	Persulfate
$H_2PO_4^-$	Dihydrogen phosphate	Sb^{3-}	Antimonide
HS^-	Hydogen sulfide or bisulfide	Se^{2-}	Selenide
HSO_3^-	Hydrogen sulfite or bisulfite	SeO_3^{2-}	Selenite
HSO_4^-	Hydrogen sulfate or bisulfate	Si^{4-}	Silicide
I^-	Iodide	SiO_4^{4-}	Orthosilicate
I_3^-	Triiodide	Te^{2-}	Telluride
IO^-	Hypoiodate		

The ability to delude yourself may be an important survival tool. Jane Wagner

Acids

Acids have been named according to old notations that the IUPAC is hesitant to change. **Binary acids** are made up of two different elements. Generally, the first element listed in an acid is hydrogen. When the acid is in aqueous solution, the prefix hydro- is used for the hydrogen ion, and the anion is modified by the suffix -ic acid (hydrochloric acid). Otherwise, the hydrogen is named first, followed by the anion, which is modified by the suffix -ide (Hydrogen chloride). **Pseudobinary acids** are named as if they were binary acids.

Common Binary and Pseudobinary Acids			
HBr	Hydrobromic acid (aq) or Hydrogen bromide	HI	Hydriodic acid (aq) or Hydrogen iodide
HCN	Hydrocyanic acid (aq) or Hydrogen cyanide	HN_3	Hydrogen azide
HCl	Hydrochloric acid (aq) or Hydrogen chloride	H_2S	Hydrogen sulfide
HF	Hydrofluoric acid (aq) or Hydrogen fluoride		

Oxoacids

Oxoacids are acids that contain oxygen. They are named in a similar fashion to oxygen-containing polyatomic compounds. One difference is that the suffix -ite is replaced by the suffix -ous; and the suffix -ate is replaced by the suffix -ic. Another naming peculiarity uses the meta- and ortho- prefixes to indicate a different water content between acids (factors of H_2O).

Common Oxoacids			
H_3AsO_3	Arsenious acid	$H_2N_2O_2$	Hyponitrous acid
H_3AsO_4	Arsenic acid	HNO_2	Nitrous acid
H_3BO_3	Orthoboric acid	HNO_3	Nitric acid
$(HBO_3)_n$	Metaboric acid	HPH_2O_2	Hypophosphorous acid
HBrO	Hypobromous acid	H_2PHO_3	Phosphorous acid
$HBrO_2$	Bromous acid	H_3PO_4	(Ortho) phosphoric acid
$HBrO_3$	Bromic acid	$H_4P_2O_5$	Diphosphorous or pyrophosphorous acid
HClO	Hypochlorous acid	$H_4P_2O_7$	Diphosphoric or pyrophosphoric acid
$HClO_2$	Chlorous acid	$H_2S_2O_4$	Dithionous acid
$HClO_3$	Chloric acid	H_2SO_3	Sulfurous acid
$HClO_4$	Perchloric acid	$H_2S_2O_3$	Thiosulfuric acid
H_2CrO_4	Chromic acid	H_2SO_4	Sulfuric acid
$H_2Cr_2O_7$	Dichromic acid	$H_2S_2O_5$	Disulfurous or pyrosulfurous acid
HIO	Hypoiodous acid	$H_2S_2O_8$	Peroxodisulfuric acid
HIO_3	Iodic acid	H_2SeO_3	Selenious acid
H_2MnO_4	Manganic acid	H_2SeO_4	Selenic acid
$HMnO_4$	Permanganic acid	H_4SiO_4	Orthosilicic acid

5.3 Organic Nomenclature

Organic chemistry is the study of molecules containing carbon. The number of known organic compounds is in the millions. Every organic molecule has a basic carbon skeleton. The key to identifying or naming an organic molecule is in the structure of its carbon skeleton. When counting the length of a carbon chain, use the following table:

Carbon Chain Names			
1 Meth-	6 Hex-	11 Undec-	16 Hexadec-
2 Eth-	7 Hept-	12 Dodec-	17 Heptadec
3 Prop-	8 Oct-	13 Tridec-	18 Octadec-
4 But-	9 Non-	14 Tetradec-	19 Nonadec-
5 Pent-	10 Dec-	15 Pentadec-	20 Icos-

Once you know a molecule's carbon framework, you can analyze the non-carbon substituents attached it. Substituents are chemical groups attached to the main carbon chain. They could be other smaller carbon chains, double, or triple bonds, or any other chemical group (such as those on the next two pages). Some basic rules when naming organic chemicals (note that these only help with small simple organic molecules):

1) Find the longest chain of carbons and count them.
2) Number the carbons on the chain to give the lowest number to substituents.
3) Identify substituents.
4) Put the substituents in alphabetical order; pay no attention to numerical prefix.
5) Put numbers in front of the substituent indicating to which carbon it is attached. If a substituent occurs more than once, indicate the number of times by using "di, tri, tetra," etc.

```
    Pentane          1,1,1-Trichloroethane       Diethyl ether

 H  H  H  H  H            Cl  H                 H  H     H  H
 |  |  |  |  |            |   |                 |  |     |  |
H-C--C--C--C--C-H      Cl-C---C-H             H-C--C--O--C--C-H
 |  |  |  |  |            |   |                 |  |     |  |
 H  H  H  H  H            Cl  H                 H  H     H  H
```

Cl Cl Cl   O

Usually, the properties of organic molecules are attributed to the different chemical groups that make up the molecular structure. These are known as **functional groups** and there is large number of them. We only have space to list the major groups and how they're usually named in a molecule. If you ever browse through an index of organic chemicals, you'll find that huge organic chemicals are not named according to rules in introductory organic chemistry courses. This is because large molecules would have unruly names. Organic chemists save time by naming the common larger organic structures. Consider cholesterol, also known as cholest-5-en-3β-ol, molecular formula $C_{27}H_{48}O$. This is not a complicated organic molecule, but try naming it.

–C–C–

Alkane. Also known as saturated hydrocarbons, they are nonpolar and dissolve in nonpolar liquids. They are relatively unreactive. Suffix -ane.

>C=C<

Alkene. Commonly known as olefins, they are nonpolar and only very slightly soluble in water. Found in unsaturated hydrocarbons such as fats and oils. Isomer prefixes cis- (same side) or trans- (opposite sides), suffix -ene.

–C≡C–

Alkyne. Common family name is acetylene. The triple bond is chemically reactive. Often used as a reactant to synthesize other organic molecules. Suffix -yne.

O M P

Arene. Aromatic hydrocarbons such as the benzene ring shown have an interesting C≃C bonding; it's like one and a half bonds. (see Resonance, Section 5.5). Prefix phenyl-, suffix -benzene (substituents have prefixes O- ortho-, M- meta-, P- para-).

–C–OH

Alcohol. The OH group is also known as the hydroxyl group. Alcohols exhibit hydrogen bonding, and are weakly basic. The OH group is widely found in biological molecules, such as carbohydrates and lipids. Suffix -ol.

O || –C–H

Aldehyde. A member of the carbonyl group (C=O). The C=O group is always found on an end of a carbon chain. Methanal (formaldehyde) HCHO is commonly used in the lab as a preservative. Suffix -al.

O || –C–C–C–

Ketone. A member of the carbonyl group (C=O). The C=O group is always found inside a carbon chain. Used in organic synthesis and as solvents. Prefix keto-, suffix -one.

O || –C–OH

Carboxylic acid. Polar, can form hydrogen bonds, and are usually weak acids. Biologically found in fatty and amino acids; sour taste (vinegar). Suffix -oic acid.

–C–O–C–

Ether. No hydrogen bonding. Ethers are highly volatile, flammable, and usually make good solvents. Groups attached to the oxygen are named in alphabetical order, then suffix -ether is added.

O || –C–O–C–

Ester. Synthesized from the dehydration of an alcohol and a carboxylic acid, they have pleasant odors, and are used in artificial flavoring. Suffix -ic (on the alcohol), and suffix -ate (on the acid).

Methanol makes you blind, propanol makes you sterile and ethanol makes you forget chemistry!

Acid anhydride. Highly reactive, and not found in nature. Used as reactants in organic synthesis. Carbon chains are listed in alphabetical order, with suffix -oic anhydride.

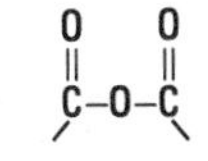

Amine. Moderately polar, can form hydrogen bonds and are typically basic. Alkaloids (nicotine, morphine, codeine) are one type of amine. Responsible for fishy smells. Prefix amino-, suffix -amine.

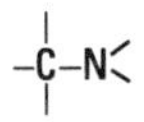

Imine. Sometimes called Schiff bases. Formed from the reaction of an amine and an aldehyde or ketone. Prefix imino-, suffix -imine.

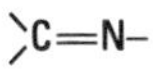

Nitrile. A highly toxic group of chemicals. Formed during organic reactions with HCN (hydrogen cyanide). Found in insecticides and rodent poisons. Prefix cyano-, suffix -nitrile.

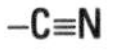

Amide. Weakly basic, form hydrogen bonds and are soluble in water (depending on molecular size). Amides may be formed during the bonding of two amino acids in a protein. This bond between two amino acids is called a peptide bond. Suffix -amide.

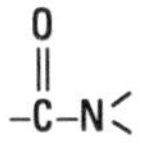

Nitro group. Formed from a reaction of nitric acid with an organic compound (usually a benzene ring). This group is found in fertilizers, disinfectants and explosives. Prefix nitro-.

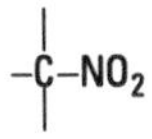

Halo alkane. Formed when a halide is bound to a carbon (usually very stable). A halide (X) may be one of F, Cl, Br, I. These compounds do not occur naturally. Many types are toxic to both humans and environment. Prefix fluoro-, chloro-, bromo- or iodo-.

Carbonyl halide. Highly reactive, does not occur in nature. Used to synthesize amides, esters, carbonyls and alcohols. Suffix -yl added to carbon chain name, followed by chloride, fluoride, bromide or iodide (X may be F, Cl, Br or I).

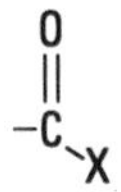

Thiol. Also known as mercaptans. Similar in many ways to alcohols. Form weak hydrogen bonds. Responsible for foul odors like skunk spray. Prefix mercapto- or sulfhydryl-, suffix- thiol.

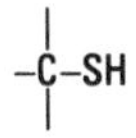

Thioether. Similar in many ways to ethers. Do not form hydrogen bonds. Carbon chains attached to the sulfur are named in alphabetical order, then given suffix -sulfide.

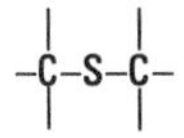

Disulfide. Disulfides are often found in biological reactions where they are converted to and from thiols. Often make important bonds (disulfide linkages) between molecules. Prefix dithio-.

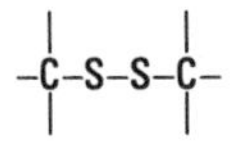

Sulfoxide. Forms a bond between two organic molecules. Found in medications such as antihypertensives, antibacterials, fungicides. Prefix sulfin- or sulfinyl-.

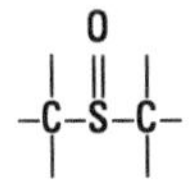

Sulfonic acid. Formed during a reaction with sulphur trioxide (SO_3) in fuming sulfuric acid (a process known as sulfonation). Prefix sulfo-, suffix -sulfonic acid.

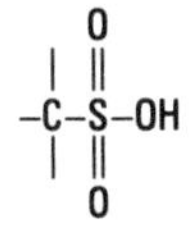

Phosphate. Very important in organisms for the storage and transfer of energy (AMP, ADP, ATP). Also found in the hydrophilic head of a phospholipid. Prefix phospho-, suffix -phosphate.

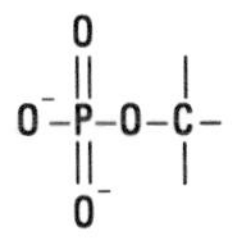

5.4 States & Properties of Matter

Useful Terms

Matter. Anything that occupies a space and has both mass and inertia.
Energy. The ability to do work; measured by amount of work done.
Volatility. The degree to which a solid or liquid can form a vapor. Volatile liquids have low boiling points and are often flammable (gasoline).

Density & Specific Gravity

Density. The amount of matter in a specific volume.
ρ = mass of substance/volume of substance

Specific gravity. A ratio reflecting the mass of a substance relative to the mass of an equal volume of water.
Sp.Gr. = mass of substance/mass of an equal volume of water at 3.98°C

Change of State

Critical pressure. The amount of pressure required to liquefy a substance at its critical temperature.

Critical temperature. Above this temperature no amount of pressure is enough to liquefy a substance.

Triple point. The temperature and pressure at which all three phases (solid, liquid, gas) are in equilibrium.

Solid $\underset{\text{Deposition}}{\overset{\text{Sublimation}}{\rightleftharpoons}}$ Gas

Solid $\underset{\text{Crystallization (Freezing)}}{\overset{\text{Fusion (melting)}}{\rightleftharpoons}}$ Liquid

Liquid $\underset{\text{Liquefaction (Condensation)}}{\overset{\text{Vaporization (Evaporation)}}{\rightleftharpoons}}$ Gas

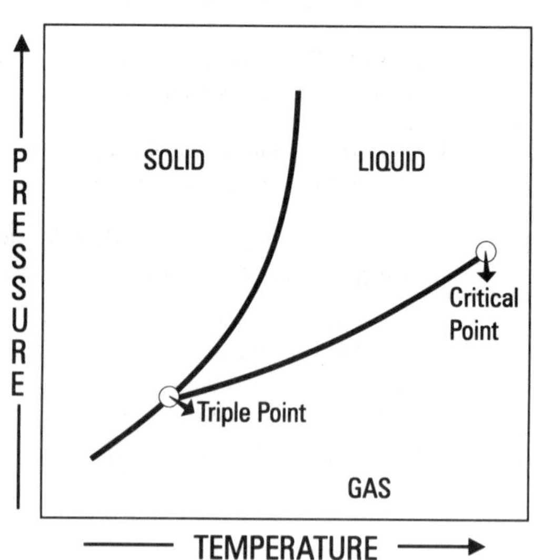

Fig. 5.4.1 Phase Diagram

Solids

Solids have constant volume, high density and definite shape; they are very difficult to compress, and expand very slightly when heated. There are two types of solids: crystalline and amorphous.

Crystalline solids have an orderly geometric structure, planar faces, definite cleavage patterns and sharp melting points.

Amorphous solids have a disordered arrangement of atoms, no definite cleavage pattern and tend to soften as temperature rises (glass, plastic).

Types of Crystalline Solids

	Metallic crystal	Ionic crystals	Molecular crystals	Network covalent crystals
Bonding	metallic bonding	ionic bonding	dipole	covalent bonding
Electrical conductivity	high	low	low	low
Hardness	soft-hard	brittle-hard	usually soft	hard
Heat conductivity	high	low	low	low
Lattice points	positive ions in an electron sea	positive/negative ions	molecules	atoms
Melting point	medium-high	high	low	low
Transparency	opaque	transparent	transparent translucent	transparent

Liquids

Liquids have constant volume, medium density, no definite shape (take the shape of the container they're in); they're difficult to compress and expand slightly when heated.

Adhesion. The molecular attraction between a liquid and another body (water sticking to the walls of a glass container).

Cohesion. The molecular attraction between molecules of a liquid. Responsible for surface tension and holding drops of liquid together.

Common Fluid Measurements

Vapor pressure. The pressure exerted above a liquid by its vapor.

Equilibrium vapor pressure. The pressure exerted above a liquid by its vapor when its rates of vaporization (evaporation) and liquefaction (condensation) are equal.

Boiling point. The temperature at which a liquid's vapor pressure equals the atmospheric pressure exerted on its surface.

Viscosity. A liquid's resistance to flow caused by internal friction.

Art is I; Science is We. Claude Bernard

Surface tension. The amount of energy required to expand the surface of a liquid by a unit area. Observable in a liquid surface acting like a membrane.

F = surface force, L = length

$\gamma = F/2L$

Poiseuille's law

P = pressure, Q = flow rate, L = length, r = radius of pipe, η = viscosity

$P_1 - P_2 = 8\,Q\eta L/\pi r^4$

Reynolds number. Calculated to predict whether a liquid's flow will be laminar (streamline) or turbulent.

η = viscosity, ρ = density, v = velocity, d = diameter

$Re = \rho v d/\eta$

$Re < 2000$ laminar flow

Re 2000-3000 flow is unpredictable

$Re > 3000$ turbulent flow

Forces of Attraction in Liquids

	H-bonded liquids	Non-H-bonded liquids	Liquid salts	Liquid metals
Particles	Molecules	Molecules	Ions	+ ions in an electron sea
Forces of attraction	H-bonding*, London forces, dipole-dipole for polar molecules	London forces, dipole-dipole for polar molecules	+ and − attraction	metallic bonding

* H-bonding = Hydrogen bonding

Gases

Gases are infinitely expandable in an open system. They have low density, no definite shape (expand to fill a container); they're easily compressed, and expand and contract during temperature changes.

Molar Volume V_m

Two sets of standard conditions are used when reporting data about gases:

Standard Temperature and Pressure (STP)
Temperature 273.15 K (0°C)
Pressure of 1 atm
V_m at STP
22.4 L/mol or 0.0224 m^3/mol

Standard Ambient Temperature and Pressure (SATP)
Temperature of 298.15 K (25°C)
Pressure of 100 kPa
V_m at SATP
24.8 L/mol or 0.0248 m^3/mol

Gas Laws

Boyle's Law

$V_1 / V_2 = P_2 / P_1$ (constant T)

Charles' and Gay-Lussac's law
$V_1/V_2 = T_1/T_2$ (constant P)

Combined gas law
$P_1V_1/T_1 = P_2V_2/T_2$

Temperature-pressure relationship
$P_1/T_1 = P_2/T_2$ (constant V)

Ideal gas law
$PV = nRT$

Avogadro's law
$V_1/n_1 = V_2/n_2$ (constant T & P)

Molar volume (L/mol)
$V_m = V/n$

Van der Waals' equation
$nRT = (P + an^2/V^2)(V - nb)$

Gas compression factor
$z = P\,V_m/RT$

Dalton's law of partial pressures
$P_{tot} = P_1 + P_2 + \ldots + P_x$

Graham's law of diffusion

$$\frac{\text{Diffusion rate of A}}{\text{Diffusion rate of B}} = \frac{\sqrt{\text{mol mass of B}}}{\sqrt{\text{mol mass of A}}}$$

Variables
V = Volume
P = Pressure
T = Temperature (K)
n = no. moles
R = Gas constant
= 8.31 J/K mol
= 8.21 m^3 atm/K mol
= 0.0821 L atm/K mol
a = Molecular attraction constant (L^2 atm/mol^2)
b = Volume constant (L/mol)

It is better to be beautiful than to be good. But it is better to be good than to be ugly. Oscar Wilde

5.5 Atoms & Molecules

The Mole

Just as a dozen is a useful measuring unit in a bakery, a mole is useful for measuring quantities of substances in the laboratory. While a dozen equals 12, a mole equals 6.0220×10^{23}. A mole of bread, then, equals 6.0220×10^{23} loaves.

Formulas

Atomic mass (approx)
u = 6.4/specific heat

Molar mass (kg/mol)
molar mass = mass/n

Moles from number of particles
n = no. of particles/N_A

Moles from mass
n = mass/molar mass

Molar volume (m^3/mol)
V_m = volume/n

Specific volume (m^3/kg)
v = volume/mass

Moles of a gas
$n = PV/RT$

Variables
n = number of moles
N_A = Avogadro's number
= 6.0220×10^{23} particles/mol
P = Pressure
V = Volume
R = Gas constant
= 8.31 J/K mol
= 8.21 m^3 atm/K mol
= 0.0821 L atm/K mol
T = Temperature (K)

The Atom

All matter is made of atoms. Atoms are made of a dense nucleus consisting of positively charged protons and neutral neutrons with negatively charged electrons surrounding this nucleus. An atom carries four pieces of information with it:

mass number →	16		2-	← ionic charge
		0		
atomic number →	8		2	← number of atoms

This information is listed in the periodic table (see Section 5.1).

Atomic Mass Unit

The atomic mass unit (amu or u) is defined so that 12 u equals the mass of the most abundant isotope of carbon (^{12}C).
1 amu = 1 dalton = 1.6605×10^{-27} kg or 1.6605×10^{-24} g

Enemies are so stimulating. Katharine Hepburn

Isotopes

Atoms of the same element with an identical number of protons but a different number of neutrons. These differences give isotopes of the same element slightly different physical properties and masses.

Electronegativity

An atom's ability to attract electrons. As a general rule electronegativity increases as you move to the right on the periodic table. It also increases as you move up the periodic table.

$F > O > Cl > N > Br > I$

Locating Electrons

The outer shells of electrons are often responsible for the physical and chemical properties of atoms. Atomic orbitals represent the electrons' most probable location around atoms. Quantum numbers are used to identify the location of these electrons. They are something like a postal code. Each electron has a different quantum number.

1) Principle quantum number, n: energy level (1, 2, 3, 4... x)
2) The subshell quantum number, ℓ: orbital type (s, p, d, f)
3) The orbital quantum number, m_ℓ: specific orbital ($-\ell$ to $+\ell$) (p_x, p_y, p_z)
4) The spin quantum number, m_s: only two electrons are allowed per orbital (+1/2, −1/2)

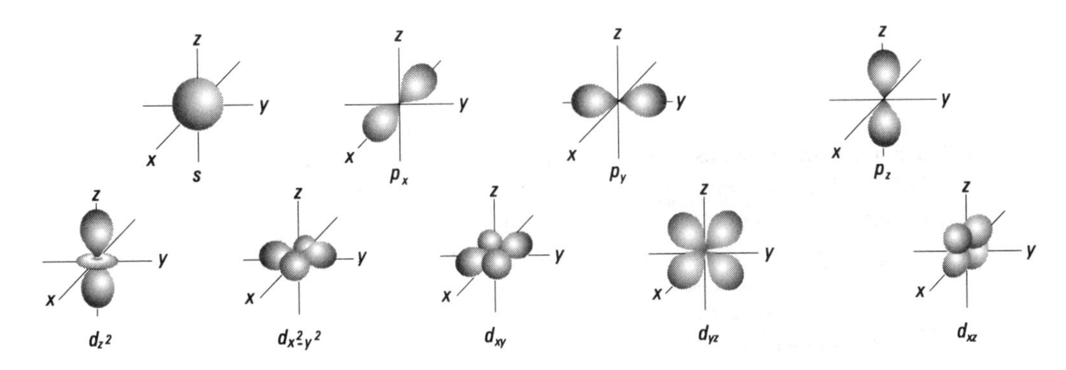

Fig. 5.5.1 Orbitals

Molecules

Types of Chemical Bonds

Electrons around one atom interact with others to form bonds.

1) **Metallic.** Positive metal ions surrounded by a sea of electrons.
2) **Ionic.** Attraction between atoms with a high difference in electronegativity.

3) **Polar covalent**. Two atoms with a moderate difference in electronegativity share electrons.

4) **Pure covalent.** Two atoms with exactly the same electronegativity share electrons.

Common Bond Lengths (Å)

Bond	Length	Bond	Length	Bond	Length
C-H	1.09	C-Br	1.93	H-H	0.74
C-C	1.54	C-I	2.14	H-F	1.1
C=C	1.34	O-H	0.96	H-Cl	1.27
C≡C	1.2	O-O	1.46	H-Br	1.5
C-O	1.43	O=O	1.25	H-I	1.7
C=O	1.2	O-N	1.39	F-F	1.42
C≡O	1.13	N-H	1.01	Cl-Cl	1.98
C-N	1.47	N-N	1.45	Br-Br	2.54
C-F	1.3	N=N	1.25	I-I	2.66
C-Cl	1.78	N≡N	1.10		

Allotropes

Some elements such as oxygen, carbon, tin, sulfur and phosphorous can exist in two or more molecular forms. Each form of the same element is an allotrope and each has different chemical and physical properties. Allotropes of oxygen are oxygen gas (O_2) and ozone (O_3). Allotropes of carbon are amorphous (charcoal), graphite and diamond.

Isomers

Compounds with the same molecular formula but different molecular structures.

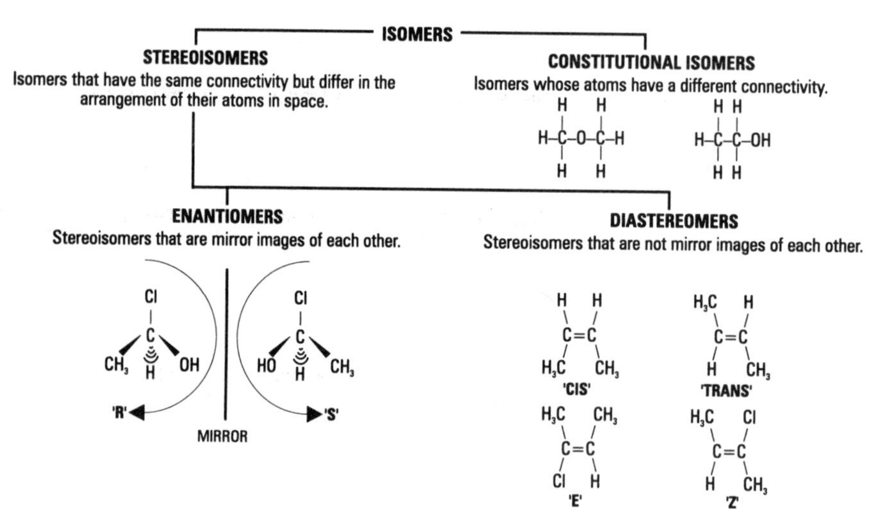

Fig 5.5.2 Isomers

Resonance

Because the arrangement of electrons is not always constant, some molecules and ions cannot be correctly represented with a single structural formula. This means that more than one Lewis structure can be written.

a) b)

Fig 5.5.3 Lewis Diagrams and Resonance

Shapes of Molecules

Hybridization. When bonding, a special mix of orbitals is often observed. For example, an "*s*" orbital and a "*p*" orbital may mix, giving two "*sp*" orbitals. Some mixed orbitals include sp, sp^2, sp^3, sp^3d, sp^3d^2.

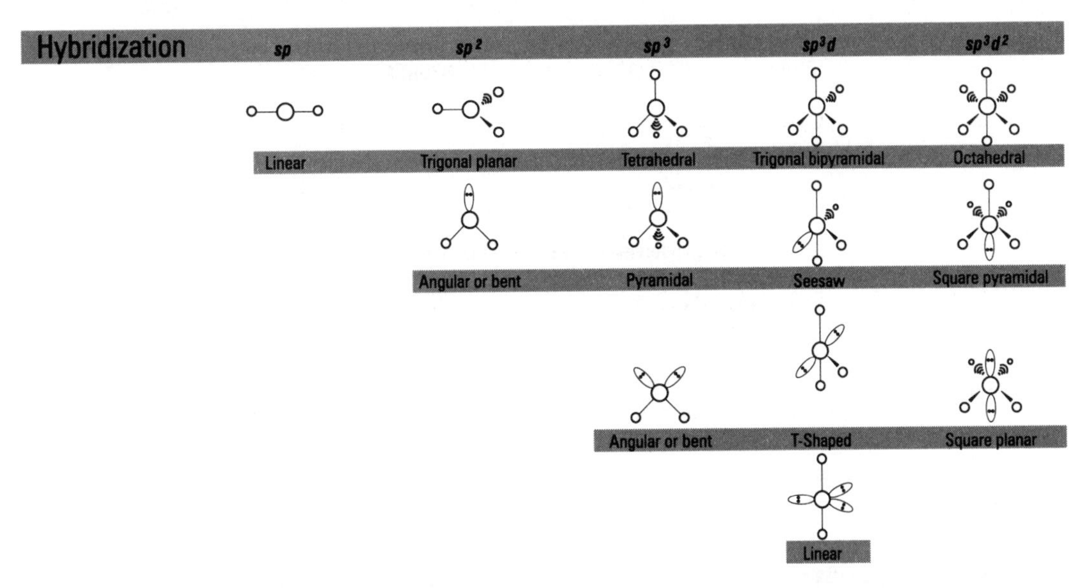

Fig 5.5.4 Shapes of Molecules

However big the fool, there is always a bigger fool to admire him. Nicolas Boileau

Intermolecular Interactions

1) **Dipole-dipole force**. A force between polar molecules.
2) **Induced dipole.** Arises when one compound or an electrical field causes a molecule to have a slightly uneven distribution of electrons.
3) **Hydrogen bonding.** A special type of bond formed when a hydrogen atom is covalently bonded to an electronegative atom and bonds to a second electronegative atom with a lone pair of electrons.
4) **Dispersion.** Forces of attraction acting on all compounds (London forces, Van der Waals forces). The force between compounds results from momentary fluctuations in the symmetry of the electron cloud around a compound.

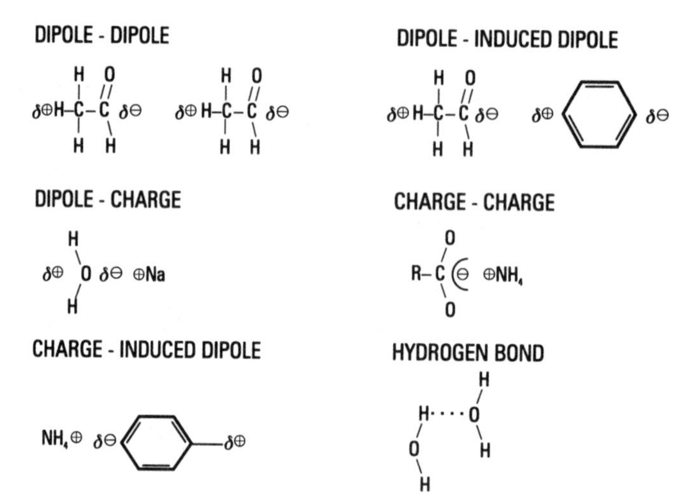

DISPERSION (London forces, Van der Waals forces)

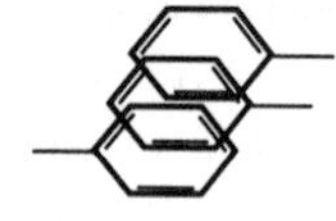

Fig. 5.5.5 Intermolecular Interactions

5.6 Chemical Reactions & Redox

Chemical Reaction

A chemical reaction is a change of bonding between atoms. Bonding changes occur because of a rearrangement or transfer of outer electrons about atoms. Chemical reactions are often accompanied by a change in energy.

Reading a Chemical Reaction

$A + B_2$	$\rightarrow$	AB_2
Reactants		Products
(starting materials)		(end materials)

Symbols Used in Chemical Equations

Symbol	Meaning	Symbol	Meaning
$+$	Mixed with	Δ	Heat energy
$\rightarrow$	Yields	$(h\nu)$	Light energy
$\not\rightarrow$	No reaction	(aq)	Aqueous, dissolved in water
$\leftrightharpoons$	Reaction favors the right	(g)	Gas
$\downarrow$	Precipitates out of solution	(s)	Solid
$\uparrow$	Gas evolves	(l)	Liquid
$\longleftrightarrow$	Resonance (see Section 5.5)		

Yields. In the laboratory, don't expect to get 100% of the product you want. The expected yield is the amount of product obtained under ideal conditions, but in reality, reactions are not perfect.

$$\% \text{ yield} = \frac{\text{Actual yield}}{\text{Expected yield}} \times 100$$

Stoichiometry

Stoichiometry is the quantitative observation of chemical reactions. It is based on the fact that each element must have the same number of atoms on either side of the arrow in an equation. This is due to the law of **conservation of matter**, which states that matter cannot be created or destroyed in a chemical reaction. With a balanced chemical equation, you can gain valuable information.

1) $A_2 + B_2 \longrightarrow AB_3$ unbalanced

2) $A_2 + 3B_2 \longrightarrow 2AB_3$ balanced

I owe nothing to Women's Lib. Margaret Thatcher

This balanced equation contains much information:

	$N_2(g)$ +	$3H_2(g)$ ⟶	$2NH_3(g)$
# moles	1 mole nitrogen gas	3 moles hydrogen gas	2 moles ammonia
# molecules	1 molecule	3 molecules	2 molecules
mass	28.01 g	6.06 g	34.07 g
molecular mass	28.01 u	6.06 u	34.07 u

The balanced equation shows the exact amount of reagents required to produce the exact amount of products. This is known as the **stoichiometric amount**. When one reagent is in excess of this stoichiometric amount, the excess will not react. The reagent present in the most limited amount is called the **limiting reagent**. It determines how much product will be formed.

Reactions in Solution

Net ionic equations show only ions and molecules that react and change. Spectator ions are not included in the net ionic equation.

Spectator ions are those ions in a solution that do not react.

Precipitates are solids produced during reactions in a solution. The solid is forced out of solution according to solubility rules (see Section 5.8).

Precipitation reactions are reactions that produce precipitates.

Solving Stoichiometry Problems

If you're taking a chemistry course, you'll probably have to solve at least one stoichiometry problem. Here are a few tips on how to solve them, along with an example. Remember, nothing beats practice.

1) Balance the equation.
2) Take all known quantities and convert to number of moles.
3) Use mole ratios to determine any unknown quantities in moles.
4) Convert your mole ratios back to whatever units you need in the answer.

Example:

Using the following unbalanced equation, calculate the mass of $NH_3(g)$ produced and determine the limiting reagent. The reaction begins with 1.00 mole $N_2(g)$ and 2.04 g of $H_2(g)$.

$$N_2 + H_2 \rightarrow NH_3$$

Quick tip – Here are some molar masses: N = 14.0 g/mol, H = 1.01 g/mol

1) Balance the equation by finding ratios for the molecules that balance the atoms on both sides of the equation. This gives you

$$N_2 + 3H_2 \rightarrow 2NH_3$$

2) The number of moles of H_2 = mass/molar mass = 2.04 g/(2 × 1.01 g/mol) = 1.01 mol. Therefore H_2 is the limiting reagent because 3 moles of it are needed to react completely with the 1 mole of N_2.
3) The number of moles NH_3 produced = 2 molecules × (1.01 mol/3 molecules) = 0.673 mol.
4) Therefore mass of NH_3 produced = molar mass x number of moles = 17.03 g/mol × 0.673 mol = 11.5 g.

To do nothing is also a good remedy. Hippocrates

Types of Reactions

Most reactions can be classified as one of the following:

1) Combination	A + B → AB	Fe + S → FeS
2) Decomposition	AB → A + B	$CaCO_3$ –Δ→ CaO + CO_2
3) Displacement	A + BC → AB + C	Mg+H_2SO_4 → H2+$MgSO_4$
4) Partner exchange	AB + CD→ AC + BD	$AgNO_3$ + NaCl → AgCl + $NaNO_3$

Chemical reactions can also be classified as **redox** and **nonredox** reactions.

Predicting Thermal Decomposition

Many compounds break down in a predictable fashion when heated. **Carbonates**, **hydrates**, **hydroxides**, **nitrates** and **sulfates** decompose according to the following rules:

1) When heated, carbonates, hydroxides, nitrates and sulfates containing Hg, Ag, Pd, Pt and Au decompose to yield the metal, *not* a metal oxide when heated.

2) When heated, **carbonates** give off carbon dioxide and form an oxide. Exception: alkali metal carbonates.

$$MgCO_3 \xrightarrow{\Delta} MgO + CO_2$$

$$K_2CO_3 \xrightarrow{\Delta} \text{No reaction}$$

Aluminum (Al) and tin (Sn) do not form carbonates.

3) When heated, **hydrates** give off water. Hydrates of **carbonates**, **hydroxides**, **nitrates** and **sulfates** will then decompose by the other rules.

$$CuSO_4 \cdot 5H_2O \xrightarrow{\Delta} CuSO_4 + 5H_2O$$

4) When heated, **hydroxides** give off water and form an oxide. Exception: alkali metal hydroxides.

$$2Al(OH)_3 \xrightarrow{\Delta} Al_2O_3 + 3H_2O$$

$$NaOH \xrightarrow{\Delta} \text{No reaction}$$

5) When heated, **nitrates** give off nitrogen dioxide, oxygen and an oxide. Exception: alkali metal nitrates will give off oxygen and form a nitrite.

$$4Fe(NO_3)_3 \xrightarrow{\Delta} 2Fe_2O_3 + 12NO_2 + 3O_2$$

$$2NaNO_3 \xrightarrow{\Delta} 2NaNO_2 + O_2$$

6) When heated, **sulfates** give off sulfur trioxide and an oxide. Exception: alkali metal and alkaline earth sulfates.

$$Al_2(SO_4)_3 \xrightarrow{\Delta} Al_2O_3 + 3SO_3$$

$$CaSO_4 \xrightarrow{\Delta} \text{No reaction}$$

The only sure thing about luck is that it will change. Henry David Thoreau

The table below is useful for predicting the outcome of reactions between metals and acids, water, oxygen gas and hydrogen gas.

Activity Series of Metals

Acid and water	Metal	Oxygen	Hydrogen
These metals liberate H_2 from water, steam or acid	Li - Lithium	These metals combine with oxygen to form oxides	These metals cannot be displaced from oxides by H_2
	K - Potassium		
	Ba - Barium		
	Sr - Strontium		
$Ca + 2H_2O \rightarrow Ca(OH)_2 + H_2(g)$	Ca - Calcium	$2Hg + O_2(g) \rightarrow 2HgO$	$Al_2O_3 + H_2(g) \rightarrow$ no reaction
	Na - Sodium		
These metals liberate H_2 from steam or acids	Mg - Magnesium		
	Be - Beryllium		
	Al - Aluminum		
	Mn - Manganese		
	Zn - Zinc		
$Mg + H_2O(g) \rightarrow MgO + H_2(g)$	Cr - Chromium		
	Cd - Cadmium		
	Fe - Iron		Oxides of these metals can be reduced with H_2
These metals liberate H_2 from acids	Ni - Nickel		
	Sn - Tin		
$Ni + H_2SO_4 \rightarrow NiSO4 + H_2(g)$	Pb - Lead		
Hydrogen's relative activity $\rightarrow$	H_2 - Hydrogen		
These metals do not react to liberate H_2 from water steam or acids.	Sb - Antimony		
	Cu - Copper		
	Hg - Mercury		
	Ag - Silver	Oxides of these metals decompose upon heating	
$Ag + H_2SO_4 \rightarrow$ no reaction	Pd - Palladium		
	Pt - Platinum		
	Au - Gold	$2HgO \xrightarrow{\Delta} 2Hg + O2(g)$	

He's not the kind of doctor that ever did anybody any good. Robert Millikan's maid

Reduction and Oxidation (Redox)

Every chemical reaction is either a redox or non-redox reaction. In every redox reaction, electrons are rearranged. In redox reactions, atoms actually lose or gain electrons. Oxidation numbers are used to keep track of electrons in redox reactions (see Section 4.6, Metabolic Cycles).

increasing oxidation number →

...,−3, −2, −1, 0, +1, +2, +3, ...

← decreasing oxidation number —

An oxidation number increases when a substance is oxidized (loses electrons). An oxidation number decreases when a substance is reduced (gains electrons). Memorize the difference using the abbreviations, "LEO" and "GER." **LEO** (**L**oss of **E**lectrons = **O**xidation) and **GER** (**G**ain of **E**lectrons = **R**eduction).

Ae^-	+	B	→	A	+	Be^-
reducing agent (electron donor)		oxidizing agent (electron acceptor)		oxidized		reduced

In any redox reaction, when an atom is oxidized, another is always reduced. Remember this rule when balancing equations with oxidation numbers.

Rules for Assigning Oxidation Numbers in Formulas

1) The oxidation number for atoms in pure metals is zero.
2) For neutral molecules the sum of positive and negative oxidation states is zero.
3) In polyatomic ions the sum of positive and negative oxidation states is equal to the charge on the ion.
4) The oxidation state of oxygen equals −2, except in peroxides (−1), and when combined with fluorine (+2).
5) The oxidation state of hydrogen is +1, except in metallic hydrides such as NaH (−1).
6) Monatomic ions such as Li^+, K^+ and Mg^{2+} have an oxidation state equal to their charge.
7) Ions of halogen atoms like F, Cl, Br and I usually have an oxidation state of −1.
8) Possible oxidation states are listed in the periodic table (Section 5.1).

The table below lists common oxidizing and reducing agents used in the laboratory.

Some Oxidizing and Reducing Agents

Oxidizing agent	Agent reduced to	Reducing agent	Agent oxidized to
Halides		**Alkali metals**	
F_2, Cl_2, Br_2,	F^-, Cl^-, Br^-	Li, Na, K	Li^+, Na^+, K^+
Metals		**Alkaline earth metals**	
Cu^+, Cu^{2+}	Cu	Mg, Ca	Mg^{2+}, Ca^{2+}
Ag^+	Ag	**Halides**	
Fe^{3+}	Fe^{2+} or Fe	Cl^-, Br^-, I^-	Cl_2, Br_2, I_2
Acids		**Other metals**	
H_2SO_4 conc.	SO_2 or H_2S or S	Fe	Fe^{2+} or Fe^{3+}
HNO_3 conc	NO_2	Fe^{2+}	Fe^{3+}
HNO_3 dil	NO or NO_2 or N_2O	Al	Al^{3+}
HNO_2	NO or N_2O	Sn^{2+}	Sn^{4+}
$HClO_3$	Cl^- or Cl_2	Sn	Sn^{4+} or Sn^{2+}
		Hydrogen	
		H_2	H^+
		H^-	H^+ or H_2
Ions		**Sulfur**	
NO_3^- in acid sol	HNO_2 or NO or N_2O	S^{2-}, H_2S (with	S (SO_4^{2-} when
ClO^- or ClO_3^-	Cl^- or Cl_2	HNO_3 conc.)	reducing agents in
BrO_3^-	Br^- or Br_2	SO_2 or SO_3^{2-}	conc. HNO_3)
IO_3^-	I^- or I_2		SO_4^{2-}
MnO_4^- (neutral or basic sol)	MnO_2	**Nitrogen**	
MnO_4^- or MnO_2 in acid sol	Mn^{2+}	NH_3	N_2
$Cr_2O_7^{2-}$ or CrO_4^{2-}	Cr^{3+}	N_2H_4	N_2
BiO_3^-	Bi^{3+}	HNO_2	HNO_3
PbO_2 in acid sol	Pb^{2+}		
Other		**Others**	
H_2O_2	H_2O	CO	CO_2
O_2	O^{2-} or H_2O	HPO_3^{2-}	H_2PO_4

Thought is the blossom; language the bud; action the fruit behind. Ralph Waldo Emerson

Predicting and Assigning Oxidation Numbers

To figure out the oxidation number of a specific atom in a compound, follow these steps:

1) Assign oxidation numbers to known atoms using the rules for assigning oxidation numbers presented earlier in this section.
2) Add the oxidation numbers.
3) In a neutral molecule, give an oxidation number to the unknown that makes the total sum of oxidation numbers equal zero.
4) In a polyatomic ion, give an oxidation number to the unknown that makes the total sum of the oxidation numbers equal the charge on the ion.

Balancing Redox Equations in Aqueous Solutions

These techniques depend on whether the chemical reaction is taking place in an acid (H^+) or a basic (OH^-) solution, so be sure to check before you start. When you know, balance the equation according to the steps below (examples on next page):

1) Split the equation into half-cell reactions.
2) Balance all the atoms except hydrogen and oxygen.
3) Balance the oxygen atoms by adding one H_2O for each oxygen atom to the side that is lacking.
4) Balance the hydrogen atoms. In acid solutions, add one H^+ for every hydrogen on the side that is lacking. In basic solutions, add one OH^- to the same side where you added your H_2O in step 3; then add one H_2O to the other side. Do this until your hydrogen and oxygen are balanced. It may sound silly, but it works.
5) Balance the charges on the ions, by adding electrons (e^-).
6) Multiply each half-cell by a number that balances the electrons. Then add the half-cells algebraically. You should now have a balanced equation.

Double check your answer!

1) The total atoms on the left side of an equation must equal the total atoms on the right side (conservation of mass).
2) Net ionic charges on the left side of an equation must equal the net ionic charges on the right (conservation of charge).
3) The total increase in oxidation numbers by reducing agents must equal the total decrease in oxidation numbers by oxidation agents.

Here's a sample of how to solve unbalanced equations in an aqueous solution.

In acid aqueous solution	In basic aqueous solution
$Mn^{2+} + BiO_3^- \rightarrow MnO_4^- + Bi^{3+}$	$Cl_2 (aq) + IO_3^- \rightarrow Cl^- + IO_4^-$
1) $Mn^{2+} \rightarrow MnO_4^-$; $BiO_3^- \rightarrow Bi^{3+}$	1) $Cl_2 \rightarrow Cl^-$; $IO_3^- \rightarrow IO_4^-$
2) $Mn^{2+} \rightarrow MnO_4^-$; $BiO_3^- \rightarrow Bi^{3+}$ (no change)	2) $Cl_2 \rightarrow 2Cl^-$; $IO_3^- \rightarrow IO_4^-$
3) $4H_2O + Mn^{2+} \rightarrow MnO_4^-$; $BiO_3^- \rightarrow Bi^{3+} + 3H_2O$	3) $Cl_2 \rightarrow 2Cl^-$; $H_2O + IO_3^- \rightarrow IO_4^-$
4) $4H_2O + Mn^{2+} \rightarrow MnO_4^- + 8H^+$; $6H^+ + BiO_3^- \rightarrow Bi^{3+} + 3H_2O$	4) $Cl_2 \rightarrow 2Cl^-$; $H_2O + 2OH^- + IO_3^- \rightarrow IO_4^- + 2H_2O$
5) $4H_2O + Mn^{2+} \rightarrow MnO_4^- + 8H^+ + 5e^-$ (x2); $6H^+ + BiO_3^- + 2e^- \rightarrow Bi^{3+} + 3H_2O$ (x5)	5) $Cl_2 + 2e^- \rightarrow 2Cl^-$ $H_2O + 2OH^- + IO_3^- \rightarrow IO_4^- + 2H_2O + 2e^-$
6) $8H_2O + 2Mn^{2-} + \rightarrow 2MnO_4^- + 16H^+ + 10e^-$; $30H^+ + 5BiO_3^- + 10e^- \rightarrow 5Bi^{3+} + 15H_2O$	6) $Cl_2 + 2e^- \rightarrow 2Cl^-$ (no change) $H_2O + 2OH^- + IO_3^- \rightarrow IO_4^- + 2H_2O + 2e^-$
The balanced equation is:	The balanced equation is:
$14H^+ + 2Mn^{2+} + 5BiO_3^- \rightarrow 2MnO_4^- + 5Bi^{3+} + 7H_2O$	$2OH^- + Cl_2 + IO_3^- \rightarrow 2Cl^- + IO_4^- + H_2O$

Formulas Used in Electrochemistry

Charge

$q = It$

Cell Potential

$W = -Eq$

Relationship Between ΔG & E

$\Delta G = -nFE$

$\Delta G° = -nFE°$

Conversion from Standard State

$\Delta G = \Delta G° + 2.303\, RT \log Q$

$\Delta G = \Delta G° + RT \ln Q$

$E = E° - 2.303\, RT/nF \log Q$

$E = E° - RT/nF \ln Q$

Variables

q = electrical charge (C)
I = current (C/s)
t = time (s)
W = work (J)
E = cell potential (V)
F = Faraday's constant
= 9.65×10^4 C/mol
n = electrons transferred (mol e^-)
T = temperature (K)
R = gas constant
= 8.31 J/K mol
= 0.0821 L atm/K mol
= 8.21 m^3 atm/K mol
° = at standard state
ΔG = Gibbs free energy (J)
Q = [products]/[reactants]

Notes on reaction spontaneity:

1) When E is positive, ΔG is negative and the reaction is spontaneous.
2) When E is negative, ΔG is positive and the reaction is not spontaneous.
3) When E and ΔG equal zero, the reaction is at equilibrium.

Judge a man by the reputation of his enemies. Arabian Proverb

All the values in the table below are given at 25°C and a pressure of 1 atm.

Electrochemical Potentials

Reaction	$E°$, V	Reaction	$E°$, V
$Ag^+ + e^- \rightarrow Ag(s)$	+0.80	$Cu^{2+} + 2e^- \rightarrow Cu(s)$	+0.34
$Ag^{2+} + e^- \rightarrow Ag^+$	+1.98	$F_2(g) + 2e^- \rightarrow 2F^-$	+2.87
$AgBr(s) + e^- \rightarrow Ag(s) + Br^-$	+0.07	$Fe^{2+} + 2e^- \rightarrow Fe(s)$	– 0.44
$AgCl(s) + e^- \rightarrow Ag(s) + Cl^-$	+0.22	$Fe^{3+} + e^- \rightarrow Fe^{2+}$	+0.77
$AgI(s) + e^- \rightarrow Ag(s) + I^-$	– 0.15	$Fe^{3+} + 3e^- \rightarrow Fe(s)$	– 0.04
$Al^3+ + 3e- \rightarrow Al(s)$	– 1.66	$[Fe(CN)_6]^{3-} + e^- \rightarrow [Fe(CN)_6]^{4-}$	+ 0.36
$Au^+ + e^- \rightarrow Au(s)$	+1.69	$Ga^{3+} + 3e^- \rightarrow Ga(s)$	– 0.53
$Au^{3+} + 3e^- \rightarrow Au(s)$	+1.50	$2H^+ + 2e^- \rightarrow H_2(g)$	0.000
$Ba^{2+} + 2e^- \rightarrow Ba(s)$	– 2.91	$2HClO(aq) + 2H^+ + 2e^- \rightarrow Cl_2(g) + 2H_2O(l)$	+ 1.63
$Be^{2+} + 2e^- \rightarrow Be(s)$	– 1.85	$HClO_2(aq) + 2H^+ + 2e^- \rightarrow HClO(aq) + H_2O(l)$	+ 1.65
$Br_2(l) + 2e^- \rightarrow 2Br^-$	+1.07	$2HNO_2(aq) + 4H^+ + 4e^- \rightarrow N_2O(g) + 3H_2O(l)$	+ 1.29
$Br_2(aq) + 2e^- \rightarrow 2Br^-$	+1.09	$H_2O_2(aq) + 2H^+ + 2e^- \rightarrow 2H_2O(l)$	+1.78
$BrO^- + H_2O(l) + 2e^- \rightarrow Br^- + 2OH^-$	+0.76	$2H_2O(l) + 2e^- \rightarrow H_2(g) + 2OH^-$	– 0.83
$BrO_3^- + 3H_2O(l) + 6e^- \rightarrow Br^- + 6OH^-$	+0.61	$H_2SO_3(aq) + 4H^+ + 4e^- \rightarrow S(s) + 3H_2O(l)$	0.45
$2BrO_3^- + 12H^+ + 10e^- \rightarrow Br_2(l) + 6H_2O(l)$	+1.52	$Hg_2^{2+} + 2e^- \rightarrow 2Hg(l)$	+0.79
$Ca^{2+} + 2e^- \rightarrow Ca(s)$	– 2.87	$2Hg^{2+} + 2e^- \rightarrow Hg_2^{2+}$	+0.92
$Cd^{2+} + 2e^- \rightarrow Cd(s)$	– 0.40	$Hg_2SO_4(s) + 2e^- \rightarrow 2Hg(l) + SO_4^{2-}$	+0.62
$Cd(OH)_2(s) + 2e^- \rightarrow Cd(s) + 2OH^-$	– 0.81	$I_2(s) + 2e^- \rightarrow 2I^-$	+0.54
$Ce^{3+} + 3e^- \rightarrow Ce(s)$	– 2.48	$IO_3^- + 3H_2O(l) + 6e^- \rightarrow I^- + 6OH^-$	+0.26
$Ce^{4+} + e^- \rightarrow Ce^{3+}$	+1.61	$In^{3+} + 3e^- \rightarrow In(s)$	– 0.34
$Cl_2(g) + 2e^- \rightarrow 2Cl^-$	+1.36	$K^+ + e^- \rightarrow K(s)$	– 2.93
$ClO^- + H_2O(l) + 2e^- \rightarrow Cl^- + 2OH^-$	+0.89	$La^{3+} + 3e^- \rightarrow La(s)$	– 2.52
$ClO_4^- + H_2O(l) + 2e^- \rightarrow ClO_3^- + 2OH^-$	+0.36	$Li^+ + e^- \rightarrow Li(s)$	– 3.05
$ClO_4^- + 2H^+ + 2e^- \rightarrow ClO_3^- + H_2O(l)$	+1.19	$Mg^{2+} + 2^{e-} \rightarrow Mg(s)$	- 2.36
$Cr^{2+} + 2e^- \rightarrow Cr(s)$	– 0.91	$Mn^{2+} + 2e^- \rightarrow Mn(s)$	– 1.19
$Cr^{3+} + e^- \rightarrow Cr^{2+}$	– 0.41	$MnO_2(s) + 2H_2O(l) + 2e^- \rightarrow Mn(OH)_2(s) + 2OH^-$	– 0.05
$Cr^{3+} + 3e^- \rightarrow Cr(s)$	– 0.74	$MnO_2(s) + 4H^+ + 2e^- \rightarrow Mn^{2+} + 2H_2O\ (l)$	+1.23
$Cr_2O_7^{2-} + 14H^+ + 6e^- \rightarrow 2Cr^{3+} + 7H_2O(l)$	+1.33	$MnO_4^- + e^- \rightarrow MnO_4^{2-}$	+0.56
$Co^{2+} + 2e^- \rightarrow Co(s)$	– 0.28	$MnO_4^- + 2H_2O(l) + 3e^- \rightarrow MnO_2(s) + 4OH^-$	+0.59
$Co^{3+} + e^- \rightarrow Co^{2+}$	+1.81	$MnO_4^- + 8H^+ + 5e^- \rightarrow Mn^{2+} + 4H_2O(l)$	+1.51
$Cu^+ + e^- \rightarrow Cu(s)$	+0.52	$Na^+ + e^- \rightarrow Na(s)$	– 2.71
$Cu^{2+} + e^- \rightarrow Cu^+$	– 0.26	$Ni^{2+} + 2e^- \rightarrow Ni(s)$	+0.15

Be good and you will be lonely. Mark Twain

Electrochemical Potentials (cont.)

Reaction	$E°$, V	Reaction	$E°$, V
$NO_3^- + H_2O(l) + 2e^- \rightarrow NO_2^- + 2OH^-$	+0.01	$SO_4^{2-} + 4H^+ + 2e^- \rightarrow H_2SO_3(aq) + H_2O(l)$	+0.17
$NO_3^- + 3H^+ + 2e^- \rightarrow HNO_2(aq) + H_2O(l)$	+0.94	$SO_4^{2-} + 8H^+ + 6e^- \rightarrow S(s) + 4H_2O(l)$	+0.36
$NO_3^- + 4H^+ + 3e^- \rightarrow NO(g) + 2H_2O(l)$	+0.96	$SO_4^{2-} + H_2O(l) + 2e^- \rightarrow SO_3^{2-} + 2OH^-$	– 0.93
$O_2(g) + 2H^+ + 2e^- \rightarrow H_2O_2(aq)$	+0.68	$S_2O_8^{2-} + 2e^- \rightarrow 2SO_4^{2-}$	+2.01
$O_2(g) + 4H^+ + 4e^- \rightarrow 2H_2O(l)$	+1.23	$Se(s) + 2H^+ + 2e^- \rightarrow H_2Se(aq)$	– 0.40
$O_2(g) + 2H_2O(l) + 4e^- \rightarrow 4OH^-$	+0.40	$Sn2+ + 2e^- \rightarrow Sn(s)$	– 0.14
$O_3(g) + 2H^+ + 2e^- \rightarrow O_2(g) + H_2O(l)$	+2.07	$Sn^{4+} + 2e^- \rightarrow Sn^{2+}$	+0.15
$O_3(g) + H_2O(l) + 2e^- \rightarrow O_2(g) + 2OH^-$	+1.24	$Sr^{2+} + 2e^- \rightarrow Sr(s)$	– 2.89
$PO_4^{3-} + 2H_2O(l) + 2e^- \rightarrow HPO_3^{2-} + 3OH^-$	– 1.12	$Te(s) + 2e^- \rightarrow Te^{2-}$	– 1.14
$PbSO_4(s) + 2e^- \rightarrow Pb(s) + SO_4^{2-}$	– 0.36	$Te(s) + 2H^+ + 2e^- \rightarrow H_2Te(aq)$	– 0.74
$Pt^{2+} + 2e^- \rightarrow Pt(s)$	+1.18	$Ti^{2+} + 2e^- \rightarrow Ti(s)$	– 1.63
$Ra^{2+} + 2e^- \rightarrow Ra(s)$	– 2.92	$Ti^{3+} + e^- \rightarrow Ti^{2+}$	– 0.37
$Rb^+ + e^- \rightarrow Rb(s)$	– 2.93	$U^{4+} + e^- \rightarrow U^{3+}$	– 0.61
$S(s) + 2e^- \rightarrow S^{2-}$	– 0.48	$V^{2+} + 2e^- \rightarrow V(s)$	– 1.18
$S(s) + 2H^+ + 2e^- \rightarrow H_2S(aq)$	+0.14	$Zn^{2+} + 2e^- \rightarrow Zn(s)$	– 0.76

A year from now you may wish you had started today. Karen Lamb

5.7 Kinetics

Useful Terms

Activation energy (*Ea*). The minimum energy required for a reaction to occur. Found by experimentation.
Intermediate (transition state, activated complex). One of the compounds produced in a reaction mechanism that reacts further, but is not seen as one of the products.
Rate determining step. The slowest step in a reaction mechanism. It determines the speed of reaction.
Reaction mechanism. A set of steps at the molecular level through which a reaction proceeds.

Reaction Rate Equations

$$aA + bB \longrightarrow cC + dD$$

Rate of Disappearance of Reactants

$$r_A = \frac{+\Delta[A]}{\Delta t}$$

$$r_B = \frac{+\Delta[B]}{\Delta t}$$

Rate of Appearance of Products

$$R_C = \frac{\Delta[C]}{\Delta t}$$

$$R_D = \frac{\Delta[D]}{\Delta t}$$

Rate Law Expression

$r = k\,[A]^m[B]^n$

Total Reaction Order

$m + n$

The **rate of reaction** is a measure of the speed at which products appear and reactants disappear.

$$\text{average rate} = \frac{\Delta\,\text{concentration}}{\Delta\,\text{time}}$$

The **rate law expression** is the mathematical relationship between the reaction rate and the concentration of reactants. k is the rate constant for a particular temperature, and is not affected by concentration. m and n are exponents to which the concentrations are raised. They are not necessarily whole numbers.

The **total reaction order** is an overall measure of how fast a reaction occurs. It is calculated by adding $m + n$. For a given chemical reaction, the values for k, m and n have to be determined experimentally.

Total Reaction Order Equations

Zero Order Equations

$r = k[A]^0 = k$

$r = k[B]^0 = k$

$r = k[A]^0[B]^0 = k$

Rate does not depend on the concentration of either reactant.

First Order Equations

$r = k[A]^1[B]^0 = k[A]$

$r = k[A]^0[B]^1 = k[B]$

Rate depends on the concentration of one reactant only.

Second Order Equations

$r = k[A]^1[B]^1$

$r = k[A]^2[B]^0 = k[A]^2$

$r = k[A]^0[B]^2 = k[B]^2$

Third Order Equations

$r = k[A]^3[B]^0 = k[A]^3$

$r = k[A]^2[B]^1$

$r = k[A]^1[B]^2$

$r = k[A]^0[B]^3 = k[B]^3$

The causes of events are never more interesting than the events themselves. Cicero

Collision Theory

The rate of reaction depends on two fundamental points:

1) The number of **collisions** between reactant particles per unit time.
2) The fraction of collisions that are successful.

Factors Affecting Reaction Rate

Catalyst. A catalyst lowers the activation energy (*Ea*) and is not destroyed during the reaction. A lower activation energy speeds up the rate of reaction hence increasing k.

Concentration. Increasing the concentration of one or more reactants increases the number of possible effective collisions and thus speeds the reaction. Decreasing the concentration of reactants decreases reaction rate. Increasing the concentration of products decreases reaction rate. Decreasing the concentration of products by removal increases reaction rate.

The nature of reactants. More complicated molecules typically react slower.

Pressure. Pressure has little effect on reaction rates involving solids and liquids. Figuring out the effect of pressure change on a gas is tricky. Think of it this way: Increasing the pressure on a gas at constant temperature has the same effect as increasing the concentration of the gas. Decreasing the pressure on a gas at constant temperature has the same effect as decreasing the concentration of the gas.

Surface area. Increasing the total area of contact between reactants increases the rate of the reaction.

Temperature. An increase in temperature increases the rate constant k, and speeds the rate of reaction. Decreasing the temperature decreases the rate constant k, and slows down the reaction rate (see the table, Stresses Affecting Equilibrium, Section 5.8).

Calculating Rate Constant

Arrhenius Equation

$$k = Ae^{-E_a/RT}$$

$$\log k = -\frac{E_a A}{2.303RT} + \log A$$

$$\log \frac{k_2}{k_1} = \frac{E_a}{2.303R} \times \frac{T_2 - T_1}{T_1 T_2}$$

Variables

A = collision frequency correction constant
e = natural log base
 = 2.303
E_a = activation energy
k = rate constant (at T)
R = ideal gas constant
 = 8.31 J/K mol
T = temperature (K)

Nothing's new, and nothing's true, and nothing matters. Sidney Morgan

5.8 Equilibrium & Solutions

Equilibrium

Equilibrium Law

$$aA + bB \rightleftharpoons cC + dD$$

The Equilibrium Constant K_{eq}

Equilibrium constant at constant temperature. This value is determined through experimentation.

$$K_{eq} = \frac{[C]^c[D]^d}{[A]^a[B]^b}$$

$K_{eq} > 1$ products are favored
$K_{eq} < 1$ reactants are favored

Types of K

K_a - Dissociation constant for an acid
K_b - Dissociation constant for a base
K_c - Gas concentration constant
K_d - Dissociation constant for complexes
K_{eq} - Equilibrium constant
K_f - Formation constant for complex ions
K_{sp} - Solubility constant for sparingly soluble salts
K_p - For partial pressures of gas
k - rate constant (see Section 5.7)

Reaction Quotient

This equation is used to predict the outcome when substances with the potential for reacting are mixed.

$$Q = \frac{[C]^c[D]^d \ldots}{[A]^a[B]^b \ldots}$$

When $Q > K_{eq}$ the concentration of reactants increases until equilibrium is reached.
When $Q < K_{eq}$ the concentration of products increases until equilibrium is reached.
When $Q = K_{eq}$ equilibrium is reached.

Changes Affecting Equilibrium

1) **Changes in temperature.** Only changes in temperature affect the K_{eq} constant.

$$3\ H_2 + N_2 \rightleftharpoons 2\ NH_3 + 5.2\ \text{kJ (exothermic reaction)}$$

   a) An increase in temperature causes exothermic reactions to absorb heat and shift the equilibrium to the left. An endothermic reaction shifts to the right.

   b) A decrease in temperature causes exothermic reactions to shift to the right and produce more heat. An endothermic reaction shifts equilibrium to the left.

2) **Changes in pressure.**

$$\text{(3 volumes of gas) } 2\ NO(g) + O_2(g) \rightleftharpoons 2\ NO_2(g) \text{ (2 volumes of gas)}$$

   a) Increases in pressure shift the equilibrium in the direction that produces a smaller volume.

   b) Decreases in pressure shift the equilibrium to produce a larger volume.

3) **Adding a catalyst**. Equilibrium is reached in a shorter period of time. Catalysts do not affect the equilibrium point.

Stresses Affecting Equilibrium

Variable	Stress	Type of reaction	Results	K_{eq}	Rate constant k	Reaction rate
Temperature	+	exothermic	less products	changes	+	+
	+	endothermic	more products	changes	+	+
	–	exothermic	more products	changes	–	–
	–	endothermic	less products	changes	–	–
Concentration reactant	+	any*	more products	no change	no change	+
	–	any*	less products	no change	no change	–
Catalyst	+	any	no change	no change	+	+
Pressure	+/–	solids	no change	no change	no change	no change
	+/–	liquids	no change	no change	no change	no change
	+	gases	equilibrium shifts to produce smaller volume	no change	no change	changes
	–	gas	equilibrium shifts to produce a large volume	no change	no change	changes

* Except for zero order reactions

Formulas

Gas Concentration Constant

$$K_C = \frac{[C]^c[D]^d}{[A]^a[B]^b}$$

Partial Pressure Constant

$$K_p = P_C{}^c P_D{}^d / P_A{}^a P_B{}^b = k_c(RT)^{\Delta n}$$

Hydrates

Hydrate. A compound with water molecules incorporated in its structure. Heating a hydrate drives off the water, leaving an anhydrous compound.

$$CuSO_4 \cdot 5H_2O \xrightarrow{\Delta} CuSO_4 + 5H_2O$$

Efflorescence. A hydrate's loss of water upon exposure to air.
Hygroscopic. A type of compound that takes up water from the air, but does not necessarily form a hydrate.
Deliquescence. Occurs when a compound absorbs enough water from moisture in the air to dissolve it completely.
Heat of hydration, ΔH_{hydr} (kJ/mol). The heat released as an ion becomes hydrated.
Heat of solution, ΔH_{sol} (kJ/mol). The enthalpy change exhibited as one mole of a substance is dissolved in a given amount of solvent.

Character is what you are in the dark. Dwight Moody

Solutions

Azeotrope. A mixture that distills without a change in composition.
Distillate. The product of distillation.
Distillation. Heating a liquid solution to boiling point, then collecting and condensing the vapor of a boiling solution.
Fractional distillation. Separating the liquid components of a solution with different boiling points.
Infinitely miscible. Liquids that mix in all proportions.
Immiscible. Liquids that do not mix, but leave distinct phases.
Miscibility. Mutual solubility of two substances in the liquid phase.
Solute. The dissolved component of a solution.
Solution. A homogeneous mixture of two or more substances.
Solvent. A medium used to dissolve solutes.
Refluxing. Preventing vapor loss while boiling a solution.

Vapor Pressure of Water at Various Temperatures

°C	mm Hg	kPa	°C	mm Hg	kPa	°C	mm Hg	kPa
0	4.58	0.61	35	42.18	5.624	70	233.7	31.16
5	6.54	0.87	40	55.32	7.375	75	289.1	38.54
10	9.20	1.23	45	71.88	9.583	80	355.1	47.34
15	12.79	1.705	50	92.51	12.33	85	433.6	57.81
20	17.54	2.338	55	118.0	15.73	90	525.8	70.10
25	23.78	3.170	60	149.4	19.92	95	633.9	84.51
30	31.82	4.242	65	187.5	25.00	100	760.0	101.3

Colligative Properties of Solutions

Colligative properties depend only on the total number of solute particles.

Vapor Pressure Lowering
$\Delta P_A = X_B P_A$

Boiling Point Elevation
$\Delta T_b = K_{bp}\, m$

Freezing Point Depression
$\Delta T_f = K_{fp}\, m$

Osmotic Pressure
$\pi = MRT$

Variables
X_B = Molar fraction of solution
P_A = Vapor pressure of pure solvent
K_{bp} = Molal boiling point depression constant
K_{fp} = Molal freezing point depression constant
m = Molality
R = Gas constant (8.31 J/mol K)
M = Molarity

Electrolytes

Compounds that yield ions in aqueous solution are called electrolytes.

1) **Strong electrolytes.** More than 50% of the solute dissociates into ions (NaCl).
2) **Weak electrolytes**. Less than 50% exist as ions in solution (acetic acid).
3) **Nonelectrolytes.** Less than 0.01% exist as ions in solution (AgCl).

Measures of Concentration

Mass percent

$$\% \text{ Mass} = \frac{\text{mass of solute}}{\text{mass of solution}} \times 100$$

Molarity (mol / L)

$$M = \frac{\text{moles of solute (mol)}}{\text{volume of solution (L)}}$$

Mole fraction

$$\text{mol fraction} = \frac{\text{moles solute}}{\text{moles solute} + \text{moles solvent}}$$

Molality (mol / kg)

$$m = \frac{\text{moles solute (mol)}}{\text{mass of solvent (kg)}}$$

Weight in Weight

$$\% w / w = \frac{\text{mass of solute (g)} \times 100}{\text{100 g of solution}}$$

Weight in Volume

$$\% w / v = \frac{\text{mass solute (g)} \times 100}{\text{100 mL of solution}}$$

Volume in Volume

$$\% v / v = \frac{\text{volume of constituent (mL)}}{\text{100 mL of solution}} \times 100$$

Solubility

Solubility product constant. K_{sp} is typically used as a measure of solubility for salts that are slightly soluble. $K_{sp} = [A^+][B^-]$

$$AB(s) \rightleftharpoons A^+ + B^- \qquad \ln K_{sp} = -\Delta G°/RT$$

$$K_{eq} = \frac{[Ag^+][Br^-]}{[AgBr]} \qquad K_{sp} = K_{eq}[AgBr] = [Ag^+][Br^-]$$

Ion product. Used to predict whether or not precipitation will occur.

$Q_i = [A^+][B^-]$ $Q_i < K_{sp}$ All ions will go into solution

$Q_i = K_{sp}$ Equilibrium

$Q_i > K_{sp}$ Precipitation will occur until $Q_i = K_{sp}$

Temperature's Effect on Solubility

Temperature usually increases solubility of solids in liquids. A few exceptions are SO_4^{2-}, SO_3^{2-}, PO_4^{3-}. The solubility of a gas will usually decreases with an increase in temperature.

Solubility of Solids and Liquids

General Solubility of Salts in Water

	Ag^+	Ba^{2+}	Ca^{2+}	Hg^+	Hg^{2+}	K^+	Li^+	Na^+	NH_4^+	Pb^{2+}	Sr^{2+}
Acetates, $CH_3CO_2^-$	M	S	S	X		S	S	S	S	S	S
Bromides, Br^-	X	S		X	M	S	S	S	S	X	S
Carbonates, CO_3^{2-}	X	X	X			S	S	S	S	X	X
Chlorides, Cl^-	X	S	S	X		S	S	S	S	SH	S
Chromates, CrO_4^{2-}	X	X	M	X		S	S	S	S	X	X
Cyanides, CN^-	X					S	S	S	S	X	X
Fluorides, F^-		X	X			S	S	S	S		X
Hydroxides, OH^-	X	M	M		X	S	S	S	S	X	M
Iodides, I^-	X	X	S	X	X	S	S	S	S	X	X
Nitrates, NO_3^-	M	S	S	S		S	S	S	S	S	S
Oxalates, $C_2O_4^{2-}$	X	X	X	X		S	X	S		S	X
Oxides, O^{2-}	X	S	M	X	X	S	S	S	S	X	M
Phosphates, PO_4^{3-}	X	X	X			S	S	S	S	X	X
Sulfates, SO_4^{2-}	M	X	M	X		S	S	S	S	X	X
Sulfides, S^{2-}	X			X	X	S	S	S	S	X	X
Sulfites, SO_3^{2-}	X	X	X			S	S	S	S	X	X

M = moderately soluble, S = soluble, SH = soluble in hot solution, X = insoluble or immiscible

Miscibility

	1	2	3	4	5	6	7	8	9	10	11	12	13	14
1. CCl_4											X			
2. n-Pentane											X			X
3. n-Hexane											X			X
4. Toluene											X			
5. Chloroform											X			
6. Diethyl ether											X			X
7. n-Butanol											X			
8. Ethanol														
9. Methanol														
10. Acetic acid(*aq*)														
11. Water	X	X	X	X	X	X	X							
12. Dioxane														
13. Acetone														
14. DMSO		X	X			X								

INCREASING DIPOLE MOMENT ↓

Common K_{sp} at 25°C

Salt	K_{sp}	Salt	K_{sp}
$AgBr$	5.35×10^{-13}	Hg_2Br_2	6.41×10^{-23}
$Ag(CH_3COO)$	1.94×10^{-3}	Hg_2Cl_2	1.45×10^{-18}
$AgCN$	5.97×10^{-17}	HgI_2	2.82×10^{-29}
$AgCO_3$	8.45×10^{-12}	Hg_2I_2	5.33×10^{-29}
$AgCl$	1.77×10^{-10}	HgS	2.00×10^{-53}
$AgCrO_4$	2.5×10^{-12}	Hg_2SO_4	7.99×10^{-7}
AgI	8.51×10^{-17}	$MgCO_3$	6.82×10^{-6}
Ag_2S	1.09×10^{-49}	MgF_2	7.42×10^{-11}
$AgSO_4$	1.20×10^{-5}	$Mg(OH)_2$	5.61×10^{-12}
$BaCO_3$	2.58×10^{-9}	$Mg_3(PO_4)_2$	9.86×10^{-25}
$BaCrO_4$	1.2×10^{-10}	$Mn(OH)_2$	2.06×10^{-13}
BaF_2	1.84×10^{-7}	$Ni(OH)_2$	5.47×10^{-16}
$Ba(OH)_2$	1.3×10^{-2}	$PbBr_2$	6.60×10^{-6}
$BaSO_4$	1.07×10^{-10}	$PbCO_3$	1.46×10^{-13}
$CaCO_3$	4.96×10^{-9}	PbC_2O_4	8.51×10^{-10}
$CaCrO_4$	6×10^{-4}	$PbCl_2$	1.17×10^{-5}
CaF_2	1.46×10^{-10}	$PbCrO_4$	2.8×10^{-13}
$Ca(OH)_2$	4.68×10^{-6}	PbI_2	8.49×10^{-9}
$Ca_3(PO_4)_2$	2.07×10^{-23}	$Pb(OH)_2$	1.42×10^{-20}
$CaSO_4$	7.10×10^{-5}	PbS	9.04×10^{-29}
$Cr(OH)_3$	6×10^{-31}	$PbSO_4$	1.82×10^{-8}
$CuBr$	6.27×10^{-9}	SnS	3.25×10^{-28}
$CuCO_3$	2.3×10^{-10}	$SrCO_3$	5.60×10^{-10}
$CuCl$	1.72×10^{-7}	$SrCrO_4$	2.2×10^{-5}
CuI	1.27×10^{-12}	SrF_2	4.33×10^{-9}
$Cu(OH)_2$	1.3×10^{-20}	$Sr(OH)_2$	6.4×10^{-3}
CuS	1.27×10^{-36}	$SrSO_4$	3.44×10^{-7}
$FeCO_3$	3.07×10^{-11}	$ZnCO_3$	1.19×10^{-10}
$Fe(OH)_2$	4.87×10^{-17}	$Zn(OH)_2$	1.2×10^{-17}
$Fe(OH)_3$	2.64×10^{-39}	ZnS	2.93×10^{-25}
FeS	1.59×10^{-19}		

Acids and Bases

Two Definitions

1) **Lewis acids** are molecules or ions capable of accepting an electron pair, an electrophile, for example. **Lewis bases** are molecules or ions capable of donating an electron pair, a nucleophile, for example.
2) **Bronsted-Lowry acids** are proton (H^+) donors. **Bases** are proton acceptors. When these acids lose a proton the negatively charged part is called its conjugate base. The stronger the acid, the weaker its conjugate base. Bases accept a proton to become conjugate acids. Bronsted-Lowry acids and bases are included in the Lewis definition.
   a) **Weak acids and base** dissociate slightly (acetic acid)

   $$H_3C\text{-}COOH \rightleftharpoons CH_3\text{-}COO^- + H^+$$

   b) **Strong acids and bases** completely dissociate (hydrochloric acid):

   $$HCl \rightarrow H^+ + Cl^-$$

   c) **Polyprotic acids** can donate more than one proton (sulphuric acid):

   $$H_2SO_4 \rightleftharpoons H^+ + SO_4^- \rightleftharpoons 2H^+ + SO_4^{2-}$$

   d) **Monoprotic acids** can donate only one proton: $HBr \rightarrow H^+ + Br^-$

Useful Terms

Amphiprotic substances can act as a base or an acid, depending on conditions.
Buffer solutions resist changes in pH. They help maintain a constant pH when acids or bases are added.
Equivalence point is reached when stoichiometric equivalents of an acid and base have reacted completely.
Endpoint is the point at which an indicator changes color.
Neutralization is the reaction between an acid and a base, yielding a salt and water.

pH	
1	0.1M HCl
2	Gastric contents
3	Vinegar
4	Tomatoes
5	Coffee
6	Rain
7	Pure water
8	Seawater
9	Borax & water
10	Milk of magnesia
11	Household ammonia
12	Washing soda
13	0.1M NaOH
14	

Acid and Base Properties

Property	Acid	Base
Taste*	sour	bitter
Litmus	red	blue
Feel*		slippery

* Do not taste or feel acids or bases.

We die before we have learned to live. Stephen Winsten

Useful Formulas for Acid-Base Chemistry

Acid and Base Ionizing Constants

$$K_a = \frac{[H^+][A^-]}{[HA]}$$

$$K_b = \frac{[BH^+][OH^-]}{[B]}$$

Hydrogen Indexing

$pH = -\log[H_3O^+]$ or $\log 1/[H_3O^+]$

$pOH = -\log[OH^-]$ or $\log 1/[OH^-]$

Percent Ionization

$$\%\ \text{Ionization} = \frac{[H^+]}{[HA]} \times 100$$

Ion Product Constant for Water

$$K_w = [H_3O^+][OH^-]$$

Equivalent Mass (g / equiv)

$$\text{Eq mass} = \frac{\text{molar mass (g / mol)}}{\text{no. of equivalents * per mole}}$$

Normality

$N = \text{equivalents} \times \text{molarity}$

Other equations

$pH + pOH = 14$

$pK_a = -\log K_a$ or $\log 1 / K_a$

Henderson – Hasselbalch

$$pH = pK_a + \log\frac{[A^-]}{[HA]}$$

*Equivalents refer to the molar ratio of titratable hydrogens. For example, HCl has 1 equivalent/mol, and H_2SO_4 has 2 equivalents/mol.

Common Bench Acids and Bases

	Formula	Molarity (M) conc/dil	% Weight conc/dil	Density conc/dil (g/mL)
Acetic acid, glacial	CH_3COOH	17.4	99.5	1.05
Acetic acid		6.27	36	1.05
Ammonium hydroxide	NH_4OH	14.8 / 5.6	28 / 10	0.90 / 0.96
Formic acid	HCOOH	23.4 / 5.75	90 / 25	1.20 / 1.06
Hydrobromic acid	HBr	8.89 / 1.3	48 / 10	1.50 / 1.08
Hydrochloric acid	HCl	11.6 / 2.9	36 / 10	1.18 / 1.05
Hydrofluoric acid	HF	32.1 / 28.8	55 / 50	1.17 / 1.16
Lactic acid	$CH_3CH(OH)COOH$	11.3	85	1.2
Nitric acid	HNO_3	15.99 / 6	71 / 30	1.42 / 1.18
Perchloric acid	$HClO_4$	11.65 / 9.2	70 / 60	1.67 / 1.54
Phosphoric acid	H_3PO_4	14.7	85	1.70
Potassium hydroxide	KOH	13.5 / 1.94	50 / 10	1.52 / 1.09
Sodium hydroxide	NaOH	19.1 / 2.75	50 / 10	1.53 / 1.11
Sodium carbonate	$NaCO_3$	1.04	10	1.10
Sulfuric acid	H_2SO_4	18.0	96	1.84

I do not believe in the collective wisdom of individual ignorance. Thomas Carlyle

Common Acids

Acid	HA	A^-	K_a
Acetic	CH_3COOH	CH_3COO^-	1.8×10^{-5}
Aluminum hydroxide	$Al(OH)_3$		4×10^{-13}
Aluminium ion	$Al(H_2O)_6^{3+}$	$Al(H_2O)_5(OH)^{2+}$	1.4×10^{-15}
Ammonium ion	NH_4^+	NH_3	5.7×10^{-10}
Arsenic	H_3AsO_4	$H_2AsO_4^-$	6×10^{-3}
	$H_2AsO_4^-$	$HAsO_4^{2-}$	1.1×10^{-7}
	$HAsO_4^{2-}$	AsO_4^{3-}	3×10^{-12}
Benzoic	C_6H_5COOH	$C_6H_5COO^-$	6.6×10^{-5}
Boric	H_3BO_3	$H_2BO_3^-$	5.8×10^{-10}
Carbonic	H_2CO_3	HCO_3^-	4.3×10^{-7}
	HCO_3^-	CO_3^{2-}	5.6×10^{-11}
Chloric	$HClO_3$	ClO_3^-	large
Chromic	H_2CrO_4	$HCrO_4^-$	1.8×10^{-1}
	$HCrO_4^-$	CrO_4^{2-}	3.2×10^{-7}
Copper(II) hydroxide	$Cu(OH)_2$		1×10^{-19}
Copper(II) ion	Cu^{2+}		1×10^{-8}
Formic	$HCOOH$	$HCOO^-$	1.8×10^{-4}
Hydrazinium ion	$NH_2NH_3^+$	NH_2NH_2	5.9×10^{-9}
Hydrobromic	HBr	Br^-	large
Hydrochloric	HCl	Cl^-	large
Hydrocyanic	HCN	CN^-	4.8×10^{-10}
Hydrofluoric	HF	F^-	3.5×10^{-4}
Hydroiodic	HI	I^-	large
Hydronium ion	H_3O^+	H_2O	1.0
Hydrophosphorous	H_3PO_2	$H_2PO_2^-$	1.2×10^{-2}
Hydrosulfuric	H_2S	HS^-	1×10^{-7}
Hypobromous	$HBrO$	BrO^-	2.1×10^{-9}
Hypochlorous	$HClO$	ClO^-	2.9×10^{-8}
Hypoiodous	HIO	IO^-	2.3×10^{-11}
Iodic	HIO_3	IO_3^-	1.7 x 10-1
Iron(II) ion	Fe^{2+}		1.2×10^{-6}
Iron(III) ion	$Fe(H_2O)_6^{3+}$	$Fe(H_2O)_5(OH)^{2+}$	4.0×10^{-3}
Lactic	$CH_3CH(OH)COOH$	$CH_3CH(OH)COO^-$	1.4×10^{-4}
Magnesium ion	Mg^{2+}		2×10^{-12}
Mercury(II) ion	Hg^{2+}		2×10^{-3}
Nitric	HNO_3	NO_3^-	large

Better to reign in Hell, than to serve in Heaven. John Milton

Common Acids (cont.)

Acid	HA	A^-	K_a
Nitrous	HNO_2	NO_2^-	7.2×10^{-4}
Oxalic	$H_2C_2O_4$	$HC_2O_4^-$	5.6×10^{-2}
	$HC_2O_4^-$	$C_2O_4^{2-}$	5.2×10^{-5}
Perchloric	$HClO_4$	ClO_4^-	large
Periodic	HIO_4	IO_4^-	2.3×10^{-2}
Permanganic	$HMnO_4$	MnO_4^-	large
Phenol	C_6H_5OH	$C_6H_5O^-$	1.3×10^{-10}
Phosphoric	H_3PO_4	$H_2PO_4^-$	7.5×10^{-3}
	$H_2PO_4^-$	HPO_4^{2-}	6.3×10^{-8}
	HPO_4^{2-}	PO_4^{3-}	1×10^{-12}
Phosphorous	H_3PO_3	$H_2PO_3^-$	3×10^{-2}
	$H_2PO_3^-$	HPO_3^{2-}	1.6×10^{-7}
Propionic	CH_3CH_2COOH	$CH_3CH_2COO^-$	1.3×10^{-5}
Sulfuric	H_2SO_4	HSO_4^-	large
	HSO_4^-	SO_4^{2-}	1.0×10^{-2}
Sulfurous	H_2SO_3	HSO_3^-	1.4×10^{-2}
	HSO_3^-	SO_3^{2-}	5.0×10^{-8}
Trichloroacetic	CCl_3COOH	CCl_3COO^-	2×10^{-1}
Water	H_2O	OH^-	1.0×10^{-14}
Zinc ion	Zn^{2+}		2.5×10^{-10}

Common Bases

Base	B	BH^+	K_b
Acetate ion	CH_3COO^-	CH_3COOH	5.7×10^{-10}
Ammonia	NH_3	NH_4^+	1.6×10^{-5}
Aniline	$C_6H_5NH_2$	$C_6H_5NH_3^+$	4.2×10^{-10}
Carbonate ion	CO_3^{2-}	HCO_3^-	2.1×10^{-4}
	HCO_3^-	H_2CO_3	2.2×10^{-8}
Chromate ion	CrO_4^{2-}	$HCrO_4^-$	3.1×10^{-8}
Nitrite ion	NO_2^-	HNO_2	1.4×10^{-11}
Sulfate ion	SO_4^{2-}	HSO_4^-	1.0×10^{-12}

Common Indicators Used in the Lab

pH interval	Indicator	Color changes (more acidic → more basic)
0.15-3.2	Methyl violet	yellow → blue → violet
0-1.8	Malachite green (acidic)	blue → blue-green
1.2-2.8	Thymol blue (acidic)	red → yellow
1.2-4.0	Benzopurpurin 4B	violet → red
1.3-3.0	Orange IV (tropeolin OO)	red → yellow
2.9-4.0	Methyl yellow	red → orange → yellow
3.0-4.6	Bromophenol blue	yellow → green → purple
3.0-5.0	Congo red	blue → red
3.1-4.4	Methyl orange	red → orange → yellow
3.8-5.4	Bromcresol green	yellow → blue
4.4-6.2	Methyl red	red → yellow
4.5-8.3	Litmus	red → pink → blue
4.8-6.8	Chlorphenol red	yellow → red
5.2-6.8	Bromcresol purple	yellow → purple
5.2-6.8	Bromphenol red	yellow → red
6.0-7.6	Bromthymol blue	yellow → green → blue
6.4-8.2	Phenol red	yellow → orange → red
7.2-8.8	Cresol red	yellow → red
7.6-9.2	m-Cresol purple	yellow → purple
8.0-9.6	Thymol blue	yellow → blue
8.3-10.0	Phenolphthalein	colorless → pink → red
9.3-10.5	Thymolphthalein	colorless → blue
10.0-12.1	Alizarin yellow R	yellow → red
11.5-14.0	Trinitrobenzene	colorless → orange
11.6-14.0	Indigo carmine	blue → yellow

Destiny. A tyrant's authority for crime and a fool's excuse for failure. Ambrose Bierce

5.9 Thermochemistry*

Exothermic reaction. A chemical reaction in which heat energy is released to its surroundings. ΔH is negative.

$$2H_2(g) + O_2(g) \rightarrow 2H_2O(l) + 571.66 \text{ kJ}$$

or

$$2H_2(g) + O_2(g) \rightarrow 2H_20(l) \quad \Delta H = -571.66 \text{ kJ}$$

Endothermic reaction. A chemical reaction in which heat is absorbed from its surroundings. ΔH is positive.

$$2HgO(s) + 181.67 \text{ kJ} \rightarrow 2Hg(l) + O_2(g)$$

or

$$2HgO(s) \rightarrow 2Hg(l) + O_2(g) \quad \Delta H = 181.67 \text{ kJ}$$

Formulas

Internal Energy Change
$\Delta E = Q + w$
ΔE (−) system loses heat
ΔE (+) system gains heat

Work
(at constant pressure)
$w = -P\Delta V$

Quantity of Heat
Heat (J)
$Q = mc\,\Delta T$
Heat of fusion (J)
$Q = ml_f$
Heat of vaporization (J)
$Q = ml_v$

Variables
c = specific heat
l_f = latent heat of fusion
l_v =latent heat of vaporization
m = mass
P = pressure
T = temperature
V = volume

Heat Capacity

Molar heat. The amount of heat needed to raise by one Kelvin the temperature of one mole of substance. Q = (# of moles) (molar heat capacity) $\Delta T = n\,C\,\Delta T$
Specific heat. The amount of heat needed to raise by one degree centigrade the temperature of one gram of a substance. Q = (mass) (specific heat) $\Delta T = m\,c\,\Delta T$

Standard State

The **standard state** of an element is the form in which it commonly exists at SATP (25°C, 100 kPa). For an element in standard state the heat of formation is zero.

* Also see Thermodynamics, Section 7.6.

It's tough to be cool with child-proof windows. Barbara Vajdik

Enthalpy

A change in enthalpy (ΔH) is a measurement of how much heat is gained or lost at constant pressure in a system.

Formulas

$H = E + PV$

$\Delta H = \Delta E + \Delta(PV)$

$\Delta H = \Delta E + P\,\Delta V$ (constant pressure)

Enthalpy Changes

Heat of reaction, ΔH. The total heat released or absorbed during a chemical reaction.

Reactants $\rightarrow$ Products

Heat of formation, ΔH_f. The amount of heat evolved or absorbed during the formation of 1 mole of a compound from its elements in their standard state.

Elements $\rightarrow$ 1 mole of substance

Heat of combustion, ΔH_c. The heat associated per unit mass or volume with the complete oxidation of 1 mole of a substance.

1 mole substance + oxygen gas $\rightarrow$ products of combustion

Heat of neutralization, $\Delta H_{neutralization}$. The quantity of heat associated with the neutralization of an acid or base.

acid + base $\rightarrow$ salt + water

Heat of solution, ΔH_{sol}. The quantity of heat evolved or absorbed when n moles of solvent dissolve 1 mole of solute.

1 mole of solute + n moles of solvent $\rightarrow$ 1 mole solute in n moles of solvent

Heat of dilution, ΔH_{dil}. The quantity of heat evolved or absorbed when a solution of 1 mole of solute in n moles of solvent is diluted by m moles of solvent.

1 mole solute in n moles of solvent + m moles of solvent $\rightarrow$
1 mole solute in $(n + m)$ moles of solvent

Heat of fusion, ΔH_{fus}. The amount of heat required to change 1 gram of a solid to a liquid.

1 gram of solid $\rightarrow$ 1 gram of liquid

$\Delta H_{fus} = -\Delta H_{crystallization}$

Heat of vaporization, ΔH_{vap}. The amount of heat required to change 1 gram of a liquid to a gas.

1 gram of liquid $\rightarrow$ 1 gram of gas

$\Delta H_{vap} = -\Delta H_{condensation}$

Heat of sublimation, ΔH_{sub}. The amount of heat required to change 1 gram of a solid to a gas.

1 gram of solid $\rightarrow$ 1 gram of gas

$\Delta H_{sub} = -\Delta H_{deposition}$

Imagination is more important than knowledge. Albert Einstein

Entropy

Entropy (S) is the quantitative measure of disorder. Changes in entropy accompany physical and chemical changes.

Physical changes that affect entropy (keeping other things constant):

1) Increasing or decreasing temperature
2) Change of state: solid, liquid, gas
3) Expanding or constricting the volume a gas occupies
4) Mixing substances

A chemical change that affects entropy:

Increasing or decreasing the number of free atoms, ions or molecules in a reaction (for example, increase in the number of moles).

Formulas

Change of State Entropy

$$\Delta S = \frac{\Delta H}{T}$$

Standard State Reaction Entropy

$$\Delta S^\circ = S^\circ_{products} - S^\circ_{reactants} = \Delta H^\circ / T$$

Gibbs Free Energy

Gibbs Free Energy (G) is a measure of the available energy capable of doing useful work; it results from a chemical or physical change.

Formulas

$H = G - PV + TS$

$\Delta H = \Delta G + T\Delta S$ (constant temperature and pressure)

$\Delta G = \Delta H - T\Delta S$ (constant temperature and pressure)

$\Delta G = \Delta G^\circ + RT \ln Q$

$\Delta G^\circ = -RT \ln K_{eq}$

$\Delta G^\circ = \Delta H^\circ - T\Delta S^\circ$

Predicting Reaction Probability at Constant Temperature and Pressure

ΔH	ΔS	ΔG	Comment
–	+	-	exothermic, work can be done, spontaneous* product formation
–	-	– (low temp)	exothermic, work can be done, spontaneous product formation
–	-	+ (high temp)	exothermic, reactants favored over products
+	+	+ (low temp)	endothermic, reactants favored over products
+	+	– (high temp)	endothermic, work can be done, spontaneous product formation
+	–	+	endothermic, reactants favored over products

* Some energy input may be required for spontaneous reactions to begin. Spontaneous reactions are not necessarily fast reactions.

Thermochemical Data

Substance	$\Delta H_f°$ (kJ/mol)	$\Delta G_f°$ (kJ/mol)	$S°$ (J/K mol)	$C_p°$ (J/K mol)
Ag	0.0	0.0	42.55	25.35
Ag_2O(*s*)	-31.05	-11.20	121.3	65.86
Ag_2S(orthorhombic)	-32.59	-40.67	144.01	76.53
AgCl(*s*)	-127.07	-109.79	96.2	50.79
$AgNO_3$(*s*)	-124.39	-33.41	140.92	93.05
Al(*s*)	0.0	0.0	28.33	24.35
Al_2O_3(α-solid)	-1675.7	-1582.3	50.92	79.04
$AlCl_3$(*s*)	-704.2	-628.8	110.67	91.84
Ar(*g*)	0.0	0.0	154.84	20.79
As(α-solid)	0.0	0.0	35.1	24.64
As(*g*)	302.5	261.0	174.21	20.79
As_4(*g*)	149.9	92.4	314	
AsH_3(*g*)	66.44	68.93	222.78	38.07
Ba(*g*)	180	146	170.24	20.79
Ba(*s*)	0.0	0.0	62.8	28.07
$BaCl_2$(*s*)	-858.6	-810.4	123.68	75.14
BaO(*s*)	-553.5	-525.1	70.42	47.78
Be(*s*)	0.0	0.0	9.5	16.44
Bi(*s*)	0.0	0.0	56.74	25.52
Br(*g*)	111.88	82.40	175.02	20.79
Br_2(*l*)	0.0	0.0	152.23	75.69
C(diamond)	1.895	2.900	2.377	6.113
C(graphite)	0.0	0.0	5.74	8.53
CCl_4(*l*)	-135.44	-65.21	216.40	131.75
CF_4(*g*)	-925	-879	261.61	61.09
CH_4(*g*)	-74.81	-50.72	186.26	35.30
C_2H_2(*g*)	226.73	209.20	200.94	43.93
C_2H_4(*g*)	52.26	68.15	219.56	43.56
C_2H_6(*g*)	-84.68	-32.82	229.60	52.63
CH_3CH_2OH(*l*)	-277.69	-174.78	160.7	111.46
CH_3COOH(*aq*)	-485.76	-396.46	178.7	
CH_3COOH(*l*)	-484.5	-389.9	159.8	124.3
CH_3OH(*l*)	-238.66	-166.27	126.8	81.6
CO(*g*)	-110.53	-137.17	197.67	29.14
CO_2(*g*)	-393.51	-394.36	213.74	37.11
CS_2(*l*)	89.70	65.27	151.34	75.7
Ca(*g*)	178.2	144.3	154.88	20.79
Ca(*s*)	0.0	0.0	41.42	25.31
$CaCl_2$(*s*)	-795.8	-748.1	104.6	72.59
$CaCO_3$(aragonite)	-1207.13	-1127.75	88.7	81.25
$CaCo_3$(calcite)	-1206.92	-1128.79	92.9	81.88
CaO(*s*)	-635.09	-604.03	39.75	42.80

Thermochemical Data

Substance	$\Delta H_f°$ (kJ/mol)	$\Delta G_f°$ (kJ/mol)	$S°$ (J/K mol)	$C_p°$ (J/K mol)
Cd(γ-solid)	0.0	0.0	51.76	25.98
Cd(*g*)	112.01	77.41	167.75	20.79
CdO(*s*)	-258.2	-228.4	54.8	43.43
Cl(*g*)	121.68	105.68	165.20	21.84
Cl_2(*g*)	0.0	0.0	223.066	33.91
Cr(*s*)	0.0	0.0	23.77	23.35
Cu(*s*)	0.0	0.0	33.15	24.44
Cu_2O(*s*)	-168.6	-146.0	93.14	63.64
$CuSO_4 \cdot 5H_2O$(*s*)	-2279.65	-1879.75	300.4	280.0
$CuSO_4$(*s*)	-771.36	-661.8	109	100.0
F(*g*)	78.99	61.91	158.75	22.74
F_2(*g*)	0.0	0.0	202.78	31.30
Fe(α-solid)	0.0	0.0	27.28	25.10
Fe_2O_3(*s*)	-824.2	-742.2	87.40	103.85
Fe_3O_4(*s*)	-1118.4	-1015.4	146.4	143.43
H(*g*)	217.97	203.25	114.71	20.78
H_2(*g*)	0.0	0.0	130.68	28.82
HBr(*g*)	-36.40	-53.45	198.70	29.14
HCl(*aq*)	-167.16	-131.23	56.5	
HCl(*g*)	-92.31	-95.30	186.91	29.12
HCN(*g*)	135.1	124.7	201.78	35.86
HF(*aq*)	-320.08	-296.82	88.7	
HF(*g*)	-271.1	-273.2	173.78	29.13
HI(*g*)	26.48	1.70	206.59	29.16
HNO_3(*aq*)	-207.36	-111.25	146.4	
HNO_3(*l*)	-174.10	-80.71	155.60	109.87
H_2O(*g*)	-241.82	-228.57	188.83	33.58
H_2O(*l*)	-285.83	-237.13	69.91	75.30
H_2O_2(*l*)	-187.78	-120.35	109.6	89.1
H_3PO_3(*s*)	-964.4			
H_3PO_4(*aq*)	-1288.34	-1142.54	158.2	
H_3PO_4(*l*)	-1266.9			
H_3PO_4(*s*)	-1279.0	-1119.1	110.50	106.06
H_2S(*g*)	-20.63	-33.56	205.79	34.23
H_2SO_4(*aq*)	-909.27	-744.53	20.1	
H_2SO_4(*l*)	-813.99	-690.00	156.90	138.91
He(*g*)	0.0	0.0	126.15	20.79
Hg(*l*)	0.0	0.0	76.02	27.98
Hg^{2+}(*aq*)	171.1	164.4	-32.2	
Hg_2^{2+}(*aq*)	172.4	153.52	84.5	
Hg_2Cl_2(*s*)	-265.22	-210.75	192.5	
$HgCl_2$(*s*)	-224.3	-178.6	146.0	

There's no place like home, after the other places close. Anonymous

Thermochemical Data

Substance	$\Delta H_f°$ (kJ/mol)	$\Delta G_f°$ (kJ/mol)	$S°$ (J/K mol)	$C_p°$ (J/K mol)
HgO(red)	-90.43	-58.54	70.29	44.06
HgO(yellow)	-90.46	-58.41	71.1	
HgS(black)	-53.6	-47.7	88.3	
HgS(red)	-58.2	-50.6	82.4	48.41
I(*g*)	106.84	70.25	180.79	20.79
I_2(*s*)	0.0	0.0	116.14	54.44
K(*g*)	89.24	60.59	160.34	20.79
K(*s*)	0.0	0.0	64.18	29.58
KBr(*s*)	-393.80	-380.66	95.60	52.30
KCl(*s*)	-436.747	-409.14	82.59	51.30
KI(*s*)	-327.90	-324.89	106.32	52.93
KOH(*aq*)	-482.37	-440.50	91.6	
KOH(*s*)	-424.76	379.08	78.9	64.9
Li(*g*)	159.37	126.66	138.77	20.79
Li(*s*)	0.0	0.0	29.12	24.77
Mg(*s*)	0.0	0.0	32.68	24.89
$MgCl_2$(*s*)	-641.32	-591.79	89.62	71.38
MgO(*s*)	-601.70	-569.43	26.94	37.15
$MgCO_3$(*s*)	-1095.8	-1012.1	65.7	75.52
N_2(*g*)	0.0	0.0	191.61	29.13
NH(*aq*)	-80.29	-26.50	111.3	
N_2H_4(*l*)	50.63	149.34	121.21	98.87
NH_3(*g*)	-46.11	-16.45	192.45	35.06
NH_4^+(*aq*)	-132.51	-79.31	113.4	79.9
NH_4Cl(*aq*)	-299.66	-210.52	169.9	
NH_4Cl(*s*)	-314.43	-202.87	94.6	84.1
NH_4NO_3(*s*)	-365.56	-183.87	151.08	139.3
NO(*g*)	90.25	86.55	210.76	29.84
NO_2(*g*)	33.18	51.31	240.06	37.20
N_2O(*g*)	82.05	104.20	219.85	38.45
N_2O_4(*g*)	9.16	97.89	304.29	77.28
N_2O_5(*g*)	11.3	115.1	355.7	84.5
N_2O_5(*s*)	-43.1	113.9	178.2	143.1
Na(*g*)	107.32	76.76	153.71	20.79
Na(*s*)	0.0	0.0	51.21	28.24
NaCl(*aq*)	-407.27	-393.13	115.5	
NaCl(*s*)	-411.15	-384.14	72.13	50.50
NaOH(*aq*)	-470.11	-419.15	48.1	
NaOH(*s*)	-425.61	-379.50	64.46	59.54
Ne(*g*)	0.0	0.0	146.33	20.79
O(*g*)	249.17	231.73	161.06	21.91
O_2(*g*)	0.0	0.0	205.14	29.36

The road to hell is paved with good intentions. Karl Marx

Thermochemical Data

Substance	$\Delta H_f°$ (kJ/mol)	$\Delta G_f°$ (kJ/mol)	$S°$ (J/K mol)	$C_p°$ (J/K mol)
$O_3(g)$	142.7	163.2	238.93	39.20
$OH^-(aq)$	-299.99	-157.24	-10.75	
$P(g)$	314.64	278.25	163.19	20.79
P(white)	0.0	0.0	41.09	23.84
P(black)	-39.3			
P(red)	-17.6	-12.1	22.80	21.21
$P_4(g)$	58.91	24.44	279.98	67.15
$PCl(s)$	-443.5			
$PCl_3(g)$	-287.0	-267.8	311.78	71.84
$PCl_3(l)$	-319.7	-272.3	217.1	
$PCl_5(g)$	-374.9	-305.0	364.58	112.80
$PH_3(g)$	5.4	13.4	210.23	37.11
$Pb(s)$	0.0	0.0	64.81	26.44
PbO(red)	-218.99	-188.93	66.5	45.81
PbO(yellow)	-217.32	-187.89	68.70	45.77
$PbO_2(s)$	-277.4	-217.33	68.6	64.64
$S(g)$	278.81	238.25	167.82	23.67
S(monoclinic)	0.33			
S(rhombic)	0.0	0.0	31.80	22.64
$SO_2(g)$	-296.83	-300.19	248.22	
$SO_3(g)$	-395.72	-371.06	256.76	50.67
$SO_3(s)$	-454.51	-374.21	70.7	
Sb(solid III)	0.0	0.0	45.69	25.23
$Si(g)$	455.6	411.3	167.97	22.25
$Si(s)$	0.0	0.0	18.83	20.00
SiO_2(α-quartz)	-910.94	-856.64	41.84	44.43
Sn(gray)	-2.09	0.13	44.14	25.77
Sn(white)	0.0	0.0	51.55	26.99
$SnO(s)$	-285.8	-256.9	56.5	44.31
$SnO_2(s)$	-580.7	-519.6	52.3	52.59
$Xe(g)$	0.0	0.0	169.68	20.79
$Zn(s)$	0.0	0.0	41.63	25.40
$ZnO(s)$	-348.28	-318.30	43.64	40.25

5.10 Nuclear Chemistry

Radiation Units

Unit	Measurement
Becquerel (Bq)	Measures activity where 1 Bq = 1 disintegration per second.
Curie (Ci)	Measures activity where 1 Ci = 3.70×10^{10} disintegrations per second.
Gray (Gy)	1 gray = 100 rad
Radiation absorbed dose (rad)	0.01 J of γ or x-ray energy absorbed per kilogram of tissue.
Roentgen equivalent man (rem)	Measures dose to humans. 1 rem has same effect as 1 Roentgen.
Roentgen (R)	Measures exposure to x-rays or γ-rays where 1 R = 3.70×10^{10} disintegrations per second.
Sievert (Sv)	1 Sv = 100 rem

Isotopes

Isotopes are forms of the same element with an identical number of protons but a different number of neutrons. These differences give isotopes of the same element slightly different physical properties and masses.

Halflives of Some Radioisotopes

Isotope	Physical halflife/ modes of decay	Biological halflife
$^{3}_{1}H$(*)	12.26 yr / β-	19 days
$^{14}_{6}C$(*)	5730 yr / β-	35 days
$^{24}_{11}Na$	15.02 h / β-	29 days
$^{40}_{19}K$(*)	1.28×10^{9} yr /EC	-
$^{42}_{19}K$	12.36 h / β-	43 days
$^{59}_{26}Fe$	44.6 days / EC	65 days
$^{60}_{27}Co$	5.27 yr / β-	-
$^{131}_{53}I$	8.04 days / β-	180 days
$^{140}_{56}Ba$	12.8 days / β-	-
$^{235}_{92}U$(*)	7.04×10^{8} yr / α	-
$^{238}_{92}U$(*)	4.47×10^{9} yr / α	-
$^{239}_{94}Pu$(*)	2400 yr/ α	-

* Naturally occurring isotopes

Formulas

Quantity of Radionuclide
$\log (q_o/q) = kt/2.303$
$\ln (q_o/q) = kt$

Activity of Sample
$\log (a_o/a) = kt/2.303$
$\ln (a_o/a) = kt$

Halflife
$t_{1/2} = 0.693/k$

Energy-Mass Relationship
$E = mc^2$

Nuclear Binding Energy
$\Delta E = \Delta mc^2$

Average Atomic Mass
(fraction of isotope A) $\times$
(mass of A) +
(fraction of isotope B) $\times$
(mass of B) + ...

Variables
k = rate constant
t = time
q = quantity at time t
q_o = initial quantity
a_o = initial activity
a = activity at time t
m = mass
E = energy
c = speed of light

Conservation of Mass Number and Nuclear Charge

The products' total sum of neutrons and protons must equal those of the reactants. The products' total charge must equal the reactants' total charge. A correctly written nuclear equation observes these rules.

Mass Defect and Nuclear Binding Energy

The actual mass of an atom does not equal the sum of all its individual components (neutrons, protons, and electrons). This difference is called the **mass defect**. The equation showing the relationship between mass and energy is $\Delta E = \Delta mc^2$ and is used to translate the missing mass into a quantity of energy. This energy is called the **nuclear binding energy**.

Nuclear Reactions

Nuclear reactions involve changes in the nuclei of atoms.

Nuclear Particles and Symbols

Particle	Symbols	Charge	Comments
Neutron	n, $^{1}_{0}n$	0	Neutral particles with greater penetrating ability in matter than β particles. More hazardous than α and β particles.
Proton	p, $^{1}_{1}P$, $^{1}_{1}H$	+1	Positively charged particles.
Beta			
Electron	β^-, $^{0}_{-1}\beta$, $^{0}_{-1}e$, e^-	–1	Beta particles are negatively charged electrons or positively charged positrons. They have a longer penetrating range in matter than α particles.
Positron	β^+, $^{0}_{+1}\beta$, $^{0}_{1}e$, $^{0}_{+1}e$, + e^+	+1	
Alpha	α, $^{4}_{2}\alpha$, $^{4}_{2}He$	+2	Positively charged helium nuclei with a very short penetrating range in matter.
Gamma rays	γ	0	Gamma rays are high energy electromagnetic waves. These waves are emitted when a high energy nucleus undergoes transition to a lower energy level.

1) **Fission** is the splitting of a heavy nucleus into two lighter ones.

   In atomic fission a portion of matter is converted into energy.

$$^{235}_{92}U + ^{1}_{0}n \rightarrow ^{139}_{56}Ba + ^{88}_{36}Kr + 9^{1}_{0} + \text{energy}$$

   When the ^{235}U sample is large enough, released neutrons collide with the sample, causing a chain reaction.

2) **Fusion** is a process that only occurs naturally in stars. Fusion releases energy by combining two light nuclei to yield a heavier one.

Reaction in stars is as follows:

$${}^{1}_{1}H + {}^{1}_{1}H \rightarrow {}^{2}_{1}H + {}^{0}_{+1}e$$
$${}^{2}_{1}H + {}^{1}_{1}H \rightarrow {}^{3}_{2}He + \gamma$$
$${}^{3}_{2}He + {}^{3}_{2}He \rightarrow {}^{4}_{2}He + 2{}^{1}_{1}H$$

Overall net reaction:

$$4{}^{1}_{1}H \rightarrow {}^{4}_{2}He + 2{}^{0}_{+1}e + 4.28 \times 10^{-12}\,J$$

3) **Spontaneous decay** of radioactive nuclei. All known nuclei above Z 83 are unstable. Unstable nuclei emit radiation in an attempt to become more stable. The process of radioactive decay cannot be influenced by temperature or pressure.

Process	Emission	Reasons
β^- decay $n \rightarrow p + e^-$	e^-	n/p ratio is too high
β^+ emission $p \rightarrow n + e^+$	e^+	n/p ratio is too low
Electron capture (EC) $e^- + p \rightarrow n$		n/p ratio is too low
α decay	${}^{4}_{2}He$	n/p ratio is too low, Z >83
γ decay (IT)** ${}^{x}_{y}A^{*x} \rightarrow {}^{x}_{y}A^{x} + \gamma$	Gamma rays	excited nuclear state
Spontaneous fission (SF)	Variable	nuclide too heavy, Z >92

** (IT) Isometric transition is a radionuclide's delayed γ-ray emission.

4) **Bombardment reactions** occur when a nucleus captures electromagnetic radiation or fast-moving particles.

a) **Alpha particle bombardment**

$${}^{27}_{13}Al + {}^{4}_{2}He \rightarrow {}^{30}_{14}Si + {}^{1}_{1}H$$
$${}^{9}_{4}Be + {}^{4}_{2}He \rightarrow {}^{12}_{6}C + {}^{1}_{0}n$$

b) **Proton bombardment**

$${}^{12}_{6}C + {}^{1}_{1}H \rightarrow {}^{13}_{7}N \text{ then } \beta\text{decay } {}^{13}_{7}N \rightarrow {}^{13}_{6}C + {}^{0}_{1}e$$

c) **Neutron bombardment**

$${}^{235}_{92}U + {}^{1}_{0}n \rightarrow {}^{94}_{40}Zn + {}^{140}_{58}Ce + 6{}^{0}_{-1}e + 2{}^{1}_{0}n$$

d) **Electron bombardment**

$${}^{9}_{4}Be + {}^{0}_{-1}e \rightarrow {}^{8}_{3}Li + {}^{1}_{0}n \text{ then } \beta\text{decay } {}^{8}_{3}Li \rightarrow {}^{8}_{4}Be + {}^{0}_{-1}e$$

If all economists were laid end to end, they would not reach a conclusion. George Bernard Shaw

5.11 IR & ^{1}H NMR Values

It's important to realize that IR and ^{1}H NMR values are not written in stone. The values shift according to the different functional groups on a molecule. Remember, the key to spectra-analysis is not only showing which functional groups are present, but also which ones are absent. Abbreviated form of intensity in the IR tables translates as follows: (s) strong, (m) medium, (w) weak and (v) variable.

Infrared Spectroscopy

Functional group	Type of vibration	Frequency (cm^{-1})	Wavelength (μ)	Intensity
Acid chloride	C=O	1800	5.56	s
Alcohols	C-O	1000-1300	7.69-10.0	s
	O-H free	3590-3650	2.74-2.78	m
	H-bonded	3200-3550	2.82-3.13	m, broad
Aldehyde	C-H	2700-2800	3.57-3.70	w
		2800-2900	3.45-3.57	w
	C=O	1690-1740	5.75-5.92	s
Alkanes	stretch C-H	2850-3000	3.33-3.51	m
	bend CH_2	1465	6.83	m
	bend CH_3	1375-1450	6.90-7.27	m
Alkenes	C-H	3000-3100	3.23-3.33	m
	out of plane bend	650-1000	10.0-15.3	s
	C=C	1600-1680	5.95-6.25	m-w
	cis (same side)	675-750	13.3-14.8	s
	trans (across)	960-975	10.2-10.4	s
Allene	C=C=C	1950-2270	4.40-5.13	s-m
Alkyl halides	C-Br	<667	>15.0	s
	C-Cl	600-800	12.5-16.7	s
	C-F	1000-1400	7.14-10.0	s
	C-I	<667	>15.0	s
Alkynes	stretch C-H	~3300	~3.03	s
	C≡C	2100-2260	4.42-4.76	m-w
Amide	stretch N-H	3100-3500	2.86-3.23	s-m
	bend N-H	1550-1640	6.10-6.45	s-m
	C=O	1630-1690	5.92-6.13	s
Amine 1° and 2°	stretch N-H	3100-3500	2.86-3.23	m
	bend N-H	1550-1640	6.10-6.45	m-s
	C-N	1000-1350	7.4-10.0	s-m

s = strong, m = medium, w = weak, v = variable

Infrared Spectroscopy (cont.)

Functional group	Type of vibration	Frequency (cm^{-1})	Wavelength (μ)	Intensity
Anhydride	C-O	1000-1300	7.69-10.0	s
	C=O	1760/1810	5.52/5.68	s
Aromatic	stretch Ar-H	3050-3150	3.17-3.28	s
	C ≃ C	1475/1600	6.25/6.78	m-w
	monosubstituted	690-710/730-770	14.1-14.5/13.0-13.7	s
	o-disubstituted	735-770	13.0-13.6	s
	m-disubstituted	680-725/750-810	13.8-14.7/12.3-13.3	s
	p-disubstituted	800-840	11.9-12.5	s
Carboxylic acid	C=O	1700-1780	5.62-5.88	s
	C-O	1000-1300	7.69-10.0	s
	O-H	2400-3400	2.94-4.17	m, broad
Ester	C=O	1730-1750	5.71-5.78	s
	C-O	1000-1300	7.69-10.0	s
Ether	C-O	1000-1300	7.69-10.0	s
Imines	C=N	1640-1690	5.91-6.10	v
Isocyantes	X=C=Y	1950-2270	4.40-5.13	s-m
Isothiocyantes	X=C=Y	1950-2270	4.40-5.13	s-m
Ketones	C=O	1680-1750	5.71-5.95	s
Nitriles	C≡N	2220-2260	4.42-4.50	m
Nitro (R-NO_2)	N ≃ O	1350/1550	6.45/7.40	s
Oximes	C=N	1640-1690	5.92-6.10	v
Sulfoxides	S=O	1050	9.52	s
Thiol	S-H	2550	3.92	w

s = strong, m = medium, w = weak, v = variable

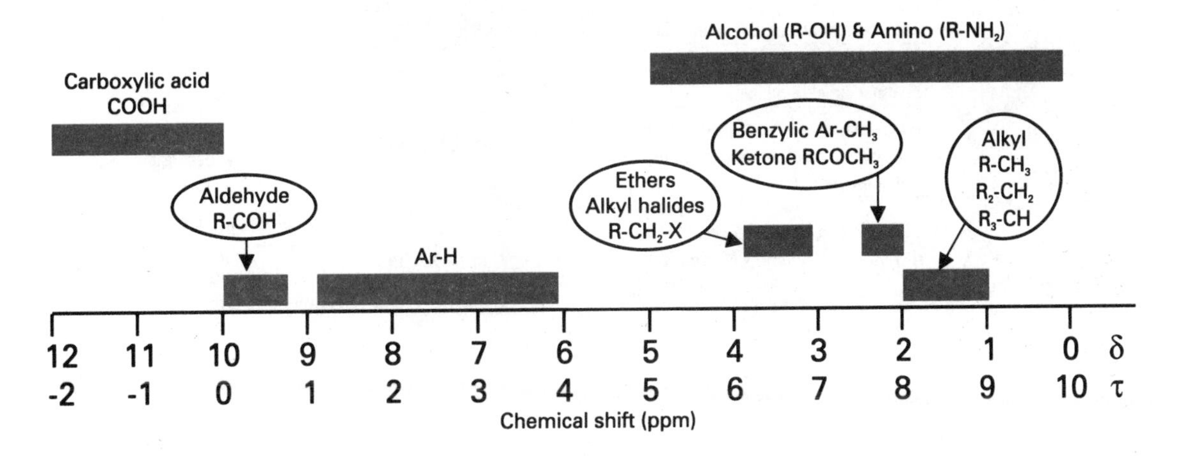

Fig. 5.11.1 Proton NMR

5.12 Chemical Index

melting point	melting point in °C at 1 atm
boiling point	boiling point in °C at 1 atm
density	density in g/cm^3 at 1 atm and 20°C
refractive index	refractive index at 20°C
solubility	sol (soluble or miscible in); insol (not soluble or miscible in)

β-Ketopropane: see acetone
Acetic acid ethyl ester: see ethyl acetate
Acetic ether: see ethyl acetate
Acetic acid anhydride: see acetic anhydride

Name	**Acetic acid, anhydrous,** *ethanoic acid*
Molecular formula	CH_3COOH
Molar mass	60.05
Melting point	16.7
Boiling point	117.9
Color	colorless
Crystal structure	rhombic
Solubility	acetone, alcohol, water, ether, CCl_4
Common uses	solvent, organic synthesis, pharmaceutical acidifier
Risks	FLAMMABLE, DO NOT INHALE FUMES, vapor and liquid irritates all tissues, burns skin, **LD_{50}**, oral, rats 3.53 g/kg
Glacial acetic acid	Conc. 17.4 M, Density: 1.05 by weight, dil 36% by weight, 6.27 M, Density: 1.05

Name	**Acetic anhydride,** *acetic acid anhydride, acetic oxide*
Molecular formula	$(CH_3CO)_2O$
Molar mass	102.09
Melting point	-73.1
Boiling point	139.6
Color	colorless
Density	1.082
Refractive index	1.3901
Solubility	alcohol, ether
Common uses	organic synthesis, dehydrating agent
Risks	FLAMMABLE, DO NOT INHALE, vapor and liquid irritates all tissues; highly reactive, avoid contact with water, **LD_{50}**, rats 1.78 g/kg

Acetic oxide: see acetic anhydride

Abbreviations: @ = at; conc = concentrated; dec = decomposes; dil = dilute; est = estimate; insol = insoluble; iv = intravenous; LC_{50} = lethal concentration 50% of the time; LD = lethal dose; LD_{50} = lethal dose 50% of the time; OES = occupational exposure standards; pet = petroleum; ppm = parts per million; sol = soluble; w/w = % weight in weight.

Name	**Acetone,** *2-propanone, dimethyl ketone, β-ketopropane, pyroacetic ether*
Molecular formula	CH_3COCH_3
Molar mass	58.08
Melting point	-94.6
Boiling point	56.4
Color	colorless
Density	0.789
Refractive index	1.359
Solubility	alcohol, ether, water
Common use	solvent
Risks	EXTREMELY FLAMMABLE, keep away from plastics and rayon, volatile; may cause dizziness; **LD_{50}**, oral, rats 10.7 mL/kg

Name	**Acetyl bromide**
Molecular formula	CH_3COBr
Molar mass	122.95
Melting point	-96
Boiling point	76
Color	yellow
Density	1.663
Solubility	acetone, chloroform, ether
Risks	REACTS WITH WATER AND ALCOHOL, irritates all tissues

Name	**Acetyl chloride**
Molecular formula	CH_3COCl
Molar mass	78.50
Melting point	-112
Boiling point	51
Color	colorless
Density	1.105
Refractive index	1.390
Solubility	acetone, ether, chloroform, petroleum ether, glacial acetic acid
Common use	acetylating agent
Risks	REACTS WITH WATER AND ALCOHOL, EXTREMELY FLAMMABLE, irritates all tissues

Name	**Acetylene,** *ethyne, ethine*. Mild odor
Molecular formula	C_2H_2
Molar mass	26.04
Melting point	-80.8
Boiling point	-84.8
Color	colorless
Solubility	acetone, chloroform
Common uses	fuel, illuminant
Risks	KEEP AWAY FROM BRASS, COPPER, SILVER. FLAMMABLE, DO NOT INHALE FUMES, explosive compound formed with copper and silver

Acetyl oxide: see acetic anhydride

Abbreviations: @ = at; conc = concentrated; dec = decomposes; dil = dilute; est = estimate; insol = insoluble; iv = intravenous; LC_{50} = lethal concentration 50% of the time; LD = lethal dose; LD_{50} = lethal dose 50% of the time; OES = occupational exposure standards; pet = petroleum; ppm = parts per million; sol = soluble; w/w = % weight in weight.

A

Alcohol: see ethanol

Name	**Acetylsalicylic acid,** *2-(Acetyloxy)benzoic acid, salicylic acid acetate*
Molecular formula	$C_9H_8O_4$
Molar mass	180.16
Melting point	135 (rapid heating)
Color	colorless
Crystal structure	monoclinic tablets or needle
Density	1.40
Solubility	ether: 1 g in 5 mL alcohol, 17 mL chloroform, 300 mLwarm water
Common uses	analgesic, antipyretic, anti-inflammatory
Risks	irritates eye, skin, respiratory tract, **LD_{50}**, oral, rats 1.5 g/kg, mice 1.1 g/kg

Name	**Aluminum,** *aluminium*
Molecular formula	Al
Molar mass	26.982
Melting point	660.5
Boiling point	2520
Color	silvery
Density	2.70
Solubility	sol in alkali, HCl, H_2SO_4; insol in conc HNO_3
Common use	electrical conductor
Risks	FLAMMABLE, do not inhale dust

Name	**Ammonia.**
Molecular formula	NH_3
Molar mass	17.03
Melting point	-77.7
Boiling point	-33.3
Color	colorless
Solubility	ether, chloroform
Common uses	solvent, manufacturing nitric acid, explosives
Risks	FUMES ARE POISONOUS, DO NOT INHALE, Nonflammable gas, pungent odor
Ammonia Water	10% ammonia in water; density: 0.957 @ 25°C; Risks: irritant to eyes, mucous membranes.
Ammonium hydroxide	28% ammonia in water; density: 0.90
Risks	REACTS VERY EXOTHERMALLY WITH MINERAL ACIDS AND H_2SO_4, keep cool in glass container

Name	**Aniline,** *phenylamine, benzenamine*
Molecular formula	$C_6H_5NH_2$
Molar mass	93.13
Melting point	-63
Boiling point	184
Color	colorless but darkens when exposed to light or air
Density	1.022
Refractive index	1.586
Solubility	alcohol, acetone
Common uses	manufacturing dyes,solvent, resins
Risks	POISONOUS, keep container closed, protect from light, **LD_{50}**, oral, rats 0.44 g/kg

Anhydrous hydrobromic acid: see hydrogen bromide
Anhydrous hydrochloric acid: see hydrogen chloride
Anhydrous hydrofluoric acid: see hydrogen fluoride

Abbreviations: @ = at; conc = concentrated; dec = decomposes; dil = dilute; est = estimate; insol = insoluble; iv = intravenous; LC_{50} = lethal concentration 50% of the time; LD = lethal dose; LD_{50} = lethal dose 50% of the time; OES = occupational exposure standards; pet = petroleum; ppm = parts per million; sol = soluble; w/w = % weight in weight.

A-B

Aspirin: see acetylsalicylic acid
Baking soda: see sodium bicarbonate

Name	**Anthracene**	Name	**Barium**
Molecular formula	$C_{14}H_{10}$	Molecular formula	Ba
Molar mass	178.23	Molar mass	137.3
Melting point	216	Melting point	729
Boiling point	342	Boiling point	1898
Color	colorless to light yellow but darkens in sunlight	Color	silvery white, yellow
Crystal structure	monoclinic plates from alcohol	Crystal structure	body centered cubic
Density	1.25	Density	3.5 @ 25°C
Solubility	1g in 70 mL methanol, 85 mL chloroform, 200 mL ether	Solubility	alcohol
Common use	dyestuff	Common use	carrier for radium
Risks	CARCINOGEN, avoid prolonged exposure	Risks	water and acid soluble barium compounds are toxic, irritates skin and eyes

Benzenamine: see aniline

Name	**Benzaldehyde,** *benzoic aldehyde.* Oxidizes in air to benzoic acid	Name	**Benzene,** *cyclohexatriene*
Molecular formula	C_6H_5CHO	Molecular formula	C_6H_6
Molar mass	106.12	Molar mass	78.11
Melting point	-26	Melting point	5.5
Boiling point	179	Boiling point	80.1
Color	colorless	Color	colorless
Density	1.04	Density	0.877
Refractive index	1.546	Refractive index	1.501
Solubility	alcohol, ether, acetone, oils	Solubility	alcohol, acetone, chloroform
Common uses	manufacture of dyes, perfumes	Common uses	organic synthesis, solvent
Risks	avoid skin contact, **LD_{50}**, oral, rats 1.3 g/kg, guinea pigs 1.0 g/kg	Risks	CARCINOGEN, FLAMMABLE, **LD_{50}**, oral, rats 3.8 mL/kg

Benzenecarboxylic acid: see benzoic acid
1,4-Benzenediol: see hydroquinone

Abbreviations: @ = at; conc = concentrated; dec = decomposes; dil = dilute; est = estimate; insol = insoluble; iv = intravenous; LC_{50} = lethal concentration 50% of the time; LD = lethal dose; LD_{50} = lethal dose 50% of the time; OES = occupational exposure standards; pet = petroleum; ppm = parts per million; sol = soluble; w/w = % weight in weight.

B-C

Name	**Benzoic acid,** *benzenecarboxylic acid, phenylformic acid, dracylic acid*
Molecular formula	$C_7H_6O_2$
Molar mass	122.12
Melting point	122.4, begins to sub. at 100
Boiling point	249.2
Crystal structure	monoclinic, plates, leaflets
Solubility	1 g in 3 mL acetone, 475 mL cool water, 345 mL warm water
Common use	preservative
Risks	irritates eyes, skin, mucous membranes

Name	**1,1'-Biphenyl,** *phenylbenzene, diphenyl, bibenzene*
Molecular formula	$C_{12}H_{10}$
Molar mass	154.2
Melting point	69-70
Boiling point	255
Color	white crystals
Solubility	alcohol, ether, insol water
Common uses	heat transfer agent, organic synthesis
Risks	causes nausea, **LD_{50}**, oral, rats 3.28 g/kg

Brimstone: see sulfur

Name	**Bromine**
Molecular formula	Br_2
Molar mass	159.808
Melting point	-7.2
Boiling point	59
Color	dark red liquid
Density	3.12
Solubility	alcohol, ether, chloroform
Common use	water disinfectant
Risks	DO NOT INHALE VAPOR, burns skin severely, keep container closed

Name	**Butane,** *n-butane*
Molecular formula	C_4H_{10}
Molar mass	58.12
Melting point	-138.4
Boiling point	-0.5
Color	colorless
Common uses	fuel, rubber synthesis
Risks	EXTREMELY FLAMMABLE

Name	**Cadmium**
Molecular formula	Cd
Molar mass	112.41
Melting point	321.1
Boiling point	767
Color	silver-white
Density	8.69 @ 25°C
Solubility	H_2SO_4; insol water
Common uses	photoelectric cells, electroplating
Risks	POSSIBLE CARCINOGENIC, DO NOT INHALE FUMES OR INGEST, irritates skin

Name	**Calcium**
Molecular formula	Ca
Molar mass	40.078
Melting point	842
Boiling point	1494
Color	silver white
Crystal structure	face centered cubic
Density	1.56 @ 25°C
Solubility	sol acid
Common use	deoxidizer in metallurgy
Risks	KEEP AWAY FROM WATER, DO NOT INHALE FUMES, REACTS WITH WATER AND ALCOHOL including moisture on skin and eyes

Abbreviations: @ = at; conc = concentrated; dec = decomposes; dil = dilute; est = estimate; insol = insoluble; iv = intravenous; LC_{50} = lethal concentration 50% of the time; LD = lethal dose; LD_{50} = lethal dose 50% of the time; OES = occupational exposure standards; pet = petroleum; ppm = parts per million; sol = soluble; w/w = % weight in weight.

C

Name	**Calcium carbonate**, exists in nature as aragonite and calcite	Name	**Camphor,** *1,7,7-trimethylbicyclo[2.2.1]heptan-2-one, 2-camphanone, 2-bornanone*
Molecular formula	$CaCO_3$	Molecular formula	$C_{10}H_{16}O$
Molar mass	100.09	Molar mass	152.24
Melting point	aragonite, dec. at 825, calcite dec. at 898.6	Melting point	179
Color	colorless or white	Boiling point	204
Crystal structure	aragonite orthorhombic, calcite hexagonal	Color	white
Density	aragonite 2.85, calcite 2.71	Crystal structure	rhombohedral crystals from alcohol
Solubility	NH_4Cl; insol water	Density	0.99 @ 25°C
Common uses	antacid, calcium supplement	Refractive index	1.546
Risks	irritates eyes and skin	Solubility	1 g in 1 mL alcohol, 0.4 mL acetone, 1 mL ether, insol water
		Common uses	manufacture of plastics, pyrotechnics, moth repellent,
		Risks	keep container closed, irritates eyes, skin, respiratory tract, **LD_{50},** oral, mice 3.0 g/kg

Carbamide: see urea
Carbolic acid: see phenol

Name	**Carbon,** allotropes are diamond, graphite and amorphous	Name	Carbon dioxide
Molecular formula	C	Molecular formula	CO_2
Molar mass	12.011	Molar mass	44.01
Melting point	4100 est	Melting point	sub. @ -78.5
Boiling point	4470 est	Color	colorless
Color	amorphous carbon is black	Common use	respiration by-product, required in photosynthesis, beverage carbonation
Density	diamond 3.51, graphite 2.25, amorphous 1.8-2.1	Risks	inhalation can cause unconsciousness, handling dry ice causes severe burns
Common uses	diamond: jewellery, cutting; graphite: lead for pencils, lubricant; amorphous: activated charcoal		

Abbreviations: @ = at; conc = concentrated; dec = decomposes; dil = dilute; est = estimate; insol = insoluble; iv = intravenous; LC_{50} = lethal concentration 50% of the time; LD = lethal dose; LD_{50} = lethal dose 50% of the time; OES = occupational exposure standards; pet = petroleum; ppm = parts per million; sol = soluble; w/w = % weight in weight.

C

Name	**Carbon monoxide,** odorless, toxic gas
Molecular formula	CO
Molar mass	28.01
Melting point	-205
Boiling point	-191.5
Color	colorless
Solubility	acetic acid, alcohol
Common use	reducing agent
Risks	FLAMMABLE, GAS IS POISONOUS, DO NOT INHALE

Name	Carbon tetrachloride
Molecular formula	CCl_4
Molar mass	153.84
Melting point	-23
Boiling point	76.7
Color	colorless
Density	1.589 @ 25°C
Refractive index	1.4607
Solubility	alcohol, chloroform, ether, insol water
Common uses	solvent, varnishes, dry cleaning and fire extinguishing
Risks	POSSIBLE CARCINOGEN, DO NOT INHALE VAPOR OR INGEST, AVOID HEATING, **LC_{50}**, mice 9528 ppm

Caustic soda: see sodium hydroxide

Name	**Chloroform,** *trichloromethane,* nonflammable liquid
Molecular formula	$CHCl_3$
Molar mass	119.39
Melting point	-63.5
Boiling point	61
Density	1.48
Refractive index	1.448
Solubility	alcohol, ether
Common use	solvent
Risks	POSSIBLE CARCINOGEN, protect from light when pure, **LD_{50}**, oral, rats 2.18 mL/kg

Name	**Chromic acid,** *chromium trioxide, chromic anhydride*
Molecular formula	CrO_3
Molar mass	99.993
Melting point	196
Boiling point	dec @ 250
Color	dark red
Crystal structure	bipyramidal
Density	2.70
Solubility	very sol water, sol alcohol, ether
Common uses	oxidizer, chromium plating
Risks	KEEP AWAY FROM COMBUSTIBLES, EXPLOSIVE REACTIONS, burns skin

Chromic anhydride: see chromic acid
Chromium trioxide: see chromic acid

Abbreviations: @ = at; conc = concentrated; dec = decomposes; dil = dilute; est = estimate; insol = insoluble; iv = intravenous; LC_{50} = lethal concentration 50% of the time; LD = lethal dose; LD_{50} = lethal dose 50% of the time; OES = occupational exposure standards; pet = petroleum; ppm = parts per million; sol = soluble; w/w = % weight in weight.

Name	Citric acid, *2-hydroxy-1,2,3-propanetricarboxylic acid, β-hydroxytricarballylic acid*	Name	Colchicine, *N-(5,6,7,9-tetrahydro-1,2,3,10-tetra-methoxy-9-oxybenzo[a]heptalen7-yl) acetamide*
Molecular formula	$C_6H_8O_7$	Molecular formula	$C_{22}H_{25}NO_6$
Molar mass	192.12	Molar mass	399.43
Melting point	anhydrous 153	Solubility	1 g in 22 mL water
Crystal structure	monohydrated, orthorhombic crystals from water solution	Common uses	genetic research, gout suppressant
Density	anhydrous 1.665 @ 25°C, monohydrate 1.542		
Risks	**LD_{50}**, rats 0.975 g/kg		

Name	Copper	Name	**Cupric sulfate,** *copper(II) sulfate* Occurs in nature as hydrocyanite
Molecular formula	Cu	Molecular formula	$CuSO_4$
Molar mass	63.546	Molar mass	159.61
Melting point	1084.9	Melting point	dec. @ 560
Boiling point	2563	Color	grayish white to greenish white
Color	copper	Crystal structure	rhombic or amorphous powder
Crystal structure	face centered cubic	Density	3.6
Density	8.96 @ 25°C	Solubility	water; insol alcohol
Solubility	HNO_3, hot H_2SO_4, insol water	Name	**Pentahydrate,** *bluestone, blue vitriol*
Common uses	making alloys bronze, brass, electrical conductor	Molecular formula	$CuSO_4 \cdot 5H_2O$
Risks	dust irritates eyes, skin, respiratory tract; **OES** short-term (dust + mist) 2 mg/m^3	Melting point	becomes anhydrous by 250°C
		Color	blue
		Crystal structure	triclinic
		Solubility	water, methanol,
		Risks	strong irritant,keep container closed to prevent hydration, **LD_{50}** oral, rats 960 mg/kg

1,4-Dehydroxybenzene: see hydroquinone
Diethylene dioxide: see 1,4-Dioxane

Abbreviations: @ = at; conc = concentrated; dec = decomposes; dil = dilute; est = estimate; insol = insoluble; iv = intravenous; LC_{50} = lethal concentration 50% of the time; LD = lethal dose; LD_{50} = lethal dose 50% of the time; OES = occupational exposure standards; pet = petroleum; ppm = parts per million; sol = soluble; w/w = % weight in weight.

C-E

Name	**Cyclohexane,** *hexahydrobenzene*
Molecular formula	C_6H_{12}
Molar mass	84.16
Melting point	6.5
Boiling point	80.7
Color	colorless
Density	0.778
Refractive index	1.426
Solubility	acetone, alcohol, ether
Common use	solvent
Risks	EXTREMELY FLAMMABLE, irritant; **OES** short-term 300 ppm

Name	**Diethyl ether,** *ethyl ether, ether*
Molecular formula	$C_4H_{10}O$
Molar mass	74.12
Melting point	-116.2
Boiling point	34.6
Color	colorless
Density	0.7134
Refractive index	1.353
Solubility	alcohol, acetone, insol water
Common uses	solvent for fats, oils, waxes
Risks	MAY EXPLODE WHEN MIXED WITH ANHYDROUS NITRIC ACID, EXTREMELY FLAMMABLE, volatile, inhalation causes dizziness

Dimethyl benzene: see xylene
Dimethyl carbinol: see isopropanol
Dimethyl ketone: see acetone
Dimethylmethane: see propane
Diphenyl: see biphenyl
Dracylic acid: see benzoic acid

Name	**1,4-Dioxane,** *1,4-diethylene dioxide*
Molecular formula	$C_4H_8O_2$
Molar mass	88.11
Melting point	11.8
Boiling point	101.1
Color	colorless
Density	1.033
Refractive index	1.418
Solubility	acetone, alcohol, water
Common use	solvent for many organic and inorganic compounds
Risks	POSSIBLE CARCINOGEN, EXTREMELY FLAMMABLE, vapor is harmful, **LD_{50}**, oral, mice 5.7 mg/kg; rats 5.2 mg/kg

Name	**Ethane,** *bimethyl, dimethyl*
Molecular formula	C_2H_6
Molar mass	30.07
Melting point	-172
Boiling point	-88.6
Color	colorless
Crystal structure	hexagonal
Solubility	insol water
Common uses	fuel
Risks	FLAMMABLE, asphyxiant

Ethanedioic acid: see oxalic acid
Ethanoic acid: see acetic acid
Ethenylbenzene: see styrene
Ether: see diethyl ether
Ethine: see acetylene

Abbreviations: @ = at; conc = concentrated; dec = decomposes; dil = dilute; est = estimate; insol = insoluble; iv = intravenous; LC_{50} = lethal concentration 50% of the time; LD = lethal dose; LD_{50} = lethal dose 50% of the time; OES = occupational exposure standards; pet = petroleum; ppm = parts per million; sol = soluble; w/w = % weight in weight.

Name	**Ethanol,** *alcohol, ethyl alcohol*
Molecular formula	CH_3CH_2OH
Molar mass	46.07
Melting point	-114.1
Boiling point	78.5
Color	colorless
Density	0.789
Refractive index	1.361
Solubility	acetone, ether, water
Common uses	alcoholic beverages, solvent, octane booster in gasoline
Risks	FLAMMABLE, causes nausea, impaired perception, **LD_{50}**, oral, rats 7.06-10.6 g/kg
	95% Alcohol, binary azeotrope with water containing 95.57% ethanol by weight bp 78.15 **Diluted alcohol** contains approx. 48.9% ethanol by volume; density: 0.936 at 15.5°C; refractive index: 0.931 @ 25°C

Name	**Ethyl acetate,** *acetic acid ethyl ester, acetic ether*
Molecular formula	$C_4H_8O_2$
Molar mass	88.1
Melting point	-83
Boiling point	77
Density	0.0902
Refractive index	1.3719
Solubility	acetone, alcohol, ether
Common uses	perfumes, solvent for nitrocellulose, photographic film
Risks	**LD_{50}**, oral, rats 113 mL/kg

Ethyl alcohol: see ethanol
Ethyl ether: see diethyl ether
Ethyne: see acetylene

Name	**Ethylenediamine,** *1,2-ethanediamine, 1,2-diaminoethane*
Molecular formula	$C_2H_8N_2$
Molar mass	60.10
Melting point	8.5, monohydrate 10
Boiling point	116, monohydrate 118
Color	colorless
Density	0.898 @ 25°C
Refractive index	1.4540 @ 26°C
Solubility	alcohol
Risks	CORROSIVE IRRITATING VAPOR, protect from atmosphere, **LD_{50}**, oral, rats 1.16 g/kg

Name	**Fluorine**
Molecular formula	F_2
Molar mass	37.996
Melting point	-219.6
Boiling point	-188.1
Color	pale yellow gas
Crystal structure	cubic
Density	1.696 @ 0°C in g/L
Risks	MANY DANGEROUS REACTIONS, DO NOT INHALE GAS, causes death, **LC_{50}**, 1 hr inhalation, rat 185 ppm; mice 150 ppm

Abbreviations: @ = at; conc = concentrated; dec = decomposes; dil = dilute; est = estimate; insol = insoluble; iv = intravenous; LC_{50} = lethal concentration 50% of the time; LD = lethal dose; LD_{50} = lethal dose 50% of the time; OES = occupational exposure standards; pet = petroleum; ppm = parts per million; sol = soluble; w/w = % weight in weight.

F-H

Name	**Formaldehyde,** *methanal, oxomethane, oxymethylene*	Name	**Formic acid,** *methanoic acid*
Molecular formula	HCHO	Molecular formula	HCOOH
Molar mass	30.03	Molar mass	46.03
Melting point	-92	Melting point	8.4
Boiling point	-19.5	Boiling point	100.5
Color	colorless	Color	colorless
Solubility	acetone, alcohol, water	Density	1.220
Risks	POSSIBLY CARCINOGENIC, very reactive, extremely irritating to mucous membranes	Refractive index	1.371
	Formaldehyde solution, *formalin,* 37% by weight formaldehyde gas in water with 10-15% methanol	Solubility	alcohol, ether, glycerol, water
Common use	embalming fluid	Common uses	reducing agent, decalcifier
Risks	**LD_{50},** oral, rats 800 mg/kg	Risks	burns skin, **LD_{50},** oral, mice 1.1 g/kg

Fuming sulfuric acid: see sulfuric acid
Glacial acetic acid: see acetic acid

Name	**Fumaric acid,** *trans-1, 2-ethylenedicarboxylic acid, E-(2)-butenedioic acid*	Name	**Hexane,** *n-hexane*
Molecular formula	$C_4H_4O_4$	Molecular formula	C_6H_{14}
Molar mass	116.1	Molar mass	86.18
Melting point	287	Melting point	-95
Color	white powder	Boiling point	69
Crystal structure	needles, monoclinic prisms or leafs from water	Color	colorless
Density	1.635	Density	0.660
Solubility	alcohol, sparingly sol water, chloroform	Refractive index	1.375
Common use	anti-oxidant	Solubility	alcohol, chloroform, ether, insol water
Risks	irritates eyes, skin, mucous membranes	Common uses	determining refractive index of minerals, constituent of pet. ether and ligroin
		Risks	EXTREMELY FLAMMABLE, inhalation harmful, **LD_{50},** rats 49.0 mL/kg

Abbreviations: @ = at; conc = concentrated; dec = decomposes; dil = dilute; est = estimate; insol = insoluble; iv = intravenous; LC_{50} = lethal concentration 50% of the time; LD = lethal dose; LD_{50} = lethal dose 50% of the time; OES = occupational exposure standards; pet = petroleum; ppm = parts per million; sol = soluble; w/w = % weight in weight.

Hydrobromic acid: see hydrogen bromide
Hydrochloric acid: see hydrogen chloride
Hydrofluoric acid: see hydrogen fluoride

Name	**Hydrazine,** *hydrazine anhydrous*
Molecular formula	H_2NNH_2
Molar mass	32.05
Melting point	2.0
Boiling point	113.5
Color	colorless liquid, white crystals
Density	1.004 @ 25°C
Solubility	alcohol, water, many organic solvents
Common uses	highly polar solvent, reducing agent for organic hydrazine derivatives, rocket fuel
Risks	POSSIBLE CARCINOGEN, DO NOT INHALE, causes burns, **LD_{50}**, oral, mice 57-59 mg/kg

Name	**Hydrogen,** isotopes are protium (^1H), deuterium (^2H), tritium (^3H)
Molecular formula	H_2
Molar mass	2.0158
Melting point	-259.3
Boiling point	-252.9
Color	colorless
Crystal structure	hexagonal
Density	0.0899 g/L @ 0°C
Common uses	fuel, hydrogenation
Risks	EXTREMELY FLAMMABLE, simple asphyxiant

Name	**Hydrogen bromide,** *anhydrous hydrobromic acid*
Molecular formula	HBr
Molar mass	80.91
Melting point	-86.9
Boiling point	-66.8
Color	colorless
Density	2.71
Solubility	alcohol, water, constant boiling azeotrope with water with approx 47% HBr by weight
Common uses	synthesis of bromides, reducing agent
Risks	CORROSIVE, causes severe burns, **LC_{50}**, inhalation, rats 2858 ppm; mice 814 ppm

Name	**Hydrobromic acid,** Hydrogen bromide gas in water. Concentrated is 48% HBr, dil 10% HBr
Color	colorless to pale yellow, darkens upon exposure to air and light
Density	conc. 1.50, dil. 1.08
Solubility	alcohol, water
Common uses	organic synthesis, analytical chemistry, sedative
Risks	causes burns, extreme irritant, protect from light

Abbreviations: @ = at; conc = concentrated; dec = decomposes; dil = dilute; est = estimate; insol = insoluble; iv = intravenous; LC_{50} = lethal concentration 50% of the time; LD = lethal dose; LD_{50} = lethal dose 50% of the time; OES = occupational exposure standards; pet = petroleum; ppm = parts per million; sol = soluble; w/w = % weight in weight.

H

Name	**Hydrogen chloride,** *anhydrous hydrochloric acid*
Molecular formula	HCl
Molar mass	36.46
Melting point	-114.2
Boiling point	-85.1
Solubility	ether, water
Risks	poisonous gas, **LC_{50}**, inhalation, rats 5666 ppm; mice 2142 ppm

Name	**Hydrochloric acid,** *muriatic acid* Hydrogen chloride gas in water. Conc. HCl(*aq*) 11.6 M is 36% HCl by weight, dil. 2.9 M is 10% HCl
Color	colorless
Density	conc. 1.18, dil 1.05
Solubility	constant boiling azeotrope with water, 20.2% HCl by weight, bp 108.6
Common uses	production of chlorides, neutralizing bases, acidifier
Risks	fumes in air and causes severe burns

Name	**Hydrogen fluoride,** *anhydrous hydrofluoric acid*
Molecular formula	HF
Molar mass	20.01
Melting point	-83.5
Boiling point	19.5
Color	colorless
Density	1.002 @ 0°C
Solubility	alcohol, water, many organic solvents
Common uses	catalyst
Risks	DO NOT INHALE FUMES, vapor and liquid irritates all tissues, severe burns, **LC_{50}**, inhalation, monkeys 1780 ppm

Name	**Hydrofluoric acid,** *fluohydric acid* Hydrogen fluoride gas in water
Color	colorless
Solubility	binary azeotrope with water 38.2% HF by weight, bp 112.2
Common uses	cleaning brass, bronze
Risks	DO NOT INHALE FUMES, vapor and liquid irritates all tissues, causes severe burns, corrosive to glass, keep in plastic

Hydrogen oxide: see water

Abbreviations: @ = at; conc = concentrated; dec = decomposes; dil = dilute; est = estimate; insol = insoluble; iv = intravenous; LC_{50} = lethal concentration 50% of the time; LD = lethal dose; LD_{50} = lethal dose 50% of the time; OES = occupational exposure standards; pet = petroleum; ppm = parts per million; sol = soluble; w/w = % weight in weight.

Name	**Hydrogen peroxide,** *hydrogen dioxide*
Molecular formula	H_2O_2
Molar mass	34.01
Melting point	-0.4
Boiling point	152
Color	colorless
Density	1.41 @ 25°C
Solubility	alcohol, ether, water
Common use	antiseptic
Risks	KEEP AWAY FROM COMBUSTIBLES, causes burns, strong oxidizer, OES short-term 2 ppm, long-term 1 ppm
	3% Hydrogen peroxide, *oxydol,* density: 1.0; **30% Hydrogen peroxide,** *superoxol,* density: 1.0

Name	**Hydroquinone,** *1,4-dehydroxy benzene, 1,4-benzenediol*
Molecular formula	$C_6H_6O_2$
Molar mass	110.11
Melting point	170-171
Boiling point	285
Color	white crystals become brown in air
Crystal structure	monoclinic prisms, needles from water
Density	1.332 @ 15°C
Solubility	acetone, alcohol, ether
Common uses	anti-oxidant, reducer and developer in photography
Risks	DO NOT INHALE VAPOR OR INGEST, keep container closed, protect from light, **LD,** injection, humans 5 g

Hydroxybenzene: see phenol

2-Hydroxybenzoic acid: see salicylic acid

2-Hydroxybenzoic acid methyl ester: see methyl salicylate

Name	**Hypochlorous acid,** found in aqueous solution only
Molecular formula	HClO
Molar mass	52.47
Color	greenish-yellow
Common use	disinfectant
Risks	DECOMPOSES TO Cl_2, O_2 AND $HClO_4$, strong oxidizing agent

Name	Hyponitrous acid
Molecular formula	$H_2N_2O_2$
Molar mass	62.03
Color	pale blue
Risks	SOLID IS EXPLOSIVE

Abbreviations: @ = at; conc = concentrated; dec = decomposes; dil = dilute; est = estimate; insol = insoluble; iv = intravenous; LC_{50} = lethal concentration 50% of the time; LD = lethal dose; LD_{50} = lethal dose 50% of the time; OES = occupational exposure standards; pet = petroleum; ppm = parts per million; sol = soluble; w/w = % weight in weight.

I-L

Name	**Indole**, *1-benzo[b]pyrrole, 2,3-benzopyrrole*
Molecular formula	C_8H_7N
Molar mass	117.14
Melting point	52
Boiling point	253-254
Color	crystalline solid is brown
Crystal structure	leafs from pet. ether or water
Density	1.22
Solubility	hot alcohol, ether
Common use	perfumes
Risks	POSSIBLE CARCINOGEN, irritates eyes, upper respiratory tract and skin, **LD_{50}**, oral, rats 1g/kg

Name	**Iodine**
Molecular formula	I_2
Molar mass	253.81
Melting point	113.5
Boiling point	185.3
Color	violet black
Crystal structure	orthorhombic
Density	4.93 @ 27°C
Solubility	chloroform, glacial acetic acid, glycerol oils, almost insol water
Common uses	topical antiseptic, reagent in analytical chemistry, synthesis of iodine compounds
Risks	DO NOT INHALE VAPOR, CORROSIVE VAPOR, **LD**, humans 2-4 g

Isopropyl alcohol: see isopropanol

Name	**Isopropanol**, *2-propanol, isopropyl alcohol, dimethyl carbinol*
Molecular formula	C_3H_8O
Molar mass	60.09
Melting point	-88.5
Boiling point	82.5
Density	0.785
Refractive index	1.377
Solubility	alcohol, water, ether
Boiling point	80.37
Density	0.834
Risks	flammable, inhalation or ingestion of vapor may cause headaches, dizziness, **LD** 100 mL

Name	**Lithium**
Molecular formula	Li
Molar mass	6.941
Melting point	180.6
Boiling point	1336
Color	silver white turns yellowish in air
Crystal structure	body centered cubic
Density	0.534 @ 25°C
Solubility	liquid ammonia
Common uses	manufacture of alloys
Risks	REACTS WITH WATER, KEEP AWAY FROM WATER, store in mineral oil

Abbreviations: @ = at; conc = concentrated; dec = decomposes; dil = dilute; est = estimate; insol = insoluble; iv = intravenous; LC_{50} = lethal concentration 50% of the time; LD = lethal dose; LD_{50} = lethal dose 50% of the time; OES = occupational exposure standards; pet = petroleum; ppm = parts per million; sol = soluble; w/w = % weight in weight.

M

Marsh gas: see methane

Name	Magnesium
Molecular form	Mg
Molar mass	24.305
Melting point	650
Boiling point	1090
Color	silver white
Crystal structure	hexagonal
Density	1.74 @ 25°C
Solubility	ammonium salts, mineral acids; insol CrO_3, water
Common use	Gringnard reagents
Risks	AVOID CONTACT WITH SKIN, FLAMMABLE, do not inhale dust

Name	**Mercury,** *hydrargyrum, quick silver, liquid silver*
Molecular formula	Hg
Molar mass	200.59
Melting point	-38.87
Boiling point	356.6
Color	silver
Density	13.53 @ 25°C
Solubility	cold dil. mineral acids; insol water
Common uses	thermometers, barometers, catalyst
Risks	DO NOT TOUCH, ABSORBED THROUGH SKIN, **OES** short-term 0.15 mg/m^3, long-term 0.05 mg/m^3

Methanal: see formaldehyde

Name	**Methane,** *marsh gas,* odorless gas
Molecular formula	CH_4
Molar mass	16.04
Melting point	-182
Boiling point	-164
Color	colorless
Solubility	alcohol, ether and other organic solvents
Common uses	fuel, organic synthesis
Risks	FLAMMABLE, simple asphyxiant

Name	**Methanol,** *methyl alcohol, wood alcohol,* poisonous liquid
Molecular formula	CH_3OH
Molar mass	32.04
Melting point	-97.8
Boiling point	64.7
Color	colorless
Density	0.7914
Refractive index	1.329
Solubility	acetone, alcohol, ether, water
Common uses	solvent, fuel, gasoline antifreeze
Risks	FLAMMABLE, consumption causes blindness and death

Methanoic acid: see formic acid
Methyl alcohol: see methanol
Methylbenzene: see toluene
Methylchloroform: see 1,1,1-Trichloroethane
Muriatic acid: see hydrogen chloride, hydrochloric acid

Abbreviations: @ = at; conc = concentrated; dec = decomposes; dil = dilute; est = estimate; insol = insoluble; iv = intravenous; LC_{50} = lethal concentration 50% of the time; LD = lethal dose; LD_{50} = lethal dose 50% of the time; OES = occupational exposure standards; pet = petroleum; ppm = parts per million; sol = soluble; w/w = % weight in weight.

M-O

Name	**Methyl salicylate,** *2-hydroxybenzoic acid methyl ester, wintergreen oil*
Molecular formula	$C_8H_8O_3$
Molar mass	152.14
Melting point	-8.6
Boiling point	220-224
Color	colorless, yellow or reddish oily liquid
Density	1.18 @ 25°C
Refractive index	1.53-1.54
Solubility	alcohol, glacial acetic acid, ether
Risks	DO NOT INGEST, POISONOUS, CAUSES DEATH, **LD**, 30 mL adults

Name	**Naphthalene**
Molecular formula	$C_{10}H_8$
Molar mass	128.17
Melting point	80.2
Boiling point	218
Color	colorless
Crystal structure	monoclinic plates from ether
Density	1.162
Refractive index	1.400
Solubility	very sol ether; sol 1 g in 13 mL alcohol, 8 mL toluene, 2 mL chloroform; insol water
Common uses	moth repellent and insecticide
Risks	poisonous vapors, absorbed through skin, **OES** short-term 15 ppm, long-term 10 ppm

Oil of vitriol: see sulfuric acid

Name	**Nitric acid**
Molecular formula	HNO_3
Molar mass	63.01
Melting point	-41.6
Boiling point	83 dec
Color	colorless to pale yellow
Density	1.5027 @ 25°C
Solubility	ether, water; concentrated nitric acid is about 70% by weight, density 1.42
Common uses	synthesis of nitrogen-containing compounds
Risks	DECOMPOSES VIOLENTLY IN ALCOHOL, CORROSIVE FUMES, burns all tissues, **OES** short-term 4 ppm, long-term 2 ppm

Name	**Oleic acid,** *(Z)-9-octadecenoic acid*
Molecular formula	$C_{18}H_{34}O_2$
Molar mass	282.45
Melting point	4
Boiling point	80-100 (dec)
Color	nearly colorless
Density	0.895 @ 25°C
Refractive index	1.4582
Solubility	acetone, alcohol, insol water
Common use	diagnostic aid
Risks	irritates eyes, skin and mucous membranes, LD_{50}, iv, mice 230 mg/kg

Oleum: see sulfuric acid, fuming
Orthophosphoric acid: see phosphoric acid

Abbreviations: @ = at; conc = concentrated; dec = decomposes; dil = dilute; est = estimate; insol = insoluble; iv = intravenous; LC_{50} = lethal concentration 50% of the time; LD = lethal dose; LD_{50} = lethal dose 50% of the time; OES = occupational exposure standards; pet = petroleum; ppm = parts per million; sol = soluble; w/w = % weight in weight.

O-P

Name	**Oxalic acid anhydrous,** *ethanedioic acid*
Molecular formula	$C_2H_2O_4$
Molar mass	90.04
Melting point	189 (dec)
Crystal structure	orthorhombic from glacial acetic acid
Density	1.90 @ 17°C
	Dihydrate
Melting point	begins dehydrating at 101
Color	colorless
Crystal structure	monoclinic tablets or prisms
Density	1.65 @ 19°C
Solubility	1 g in 2.5 mL alcohol, 7 mL water, 2 mL hot water
Common uses	analytical reagent, reducing agent
Risks	poisonous, burns skin, **LD_{50}**, oral, rats 10% solution, 4.75ml/kg; **OES** short-term 2 mg/m^3

Name	**Oxygen,** two allotropes oxygen (O_2) and ozone (O_3)
Molecular formula	O_2
Molar mass	31.999
Melting point	-218.4
Boiling point	-183
Color	colorless
Crystal structure	cubic
Solubility	1 mL in 32 mL of warm water
Common uses	fuel for welding, rocket propellant
Name	**Ozone,** *triatomic oxygen*
Molecular formula	O_3
Molar mass	48.00
Melting point	-193
Boiling point	-112
Color	pale blue gas, dark blue liquid
Solubility	alkali solutions, oils
Common uses	air disinfectant, organic synthesis
Risks	very toxic, powerful oxidizing agent, **OES** short-term 0.3 ppm, long-term 0.1 ppm

Oxybenzene: see phenol
Ozone: see oxygen, ozone

Name	**Paraffin wax,** a mixture of hydrocarbons
Molecular formula	C_nH_{2n+2}
Melting point	50-57
Color	colorless or white
Density	approx. 0.90
Solubility	chloroform, ether, insol water
Common uses	manufacture of paraffin paper, candles
Risks	**OES** short-term 6 mg/m^3, long-term 2 mg/m^3

Name	**Pentane,** *n-pentane*
Molecular formula	C_5H_{12}
Molar mass	72.15
Melting point	-129.7
Boiling point	36.1
Density	0.6264
Refractive index	1.358
Solubility	alcohol, ether
Risks	FLAMMABLE, **LC_{50}**, inhalation, mice 128-200 ppm

Perchloromethane: see carbon tetrachloride

Abbreviations: @ = at; conc = concentrated; dec = decomposes; dil = dilute; est = estimate; insol = insoluble; iv = intravenous; LC_{50} = lethal concentration 50% of the time; LD = lethal dose; LD_{50} = lethal dose 50% of the time; OES = occupational exposure standards; pet = petroleum; ppm = parts per million; sol = soluble; w/w = % weight in weight.

P

Name	Perchloric acid
Molecular formula	$HClO_4$
Molar mass	100.5
Melting point	-112
Color	colorless
Density	1.76
Solubility	water
Common use	oxidizer in analytical chemistry
Risks	volatile, hygroscopic, corrosive

Name	Periodic acid
Molecular formula	H_5IO_6, $HIO_4 \cdot 2H_2O$
Molar mass	227.94
Melting point	122
Boiling point	140 dec
Color	white
Crystal structure	monoclinic
Solubility	freely sol in water, sol alcohol, slightly sol ether
Common use	oxidizes organic molecules in organic synthesis
Risks	oxidizes organic compounds

Petrohol: see isopropanol
Phenyl hydroxide: see phenol
Phenyl amine: see aniline
Phenyl benzene: see biphenyl
Phenylethylene: see styrene
Phenylformic acid: see benzoic acid

Name	**Phenol,** *carbolic acid, phenyl hydroxide, hydroxybenzene*
Molecular formula	C_6H_5OH
Molar mass	94.11
Melting point	40.9
Boiling point	182
Color	colorless, reddens on exposure to air and light
Density	1.07
Refractive index	1.54
Solubility	very sol alcohol, chloroform ether, glycerol, sol 1 g in 15 mL water
Common use	antiseptic
Risks	poisonous, burns skin, protect from light, **LD**, 1-15 g in humans

Name	**Phosphoric acid,** *orthophosphoric acid*
Molecular formula	H_3PO_4
Molar mass	98.00
Melting point	42.35
Boiling point	approx 200 dec
Color	colorless
Solubility	alcohol, water
Common uses	fertilizer manufacture, acid catalyst, analytical chemistry
Risks	irritates eyes, skin, and mucous membranes

Abbreviations: @ = at; conc = concentrated; dec = decomposes; dil = dilute; est = estimate; insol = insoluble; iv = intravenous; LC_{50} = lethal concentration 50% of the time; LD = lethal dose; LD_{50} = lethal dose 50% of the time; OES = occupational exposure standards; pet = petroleum; ppm = parts per million; sol = soluble; w/w = % weight in weight.

P

Name	**Phosphorous,** allotropes are black, red and white
	Phosphorous black
Molecular formula	P (polymorphic)
Color	black
Crystal structure	orthorhombic
Solubility	insol organic solvents
	Phosphorous red
Molecular formula	P_4
Molar mass	123.89
Melting point	416 sub, ignites at about 200 in air
Color	red to violet
Crystal structure	cubic or amorphous
Density	2.34
Solubility	alcohol; insol ether, NH_3
Common uses	pyrotechnics, organic synthesis
Risks	KEEP AWAY FROM OXIDIZING AGENTS, MAY EXPLODE WITH CONTACT OR FRICTION

	Phosphorous white
Molecular formula	P_4
Molar mass	123.89
Melting point	44.1
Boiling point	280
Color	white
Crystal structure	cubic
Density	1.82
Refractive index	2.144
Solubility	alkane, chloroform
Common uses	rat poisons, smoke screen production, analysis of gases
Risks	IGNITES IN MOIST AIR @ 30°C, DO NOT TOUCH, handle with forceps, **LD** 50-100 mg

Name	**Platinum**
Molecular formula	Pt
Molar mass	195.08
Melting point	1773.5
Boiling point	3827
Color	silver
Crystal structure	face centered cubic
Density	21.5
Common uses	oxidation catalyst, electrodes, jewellery

Name	**Potassium,** *kalium*
Molecular formula	K
Molar mass	39.098
Melting point	63.7
Boiling point	759
Color	silver white
Crystal structure	body centered cubic
Density	0.856
Solubility	acids
Common uses	synthesis of inorganic and organic compounds
Risks	REACTS WITH OXYGEN, WATER, ACIDS

Abbreviations: @ = at; conc = concentrated; dec = decomposes; dil = dilute; est = estimate; insol = insoluble; iv = intravenous; LC_{50} = lethal concentration 50% of the time; LD = lethal dose; LD_{50} = lethal dose 50% of the time; OES = occupational exposure standards; pet = petroleum; ppm = parts per million; sol = soluble; w/w = % weight in weight.

P

Name	**Potassium permanganate,** *permanganic acid potassium salt, chameleon mineral*
Molecular formula	$KMnO_4$
Molar mass	158.03
Melting point	240 dec
Color	dark purple
Crystal structure	rhombic
Density	2.703
Refractive index	1.59
Solubility	very sol acetone, methanol; sol H_2SO_4, 1 g in 16 mL cold water
Common uses	organic synthesis, analytical reagent
Risks	CAUTION: STRONG OXIDIZER, burns skin, **LD_{50}**, oral, rats 1.09 g/kg

Name	**Potassium iodide**
Molecular formula	KI
Molar mass	166.02
Melting point	680
Color	colorless-white crystals
Crystal structure	cubic
Density	3.12
Solubility	alcohol, water
Risks	**LD_{50}**, iv, 285 mg/kg

Name	**Propane,** *dimethylmethane, propyl hydride*
Molecular formula	$CH_3CH_2CH_3$
Molar mass	44.10
Melting point	-189.7
Boiling point	-42.1
Solubility	alcohol,chloroform
Common uses	fuel, organic synthesis, refrigerant
Risks	FLAMMABLE

Name	**Pyridine**
Molecular formula	C_5H_5N
Molar mass	79.10
Melting point	-42
Boiling point	115
Color	colorless
Density	0.982
Refractive index	1.509
Solubility	alcohol, ether, pet. ether
Common uses	solvent, organic synthesis, weak base
Risks	FLAMMABLE, irritates skin and respiratory tract, **LD_{50}**, oral, rats 1.58 g/kg

2-Propanol: see isopropanol
2-Propanone: see acetone
Propyl hydride: see propane

Abbreviations: @ = at; conc = concentrated; dec = decomposes; dil = dilute; est = estimate; insol = insoluble; iv = intravenous; LC_{50} = lethal concentration 50% of the time; LD = lethal dose; LD_{50} = lethal dose 50% of the time; OES = occupational exposure standards; pet = petroleum; ppm = parts per million; sol = soluble; w/w = % weight in weight.

S

Name	**Salicylic acid,** *2-hydroxybenzoic acid*
Molecular formula	$C_7H_6O_3$
Molar mass	138.12
Melting point	157-159
Boiling point	211 sub
Color	white, discolors in light
Crystal structure	monoclinic
Density	1.44
Refractive index	1.565
Solubility	1 g in 3 mL acetone, 2.7 mL alcohol, 3 mL ether
Common uses	organic synthesis, dye manufacture, analytical chemistry
Risks	protect from light, may irritate skin, ingestion of large quantities may cause vomiting, **LD_{50},** iv, mice 0.5 g/kg

Name	**Selenium chloride**
Molecular formula	$SeCl_2$
Molar mass	228.83
Melting point	-85
Color	deep red
Density	2.77 @ 25°C
Solubility	chloroform, CCl_4
Risks	decomposes in water

Salicylic acid acetate: see acetyl salicylic acid
Salt: see sodium chloride
SDS: see sodium dodecyl sulfate
Silicic anhydride: see silica
Silicon dioxide: see silica

Name	**Silica,** *silicon dioxide, silicic anhydride* nonflammable gas, suffocating odor
Molecular formula	SiO_2
Molar mass	60.09
Solubility	HF; insol water
Common use	manufacture of glass
Risks	poisonous, irritating to eyes

Name	**Silver**
Molecular formula	Ag
Molar mass	107.868
Melting point	960.5
Boiling point	2163
Color	silver
Crystal structure	face centered cubic
Density	10.5 @ 25°C
Solubility	insol water
Common uses	catalyst for hydrogenation and oxidation, toxic to bacteria and lower forms of life, inert to most acids
Risks	REACTS WITH DIL. NITRIC ACID AND HOT CONC. SULFURIC ACID AND HCl

Abbreviations: @ = at; conc = concentrated; dec = decomposes; dil = dilute; est = estimate; insol = insoluble; iv = intravenous; LC_{50} = lethal concentration 50% of the time; LD = lethal dose; LD_{50} = lethal dose 50% of the time; OES = occupational exposure standards; pet = petroleum; ppm = parts per million; sol = soluble; w/w = % weight in weight.

S

Name	**Silver chloride**
Molecular formula	AgCl
Molar mass	143.34
Melting point	455
Boiling point	1550
Color	white powder, darkens when exposed to light
Solubility	ammonia, about 1 mg in 500 mL water, insol dil acids
Comments	photo decomposition, protect from light

Name	**Silver nitrate**
Molecular formula	$AgNO_3$
Molar mass	169.89
Melting point	212
Boiling point	440 dec
Color	colorless
Crystal structure	rhombohedral
Density	4.35
Refractive index	1.729
Solubility	1 g in 235 mL acetone, 30 mL alcohol, 6.5 mL hot alcohol
Common uses	photography, mirror manufacture, production of other silver salts
Risks	POISONOUS, causes burns

Silver dioxide: see silica
Soda lye: see sodium hydroxide
Sodium acid sulfite: see sodium bisulfite

Name	**Sodium bicarbonate,** *sodium hydrogen carbonate, baking soda*
Molecular formula	$NaCHO_3$
Molar mass	84.00
Melting point	50 dec
Color	white
Crystal structure	monoclinic
Density	2.159
Solubility	6.9 g in 100 mL cold water, 16.4 g in 100 mL hot water, slightly sol alcohol
Common uses	source of CO_2, antacid, production of other sodium salts

Name	**Sodium bisulfate,** *sodium hydrogen sulfate*
Molecular formula	$NaHSO_4$
Molar mass	120.07
Melting point	315
Boiling point	dec
Color	colorless
Crystal structure	triclinic
Density	2.435
Solubility	28.6 g in 100 mL cold water, 100 g in 100 mL hot water, slightly sol alcohol, insol NH_3
Risks	keep container closed, decomposed by alcohol

Abbreviations: @ = at; conc = concentrated; dec = decomposes; dil = dilute; est = estimate; insol = insoluble; iv = intravenous; LC_{50} = lethal concentration 50% of the time; LD = lethal dose; LD_{50} = lethal dose 50% of the time; OES = occupational exposure standards; pet = petroleum; ppm = parts per million; sol = soluble; w/w = % weight in weight.

S

Name	**Sodium bisulfite,** *sodium acid sulfite, sodium hydrogen sulfite*
Molecular formula	$NaHSO_3$
Molar mass	104.07
Melting point	dec
Color	white, yellow in solution
Crystal structure	monoclinic
Density	1.48
Refractive index	1.526
Solubility	water; slightly sol alcohol
Common uses	disinfectant, bleach
Risks	conc. solutions are irritants to skin and mucous membranes, **LD_{50}**, iv, rats 115 mg/kg

Name	**Sodium chloride,** *salt*
Molecular formula	NaCl
Molar mass	58.45
Melting point	804
Density	2.17
Color	colorless, white
Crystal structure	cubic
Solubility	1 g in 2.8 mL warm water, 2.6 mL hot water
Risks	**LD_{50}**, oral, rats 3.75g/kg

Name	**Sodium dodecyl sulfate,** *SDS, sodium lauryl sulfate, sulfuric acid monododecyl ester sodium salt*
Molecular formula	$C_{12}H_{25}NaO_4S$
Molar mass	288.38
Color	white or cream crystals, flakes or powder
Common uses	anionic detergent, toothpaste, wetting agent, protein and lipid separation
Risks	**LD_{50}**, oral, rats 1.29 mg/kg

Name	**Sodium hydroxide,** *soda lye, sodium hydrate, caustic soda*
Molecular formula	NaOH
Molar mass	40.01
Melting point	318.4
Boiling point	1390
Color	white
Density	2.13 @ 25°C
Refractive index	1.3576
Solubility	very sol alcohol, 1 g in 0.9 mL water, 0.3 mL hot water; insol acetone, ether. Conc. is 50% by weight, 19.1 M
Common uses	neutralizing acids, acid-base titration
Risks	CAUTION: VERY EXOTHERMIC WHEN DISSOLVING IN WATER, very corrosive to all tissues, **LD_{50}**, oral, rabbits 0.5 g/kg

Sodium hydrate: see sodium hydroxide
Sodium hydrogen carbonate: see sodium bicarbonate
Sodium hydrogen sulfate: see sodium bisulfate
Sodium hydrogen sulfite: see sodium bisulfite
Sodium lauryl sulfate: see sodium dodecyl sulfate

Abbreviations: @ = at; conc = concentrated; dec = decomposes; dil = dilute; est = estimate; insol = insoluble; iv = intravenous; LC_{50} = lethal concentration 50% of the time; LD = lethal dose; LD_{50} = lethal dose 50% of the time; OES = occupational exposure standards; pet = petroleum; ppm = parts per million; sol = soluble; w/w = % weight in weight.

S

Name	**Sodium metabisulfite,** *sodium pyrosulfite*
Molecular formula	$Na_2O_5S_2$
Molar mass	190.10
Melting point	150 dec
Color	white
Density	1.4
Solubility	freely sol water, glycerol, slightly sol alcohol
Common use	antioxidant

Name	**Styrene,** *ethenylbenzene, cinnamene, vinylbenzene*
Molecular formula	$C_6H_5CH{=}CH_2$
Molar mass	104.14
Melting point	-30.6
Boiling point	145-146
Color	colorless to yellow
Density	0.906
Refractive index	1.547
Common uses	converted to polymer polystyrene, synthetic rubber, insulators
Risks	irritates eyes and mucous membranes, **LD_{50}**, iv, mice 90 mg/kg

Sodium pyrosulfite: see sodium metabisulfite

Name	**Sulfur,** *sulphur, brimstone*
Molecular formula	S
Molar mass	32.06
Melting point	β 115, γ 107
Color	α yellow, β pale yellow, γ pale yellow
Crystal structure	α = rhombic, β = monoclinic, γ = amorphous
Density	α 2.07, β 1.96, γ 1.92
Refractive index	α 1.957
Solubility	CCl_4, slightly sol alcohol, ether liquid NH_3, insol water
Common uses	production of sulfuric acid, carbon disulfide, gun powder
Risks	irritates skin and mucous membranes

Name	**Sulfuric acid,** *oil of vitriol*
Molecular formula	H_2SO_4
Molar mass	98.08
Melting point	-38.9
Boiling point	167
Color	colorless
Density	1.841
Refractive index	1.405
Solubility	water
Common uses	production of fertilizers, explosives, common lab acid
Risks	VERY EXOTHERMIC WHEN DISSOLVING IN WATER, VERY CORROSIVE, corrosive to all tissues, **LD_{50}**, oral, rats 2.14 g/kg
	Fuming sulfuric acid, *oleum* Concentrated sulfuric acid with free SO_3 is 18 M, 96% by weight
Molecular formula	H_2SO_4 with free SO_3
Melting point	10
Boiling point	290
Color	colorless
Density	1.84
Risks	VERY CORROSIVE, DO NOT INHALE FUMES, keep closed in glass container, very exothermic when dissolving in water

Abbreviations: @ = at; conc = concentrated; dec = decomposes; dil = dilute; est = estimate; insol = insoluble; iv = intravenous; LC_{50} = lethal concentration 50% of the time; LD = lethal dose; LD_{50} = lethal dose 50% of the time; OES = occupational exposure standards; pet = petroleum; ppm = parts per million; sol = soluble; w/w = % weight in weight.

S-T

Sulfuric anhydride: see sulfur trioxide
Sulfurous oxychloride: see thionyl chloride

Name	**Sulfur trioxide,** *sulfuric anhydride,* component of fuming sulfuric acid
Molecular formula	α SO_3, β $(SO_3)_2$
Melting point	α 62.3, β 32.5, γ 16.8
Boiling point	α 44.8, β 50 sub, γ 44.8
Density	α 1.97, γ 1.92
Common uses	intermediate during production of sulfuric acid, production of explosives
Risks	corrosive, irritates mucous membranes

Name	**Thionyl chloride,** *sulfurous oxychloride*
Molecular formula	$SOCl_2$
Molar mass	118.98
Melting point	-104.5
Boiling point	76
Color	colorless, pale yellow or reddish
Density	1.638
Refractive index	1.517
Solubility	chloroform, CCl_4, dec in water
Common use	production of acyl chlorides
Risks	DO NOT INHALE VAPOR, corrosive to eyes, skin and mucous membranes

Sulphur: see sulfur
Tetrachloromethane: see carbon tetrachloride
TNT: see 2,4,6-Trinitrotoluene

Name	**Toluene,** *methylbenzene*
Molecular formula	$C_6H_5CH_3$
Molar mass	92.13
Melting point	-95
Boiling point	110.6
Density	0.867
Refractive index	1.496
Solubility	acetone, alcohol
Common uses	solvent, benzene substitute
Risks	FLAMMABLE, **LD_{50}**, oral, rats 7.53 g/kg

Name	**1,1,1-Trichloroethane,** *methylchloroform*, nonflammable
Molecular formula	$C_2H_3Cl_3$
Molar mass	133.42
Melting point	-32.5
Boiling point	74.1
Density	1.338
Refractive index	1.4384
Solubility	acetone, CCl_4, insol water
Risks	irritates eyes, mucous membranes

Triatomic oxygen: see ozone
Trichloromethane: see chloroform

Abbreviations: @ = at; conc = concentrated; dec = decomposes; dil = dilute; est = estimate; insol = insoluble; iv = intravenous; LC_{50} = lethal concentration 50% of the time; LD = lethal dose; LD_{50} = lethal dose 50% of the time; OES = occupational exposure standards; pet = petroleum; ppm = parts per million; sol = soluble; w/w = % weight in weight.

T-W

Name	2,4,6-Trinitrotoluene, *TNT*
Molecular formula	$C_7H_5N_3O_6$
Molar mass	227.13
Melting point	80.1
Boiling point	240 EXPLODES
Color	yellow
Crystal structure	monoclinic
Density	1.65
Solubility	acetone, ether, sparingly sol in water
Common use	explosives
Risks	EXPLOSIVE, absorbed through skin

Name	Turpentine
Color	colorless
Density	0.854-0.868 @ 25°C
Refractive index	1.468-1.478
Solubility	alcohol, chloroform, ether, insol water
Common uses	solvent, polishes
Risks	irritates skin and mucous membranes, absorbed through skin, lungs

Vinylbenzene: see styrene

Name	Urea, *carbamide*
Molecular formula	H_2NCONH_2
Molar mass	60.06
Melting point	133
Boiling point	dec
Crystal structure	tetragonal prisms
Density	1.323
Refractive index	1.484
Solubility	1 g in 1 mL of water
Common uses	fertilizer, production of plastics and resins

Name	Water, *hydrogen oxide, odorless liquid*
Molecular formula	H_2O
Molar mass	18.016
Melting point	0.
Boiling point	100.
Density	0.917 g/cm^3 @ 0°C (ice), 1.000000 g/mL @ 3.98°C, 0.997 g/mL @ 25°C
Refractive index	1.3330
Vapor pressure	23.756 mm Hg @ 25°C
Surface tension	0.07197 N/m
Heat capacity	75.2 J/K mol @ 25°C
Dipole moment	1.84 D
Triple point	0.0100°C @ 4.58 mm Hg
Thermochemical data	$\Delta H°$ of liquid, -286 kJ/mol @ 25°C $\Delta H°_{fus}$ 6.02 kJ/mol, $\Delta H°_{vap}$ 40.7 kJ/mol @ 100°C

Wintergreen oil: see methyl salicylate
Wood alcohol: see methanol

Abbreviations: @ = at; conc = concentrated; dec = decomposes; dil = dilute; est = estimate; insol = insoluble; iv = intravenous; LC_{50} = lethal concentration 50% of the time; LD = lethal dose; LD_{50} = lethal dose 50% of the time; OES = occupational exposure standards; pet = petroleum; ppm = parts per million; sol = soluble; w/w = % weight in weight.

X-Z

Name	**Xylene,** dimethylbenzene
Molecular formula	C_8H_{10}
Molar mass	106.16
Melting point	meta -47.4, ortho -25, para 13
Boiling point	meta 139.3, ortho 144, para 137
Density	meta 0.8684 @ 15°C, ortho 0.8801, para 0.8610
Refractive index	meta 1.497, ortho 1.506, para 1.496
Solubility	acetone, alcohol, ether, insol in water
Common use	solvent
Risks	avoid inhalation, **LD_{50}**, meta: oral, rats 7.71 mL/kg

Name	**Zinc**
Molecular formula	Zn
Molar mass	65.38
Melting point	419.6
Boiling point	908
Color	bluish-white
Crystal structure	hexagonal
Density	7.14 @ 25°C
Solubility	acid, alkane, acetic acid, insol water
Common uses	in alloys, batteries, reducing agent

Name	**Zinc chloride**
Molecular formula	$ZnCl_2$
Molar mass	136.29
Melting point	290
Boiling point	732
Color	white
Density	2.91 @ 25°C
Solubility	freely sol acetone
Risks	irritates skin, mucous membranes

Abbreviations: @ = at; conc = concentrated; dec = decomposes; dil = dilute; est = estimate; insol = insoluble; iv = intravenous; LC_{50} = lethal concentration 50% of the time; LD = lethal dose; LD_{50} = lethal dose 50% of the time; OES = occupational exposure standards; pet = petroleum; ppm = parts per million; sol = soluble; w/w = % weight in weight.

GEOLOGY

6

6.1 Geological Time Scale

Phanerozoic Time Scale

Eon	Era	Period	Epoch	Millions of years ago
PHANEROZOIC	CENOZOIC	Quaternary	Holocene	0.01
			Pleistocene	2
		Tertiary	Pliocene	5
			Miocene	24
			Oligocene	37
			Eocene	58
			Paleocene	66
	MESOZOIC	Cretaceous		144
		Jurassic		208
		Triassic		245
	PALEOZOIC	Permian		286
		Carboniferous	Pennsylvanian	320
			Mississippian	360
		Devonian		408
		Silurian		438
		Ordovician		505
		Cambrian		570

Precambrian Time Scale

Eon		Era	Millions of years ago*
PRECAMBRIAN	Proterozoic	Late	900
		Middle	1600
		Early	2500
	Archean	Late	3000
		Middle	3400
		Early	4600

*Dates in table represent end of each period.

Geology is the science of earthly things. It studies the composition, history and processes of the Earth. The Earth began to form about 5 billion years ago from a cloud of rock fragments in the early solar system. The time scale on the left is based on fossil records and radiometric dating techniques. The dates in the table represent the end of each time frame.

Precambrian time spans over 4 billion years. Rocks formed in the Precambrian eon are generally complex in structure, and usually lack fossils. Very few fossils have been traced back to the Precambrian eon. The Phanerozoic eon, on the other hand, includes rocks with richly diverse fossils. The Phanerozoic eon spans the Paleozoic, Mesozoic and Cenozoic eras.

6.2 The Earth

Layers of the atmosphere	The standard atmosphere (US)			Elements in sea water	
	Name	Formula	Percent volume (%)	Name	Concentration (g/L)
Magnetosphere	Nitrogen	N_2	78.084	Oxygen	857
Exosphere	Oxygen	O_2	20.9476	Hydrogen	108
Thermosphere	Argon	Ar	0.934	Chlorine	19.0
Ionosphere	Carbon dioxide	CO_2	0.0314	Sodium	10.5
Mesosphere	Neon	Ne	0.001818	Magnesium	1.35
Stratosphere	Helium	He	0.000524	Sulfur	0.89
Troposphere	Methane	CH_4	0.0002	Calcium	0.40
	Krypton	Kr	0.000114	Potassium	0.38
	Hydrogen	H_2	0.00005	Bromine	0.07
	Xenon	Xe	0.0000087	Carbon	0.03
				Others	2.39

Elements in the Earth's crust	
Element	Percentage by weight
Oxygen	46.6
Silicon	27.7
Aluminum	8.1
Iron	5.0
Calcium	3.6
Sodium	2.8
Potassium	2.6
Magnesium	2.1
All others	1.5

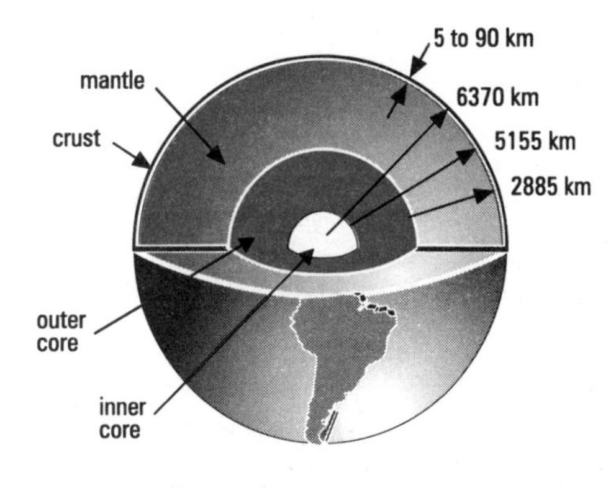

Fig. 6.2.1 Earth's Layers

Layers in the Earth						
Name	Begins (km)	Ends (km)	Thickness (km)	Density (kg/m³)	Phase	Temperature (°C)
Crust-continental	0	25-90	25-90	2500	solid	0-700
-oceanic	0 (below ocean)	5-10	5-10	3000	solid	0-700
Mantle-upper	5-90	400	310-395	3000	solid*	700-1300
-transition zone	400	1050	650	3380-4600	solid	1300-1800
-lower	1050	2885	1835	4600-5400	solid	1800-2800
Outer core	2885	5155	2270	9900-12 200	liquid	3200
Inner core	5155	6370	1215	13 000	solid	8100
* Except for a softer molten layer						

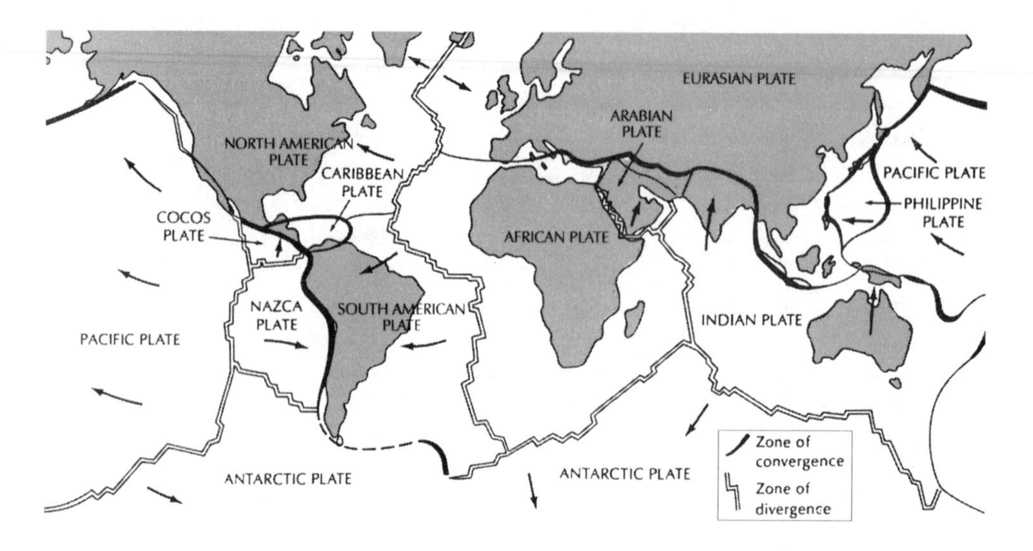

Fig. 6.2.2a Earth's Lithospheric Plates, Zones of Convergence and Divergence

The Changing Surface of the Earth

The theory of **plate tectonics** describes the surface of the Earth as constantly changing. The theory is based on observations suggesting that over time, the continents drift about the surface of the Earth. Studies of ocean basins support the theory. Plate tectonics theory proposes that the Earth is covered with large plates. As these plates move, they change the physical features of the planet. These plates are made up of the crust and upper mantle; together they're called the **lithosphere**. The lithosphere moves about on top of a softer, hotter layer known as the **asthenosphere**. There are seven major **lithospheric plates** along with several smaller platelets. **Divergence zones** exist where plates move away from each other. **Lateral movement zones** exist where plates slide against each other. **Convergence zones** exist where plates push into each other.

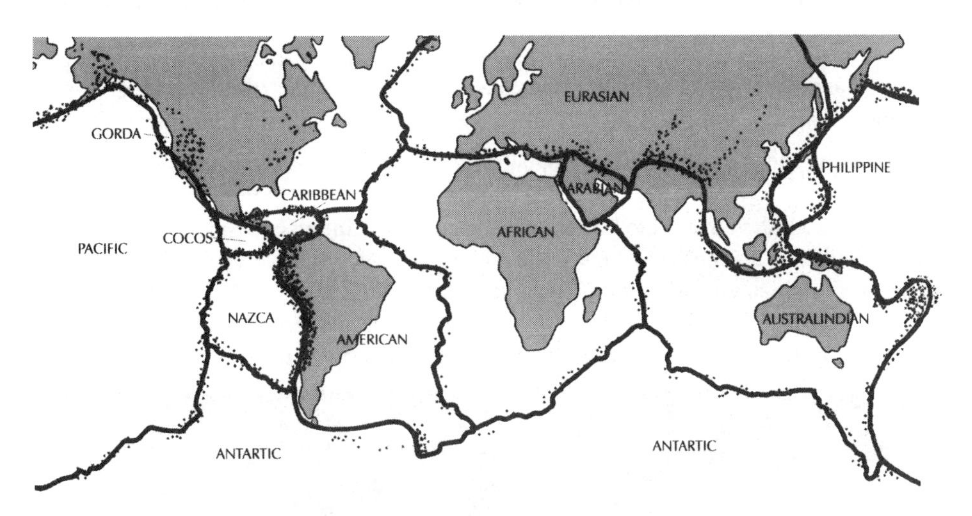

Fig. 6.2.2b Concentration of Earthquakes Corresponding to Boundaries of Lithospheric Plates

The table below lists some useful data about our planet Earth.

Earth Data

Quantity	Symbol	Value	Quantity	Symbol	Value
Mass of the Earth	M_e	5.9763×10^{24} kg	Mean density	δ	5517.0 kg/m^3
Mean rotational velocity	U_{orb}	2.98 km/s	Sidereal rotational period	P	86 164.09 s
Sidereal rotational period about sun	P_{orb}	3.1558×10^7s	Angular rotational velocity	ω	7.29212×10^{-5} rad/s
Mean equatorial radius	R_{eq}	6 378 km	Mean equatorial rotational velocity	v	0.465 km/s
Average acceleration of gravity at equator	g_{eq}	9.78057 m/s^2	Moment of inertia	I	8.070×10^{42} kg m^2
Average acceleration of gravity at poles	g_p	9.83225 m/s^2	Moment of rotation	L	5.885×10^{33} kg m^2/s
Average acceleration of gravity on Earth	g	9.80665 m/s^2	Area of continents	S_c	1.49×10^8 km^2
Mean radius	R_e	6 370.949 km	Area of oceans	S_o	3.61×10^8 km^2
Area of surface	S_e	5.1005×10^8 km^2	Greatest height (Mount Everest)		8 840 m
Volume of the Earth	V_e	1.0832×10^{12} km^3	Greatest sea depth		10 924 m

There is a remedy for everything: it is called death. Portuguese proverb

6.3 Rock Types

Rocks are mixtures of minerals. Just as biologists classify living things, geologists classify rocks into groups. Rocks are classified on the basis of their origins (how they were formed). There are three major types: **igneous, sedimentary** and **metamorphic** rock.

Igneous rocks are made from molten magma that has cooled and solidified. Igneous rocks formed above the Earth's surface are called **extrusive**; igneous rocks that have solidified under the Earth's surface are called **intrusive**. Since liquid magma is rich in silica, igneous rocks are further subdivided according to their silica content (SiO_2). Classification based on silica content: >66% = acid; 52-66% = intermediate; 45-52% = basic; <45% = ultrabasic. Geologists' use of the terms "acidic" and "basic" do not refer to chemical properties, but to the archaic term "silicious acid."

Igneous Rock Types

	Acidic (Granite)	Intermediate (Andesitic)	Basic (Basaltic)	Ultrabasic (Ultramafic)
Extrusive (volcanic)	Rhyolite	Andesite	Basalt	*
Intrusive (pultonic)	Granite	Diorite	Gabbro	Peridotite
% Silica	>66%	52-66%	45-52%	<45%
Color	light	→ Increasing darkness		→ dark
Density	2.7	→ Increasing density		→ 3.3

* Extrusive ultrabasic rock is rarely found

Coarse-grained rocks:
1 Syenite **2** Granite
3 Granodiorite **4** Diorite
5 Gabbro **6** Peridotite

Pale and Dark minerals:
a Orthoclase feldspar
b Quartz
c Plagioclase feldspar
d Biotite **e** Amphibole
f Pyroxene **g** Olivine

Fine-grained rocks:
7 Trachyte **8** Rhyolite
9 Rhyodacite
10 Andesite **11** Basalt

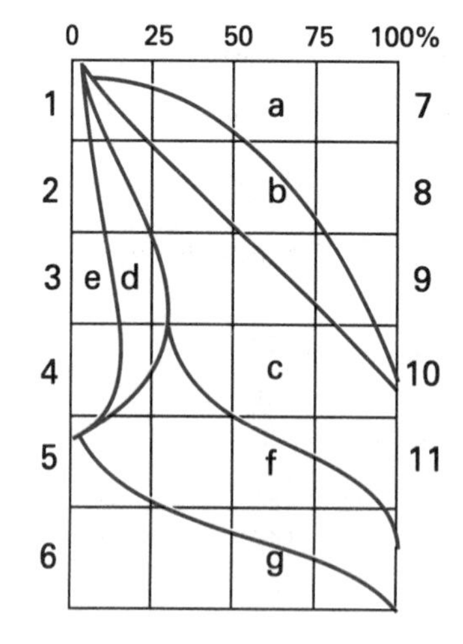

Fig. 6.3.1 Percentages of Minerals in Igneous Rocks

Silicates (minerals containing silica) make up igneous rock in different proportions. Common silcates are quartz, orthoclase feldspar, plagioclase feldspar, biotite, amphibole (hornblende), pyroxine (augite) and olivine.

Igneous rocks have a general composition of silicate minerals. For example, granite is typically 40% orthoclase feldspar, 25% quartz, 15% plagioclase feldspar, 15% biotite and 5% amphibole. Granite is also classified as acidic because the SiO_2 content of its minerals is greater than 66%.

Sedimentary rock covers approximately 3/4 of all land; it is formed from the deposits of many kinds of material. There are three basic types of sedimentary rock: **clastic** sediment, **precipitated chemical** sediment and **organic** sediment. In nature, most sedimentary rocks are mixtures of these three basic types.

Clastic (or detrital) sediment is made up of rock fragments that have become cemented firmly together. Geologists classify the size of the fragments according to the table on the right. **Precipitated chemical** sediment is made of inorganic chemical materials that were originally dissolved in water, but left the solution to form solid rock. **Organic** sediment is made up of the solidified remains and secretions of living organisms.

Rock Fragment Sizes

Name	Subdivisions	Particle size (mm)
Mud	Clay	smaller than 1/256
	Silt	1/256-1/16
Sand	Sand	1/16-2
Gravel	Granule	2-4
	Pebble	4-64
	Cobble	64-256
	Boulder	larger than 256

Divisions of Sedimentary Rock

Sediment type	Rock types	Rock subtypes	Information
Clastic	Conglomerate		gravel-size rounded rock fragments
	Breccia		gravel-size angular rock fragments
	Sandstone	Arkose	sand-size fragments of quartz and feldspar
		Graywacke	sand-size fragments with a dark color containing quartz and feldspar
		Quartz arenite	sand-size fragments of quartz
	Shale		mud-size fragments that typically split into thin layers
	Mudstone		mud-size fragments that break into clumps
Precipated	Evaporites	Halite	sodium chloride, $NaCl$
Chemical		Gypsum	calcium sulfate dihydrate, $CaSO_4 \cdot 2H_2O$
		Anhydrite	calcium sulfate, $CaSO_4$
		Borax	sodium borate decahydrate, $Na_2B_4O_7 \cdot 10H_2O$
		Potash	potassium chloride, KCl
	Carbonates	Limestone	calcium carbonate, $CaCO_3$
		Dolomite	dolomite, $CaMg(CO_3)_2$
Organic	Limestone		shell debris and fossil fragments
	Coal	Peat	plant remains that have become brown and soft
		Lignite	brown material formed when peat is subjected to heat and pressure
		Bituminous coal	black with few plant remains left visible
		Anthracite coal	high quality coal, black with a homogeneous composition
	Chert	Flint	gray rock formed from siliceous microscopic skeletons
		Jasper	reddish rock formed from siliceous microscopic skeletons and Fe_2O_3

Metamorphic rock is formed when environmental conditions cause changes in igneous, sedimentary or other metamorphic rocks. Typically, these changes are in grain size and/or mineral content, while the rock remains in its solid state. **Grade** is the term used to indicate the degree of change.

Factors causing metamorphism include temperature, pressure, and chemical environment. The classes of metamorphism depend on environmental factors. **Contact metamorphism** is caused by heat and chemical factors. **Regional metamorphism** is caused by a combination of heat, pressure and chemical factors.

Rock texture is the common method of classifying metamorphic rocks. **Foliated** rock is layered. **Nonfoliated** rock does not exhibit layers.

Common Metamorphic Rock Types

Rock texture	Rock name	Environment	Typical grade	Parent rock
Foliated	amphibole	regional	medium-high	basalt
	gneiss	regional	high	igneous or sedimentary
	phyllite	regional	medium	slate
	schist	regional	low-high	igneous or sedimentary
	slate	regional or contact	low	mudstone or shale
Nonfoliated	hornfels	contact	low-high	igneous or sedimentary
	quartzite	contact	low-high	sandstone
	marble	regional or contact	low-high	limestone or dolomite

The Rock Cycle

Under the right conditions, rocks change between igneous, sedimentary and metamorphic states in a cyclic fashion. The diagram below shows this cycle, as well as short cuts that also take place.

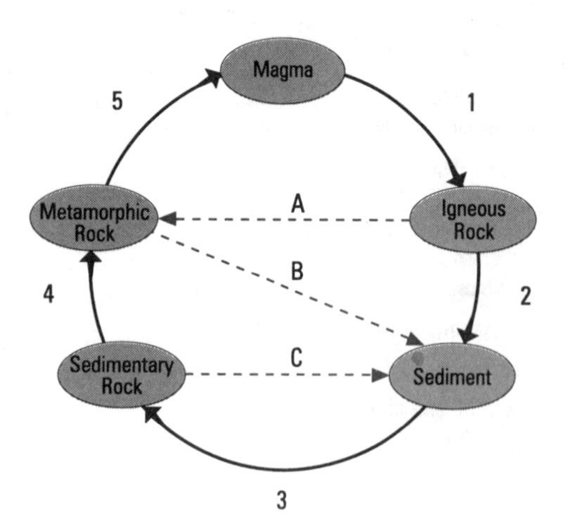

1) Molten magma cools and solidifies into igneous rocks.
2) Igneous rocks wear away into fragments after weathering and erosion. They are then transported and deposited into sediments.
3) Sediments become buried and harden into sedimentary rock (lithification).
4) High temperature and pressure cause sedimentary rock to change into metamorphic rock.
5) High temperature and pressure liquefy the metamorphic rock, reforming magma.

The Short Cuts

A) High heat and pressure convert igneous rock directly to metamorphic rock.
B) Metamorphic rock wears away into fragments and is transported and deposited into sediment.
C) Sedimentary rock wears away into fragments and is transported and deposited back into sediment.

I have no right to complain that God wished to sustain me in the world as a person who is not the principal and most perfect of all. René Descartes

6.4 A Key to Minerals

Minerals are combinations of chemical elements. They are the building blocks in the formation of rocks. Geologists use various methods to classify and identify minerals: chemical formula, color, cleavage (how they split), hardness (H), specific gravity (SG) and crystal structure (CS).

Key to crystal structure			
Isometric	1	Orthorhombic	4
Tetragonal	2	Monoclinic	5
Hexagonal	3	Triclinic	6
Amorphous	A		

Mohs hardness scale			
Hardness	Mineral	Hardness	Mineral
1	Talc	6	Orthoclase
2	Gypsum	7	Quartz
3	Calcite	8	Topaz
4	Fluorite	9	Corundum
5	Apatite	10	Diamond

Common Minerals

Mineral	Chemical formula	Cleavage	Color	SG	H	CS
Albite	See *Plagioclase Feldspar*					
Amphibole group	Silicates of Ca, Na, Mg, Fe, Al	two at 56° and 124°	black, green, brown	3.0-3.4	5-6	5
Andalusite	Al_2SiO_5	two fair at 89° and 91° uneven subchonchoidal fracture	white, gray, green, brown	3.1-3.2	7.5	4
Anorthite	See *Plagioclase Feldspar*					
Apatite	$Ca_5(PO_4)_3(F,Cl,OH)$	poor in one direction	colorless, white, various colors	3.1-3.2	5	3
Augite	See *Pyroxene group*					
Bauxite	Hydrated aluminun oxides	earthy fracture	white to dark reddish-brown	2.4-2.6	1-3	A
Beryl	$Be_3Al_2Si_6O_{18}$	poor basal cleavage	emerald green, pink, yellow, blue	2.6-2.8	7.5-8	3
Biotite	$K(Mg,Fe)_3(AlSi_3O_{10})(OH,F)_2$	in one direction	black, dark brown or dark green	2.8-3.4	2.5-3	5
Bornite	Cu_5FeS_4	uneven conchoidal fracture	red to brown	4.9-5.1	3	1
Calcite	$CaCO_3$	rhombohedral, three cleavages at 75°	colorless or white	2.7	3	3
Cassiterite	SnO_2	uneven fracture	black or brown	6.8-7.1	6-7	2
Chalcedony	SiO_2 (variety of quartz)	conchoidal fracture	white, gray, blue, red, brown, black	2.6	7	A
Chalcopyrite	$CuFeS_2$	poor in one direction	yellow	4.1-4.3	3.5-4	2
Chlorite	$(Mg,Fe)_6(AlSi_3)O_{10}(OH)_8$	in one direction	light to dark green	2.6-3.3	2-2.5	5

Common Minerals

Mineral	Chemical formula	Cleavage	Color	SG	H	CS
Chromite	$FeCr_2O_4$	uneven chonchoidal fracture	iron black	4.3-5.0	5.5	1
Chrysoberyl	$BeAl_2O_4$	distinct in one direction, poor in two others	yellow-green	3.5-3.8	8.5	4
Cinnabar	HgS	in three directions at 60° and 120°	vermillion red	8.0-8.2	2.5	3
Clay Minerals	See *Kaolinite*					
Corundum	Al_2O_3	no cleavage, but will part along rhombohedral planes	ruby: red; saphire: all others	3.9-4.1	9	3
Diamond	C	octahedral cleavage	colorless, white, pale shades	3.5	10	1
Dolomite	$CaMg(CO_3)_2$	rhombohedral, three cleavages at 75°	colorless, white, various pale colors	2.8-2.9	3.5-4.0	3
Epidote	Ca, Al and Fe silicates	in one direction lengthwise	yellow-green to dark brown	3.3-3.6	6-7	5
Feldspar	See *Orthoclase feldspar* and *Plagioclase feldspar*					
Fluorite	CaF_2	octahedral, in four directions	colorless or varied colors	3.0-3.2	4	1
Galena	PbS	three cleavages at 90°	dark gray	7.4-7.6	2.5	1
Garnet	group of related silicate minerals with Al, Ca, Cr, Fe, Mg, Mn, Ti	uneven conchoidal fracture	deep red, brown, yellow, green, black	3.5-4.3	6.5-7.5	1
Graphite	C	good in one direction	steel gray, black	2.3	1-2	3
Gypsum	$CaSO_4{\cdot}2H_2O$	in one direction, poor in two others	colorless, white, or light tints	2.3-2.4	2	5
Hematite	Fe_2O_3	uneven fracture	steel gray, red, black	4.9-5.3	5.0-6.5	3
Hornblende	See *Amphibole group*					
Kaolinite	$Al_2SiO_5(OH)_4$	earthy fracture	colorless, white, gray	2.6	2-2.5	5
Kyanite	Al_2SiO_5	pinacoidal cleavage	white to light blue	3.5-3.7	5-7	6
Labradorite	See *Plagioclase feldspar*					
Limonite	$FeO(OH){\cdot}n(H_2O)$	earthy or chonchoidal fracture	yellow to brown or black	2.7-4.3	4.0-5.5	A
Magnetite	Fe_3O_4	uneven subchonchoidal fracture	iron black	4.9-5.2	5.5-6.5	1
Malachite	$Cu_2CO_3(OH)_2$	in one direction	bright to dark green	3.9-4.0	3.5-4.0	5
Mica group	See *Biotite and Muscovite*					
Molybdenite	MoS_2	in one direction	bluish lead-gray	4.7-4.8	1.0-1.5	3
Muscovite	$KAl_3Si_3O_{10}(OH)_2$	in one direction	colorless, white, multi-colored	2.7-3.0	2-2.5	5

Thou call'dst me dog before thou hadst cause, but since I am a dog, beware my fangs. William Shakespeare

Common Minerals

Mineral	Chemical formula	Cleavage	Color	SG	H	CS
Olivine	$(Mg,Fe)_2SiO_4$	poor in two directions at 90° conchoidal fracture	yellowish green	3.2-4.3	6.5-7	4
Opal	$SiO_2 \cdot nH_2O$	chonchoidal fracture	colorless, white, red, yellow, green	1.9-2.2	5.5-6.5	A
Orthoclase feldspar	$KAlSi_3O_8$	two directions at 90°	colorless, white, red, yellow, green	2.5-2.6	6-6.5	5
Peridot	$(Mg,Fe)_2SiO_4$	poor in two directions at 90°	yellowish-green	3.2-3.5	7	4
Plagioclase feldspar	$NaAlSi_3O_8$ (Albite) $CaAl_2Si_2O_8$ (Anorthite)	two at 86°, poor in a third direction	white, colorless, gray	2.6-2.8	6	6
Pyrite	FeS_2	uneven fracture	light yellow	4.9-5.2	6-6.5	1
Pyroxene	silicates of Fe, Mg, Na, Ca, and Al	in two directions at nearly right angles	green, brown, black	3.2-3.6	5-6	5
Pyrrhotite	$Fe_{1-x}S(x=0\text{-}0.2)$	uneven fracture	yellow to brown	4.5-4.6	3.5-4.5	4
Quartz	SiO_2	rhombohedral cleavage	colorless, white, brown, violet, or rose	2.6	7	3
Ruby	See *Corundum*					
Saphire	See *Corundum*					
Serpentine	$(Mg,Fe)_3Si_2O_5(OH)_4$	splintery fracture	green, yellow, brown	2.5-2.6	3-5	5
Sillimanite	Al_2SiO_5	in one direction lengthwise	white, gray, brown	3.2-3.3	6-7	4
Sphalerite	ZnS	six directions at 60°	yellow, red, brown	3.9-4.1	3.5-4.0	1
Spinel	$MgAl_2O_4$	chonchoidal fracture	blue, brown, green, red, black	3.6-4.0	7.5-8	1
Staurolite	$Fe_2Al_9Si_4O_{22}(OH)_2$	poor in one direction lengthwise	yellowish-brown to reddish-brown	3.7-3.8	7	5
Sulfur	S	poor in two directions	light yellow	2.0-2.1	1.5-2.5	4
Sylvite	KCl	cubic cleavage	colorless, white, reddish blue	2.0	2	1
Talc	$Mg_3Si_4O_{10}(OH)_2$	in one direction	white to pale green	2.7-2.8	1	5
Topaz	$Al_2SiO_4(OH,F)_2$	perfect basal cleavage in one direction	colorless, white, yellow, violet, blue	3.4-3.6	8	4
Tourmaline	$Na(Mg,Fe)_3Al_6(BO_3)_3(Si_6O_{18})(OH,F)_4$	uneven conchoidal fracture	black, green, red, pink or blue	3.0-3.3	7-7.5	3
Turquoise	$CuAl_6(PO)_4(OH)_8 \cdot (4\text{-}5)H_2O$	chonchoidal fracture	blue, green	2.6-2.8	5-6	6
Wollastonite	$CaSiO_3$	two at nearly right angles	colorless, white, gray	2.8-2.9	4.5-5	6
Zircon	$ZrSiO_4$	two poor cleavages	colorless, gray, red	4.6-4.7	7	2

6.5 Crystal Systems

All crystals fit into one of six crystal systems, regardless of the infinite range of shapes possible. Each system has its own characteristics but may share some common forms.

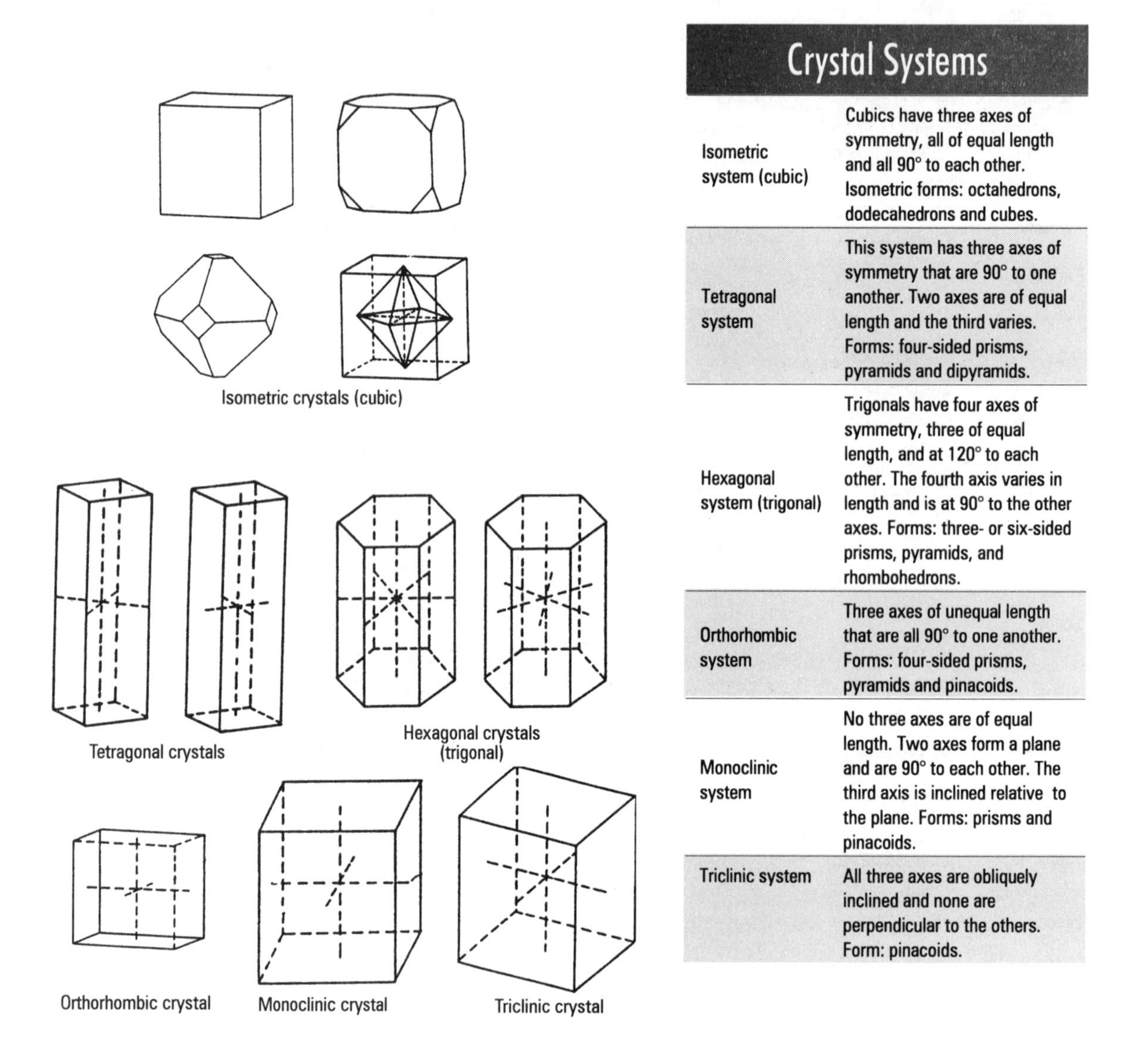

Crystal Systems	
Isometric system (cubic)	Cubics have three axes of symmetry, all of equal length and all 90° to each other. Isometric forms: octahedrons, dodecahedrons and cubes.
Tetragonal system	This system has three axes of symmetry that are 90° to one another. Two axes are of equal length and the third varies. Forms: four-sided prisms, pyramids and dipyramids.
Hexagonal system (trigonal)	Trigonals have four axes of symmetry, three of equal length, and at 120° to each other. The fourth axis varies in length and is at 90° to the other axes. Forms: three- or six-sided prisms, pyramids, and rhombohedrons.
Orthorhombic system	Three axes of unequal length that are all 90° to one another. Forms: four-sided prisms, pyramids and pinacoids.
Monoclinic system	No three axes are of equal length. Two axes form a plane and are 90° to each other. The third axis is inclined relative to the plane. Forms: prisms and pinacoids.
Triclinic system	All three axes are obliquely inclined and none are perpendicular to the others. Form: pinacoids.

Fig. 6.5.1 Crystal Systems

PHYSICS

7

7.1 Symbols in Physics

One thing you'll find when studying physics is that the same symbols are used to represent many different values. Use the following table as a quick reference. We do not include the subscripts that are used in the sections that follow. For a more complete list, see Section 2.4, SI Units, and Unit 1, Constants for Science.

Symbols in Physics	
'	final value
0	initial value
θ	angle
α	angular acceleration
δ	phase angle
ε_0	permittivity of free space (8.85×10^{-12} $C^2/N\ m^2$)
$\mathscr{E}_{mf}$	electromotive force, emf
ϕ	electric flux, magnetic flux
λ	wavelength, line charge density
μ	mass per unit length, magnetic moment, permeability of free space (1.26×10^{-26} N/A^2)
π	3.14
ρ	density, volume charge density
σ	conductivity constant, surface charge density, Stefan's constant (5.67×10^{-8} W/m^2K^4)
τ	torque
ω	angular velocity
A	amplitude, area
a	acceleration
a_c	centripetal acceleration
a_t	tangential acceleration
B	magnetic field, bulk modulus
C	capacitance, drag coefficient
c	specific heat, speed of light in a vacuum (3.00×10^8 m/s)
d	distance
E	electric field, energy
e	2.71, emissivity
F	force
f	focal length, frequency
G	gravitational constant (6.67×10^{-11} $N\ m^2/kg^2$)
g	average acceleration due to gravity on Earth ($9.81\ m/s^2$)

Genius is eternal patience. Michelangelo

Symbols in Physics (cont'd.)

H	enthalpy, rate of heat transfer
h	height, Planck's constant (6.63 x 10^{-34} J s)
I	moment of inertia, impulse, electric current, intensity
i	unit vector in x axis, image distance
j	unit vector in y axis
J	current density
k	unit vector in z axis, Boltzmann's constant (1.38 x 10^{-23} J/K), Coulomb's constant (9.0 x 10^9 N m^2/C^2), Biot-Savart constant (10^{-7} T m/A), spring constant, dielectric constant
L	angular momentum, length, inductance
l	length
M	total mass
m	mass
o	object distance
P	power, pressure, power of a lens in diopters
p	linear momentum, rate of heat loss by radiation
Q	electric point charge, heat
q	electric point charge
R	resistance, gas constant (8.31 J/K mol)
r	radius, distance separating points
S	entropy
s	displacement, arclength
T	temperature, period
t	time
U	internal energy
V	volume, voltage
v	velocity
W	work
w	weight

7.2 Motion

Symbols

$_0$ = initial value
θ = angle
θ = unit vector tangent to circular motion
ϕ = angle
α = angular acceleration
τ = torque
ω = angular velocity
a = acceleration
$\mathbf{a_c}$ = centripetal acceleration
$\mathbf{a_t}$ = tangential acceleration
F = force
g = acceleration of gravity (9.81 m/s²)
I = moment of inertia
i = unit vector in *x* axis
j = unit vector in *y* axis
k = unit vector in *z axis*
L = angular momentum
m = mass
M = total mass
p = linear momentum
r = unit vector pointing away from center
r_c = position of center of mass
s = displacement
t = time
v = velocity
v_c = velocity of center of mass
x = position on *x* axis

Motion in One Dimension

Velocity (m / s)

Speed (the magnitude of velocity)
average speed = $\Delta s / \Delta t$

Average Velocity

$\bar{v} = \Delta x / \Delta t$

$\bar{v} = 1/2(v_o + v)$

Velocity with Constant Acceleration

$v = v_0 + at$

$v^2 = v_0^2 + 2a\Delta x$

$v = \sqrt{v_0^2 + 2a\Delta x}$

Instantaneous Velocity

$$v = \lim_{\Delta t \to 0} \frac{\Delta x}{\Delta t} = \frac{dx}{dt}$$

Velocity Integral

$$v = v_0 + \int a\,dt$$

Acceleration (m / s²)

Average Acceleration

$\bar{a} = \Delta v / \Delta t$

Instantaneous Acceleration

$$a = \lim_{\Delta t \to 0} \frac{\Delta v}{\Delta t} = \frac{dv}{dt} = \frac{d^2 x}{dt^2}$$

Displacement (m)

Displacement with Constant Acceleration

$x = x_0 + v_0 t + 1/2\,at^2$

$x = x_0 + 1/2(v_0 + v)t$

$\Delta x = v_0 t + 1/2\,at^2$

$\Delta x = 1/2(v_0 + v)t$

$$\Delta x = \frac{v^2 - v_0^2}{2a}$$

Displacement Integral

$$x = x_0 + \int v\,dt = x_0 + v_0 t + \int at\,dt$$

Refresher for Vectors in Two Dimensions

Components of Vector *A*

$A = (A_x, A_y)$

$A_x = A\cos\theta$

$A_y = A\sin\theta$

Magnitude of Vector A

$A = \sqrt{A_x^2 + A_y^2}$

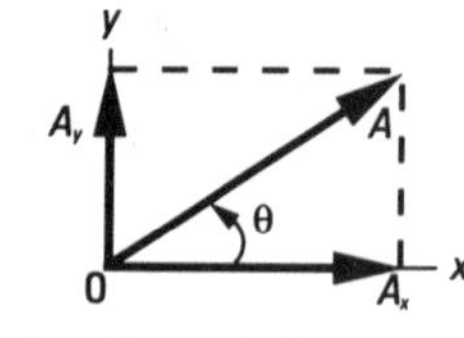

Fig. 7.2.1 Vectors in Two Dimensions

Angle Between *A* and *x* Axis

$\tan\theta = A_y / A_x$

$\theta = \arctan A_y / A_x$

Motion in Two Dimensions

Angle Determination
displacement $\tan\theta_s = y/x$
velocity $\tan\theta_v = v_y / v_x$
acceleration $\tan\theta_a = a_y / a_x$

Velocity (m/s)
$\boldsymbol{v} = (v_x, v_y) = v_x\mathbf{i} + v_y\mathbf{j}$

Average Velocity
$\bar{\boldsymbol{v}} = \Delta\boldsymbol{s} / \Delta t$

Magnitude of Velocity
$v = \sqrt{v_x^2 + v_y^2}$

Instantaneous Velocity
$$\boldsymbol{v} = \lim_{\Delta t \to 0} \frac{\Delta \boldsymbol{s}}{\Delta t} = \frac{d\boldsymbol{s}}{dt}$$

Velocity Integral
$$\boldsymbol{v} = \boldsymbol{v}_0 + \int \boldsymbol{a}\, dt$$

Velocity (constant acceleration)
$\boldsymbol{v} = \boldsymbol{v}_0 + \boldsymbol{a}t$

Acceleration (m/s^2)
$\boldsymbol{a} = (a_x, a_y) = a_x\mathbf{i} + a_y\mathbf{j}$

Average Acceleration
$\bar{\boldsymbol{a}} = \Delta\boldsymbol{v}/\Delta t$

Instantaneous Acceleration
$$\boldsymbol{a} = \lim_{\Delta \to 0} \frac{\Delta v}{\Delta t} = \frac{dv}{dt} = \frac{d^2\boldsymbol{s}}{dt^2}$$

Magnitude of Acceleration
$a = \sqrt{a_x^2 + a_y^2}$

Displacement (m)
$\boldsymbol{s} = (x, y) = x\mathbf{i} + y\mathbf{j}$

Components of Displacement
(constant acceleration)
$x = x_0 + v_{x0}t + \frac{1}{2}a_x t^2$
$y = y_0 + v_{y0}t + \frac{1}{2}a_y t^2$

Displacement Integral
$$\boldsymbol{s} = \boldsymbol{s}_0 + \int \boldsymbol{v}\, dt = s_0 + \boldsymbol{v}_0 t + \int \boldsymbol{a}t\, dt$$

Displacement (constant acceleration)
$\boldsymbol{s} = \boldsymbol{s}_0 + \boldsymbol{v}_0 t + \frac{1}{2}\boldsymbol{a}t^2$

Projectile Motion

Vertical Displacement (m)
$y = y_0 + v_0(\sin\theta_0)t - \frac{1}{2}gt^2$
$y = y_0 + v_{y_0}t - \frac{1}{2}gt^2$

Projected From the Surface $(y_0 = 0)$
$y = v_0(\sin\theta_0)t - \frac{1}{2}gt^2$
$y = v_{y_0}t - \frac{1}{2}gt^2$

Dropped Object
$\Delta y = -\frac{1}{2}gt^2$

Maximum Height
$$h = \frac{v_0^2 \sin^2\theta_0}{2g}$$

Horizontal Displacement (m)
$x = v_{x_0}t = v_0(\cos\theta_0)\,t$

Maximum Range
$R_x = v_0(\cos\theta_0)t_{max}$
$$R_x = \frac{v_0^2 \sin 2\theta_0}{g}$$
$$R_x = \frac{2v_0^2 \sin\theta_0 \cos\theta_0}{g}$$

Velocity (m/s)
$v_x = v_{x0} = v_0 \cos\theta_0$
$v_y = v_{y0} - gt = v_0(\sin\theta_0) - gt$

Time(s)

Total Time
$$t_{max} = \frac{2v_0 \sin\theta_0}{g}$$

Time to Maximum Height
$$t_{1/2} = \frac{v_0 \sin\theta_0}{g}$$

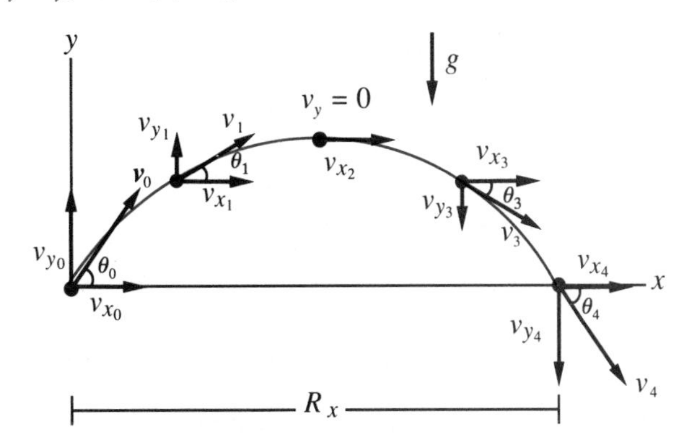

Fig 7.2.2 Projectile Vectors

I did not attend his funeral; but I wrote a nice letter saying I approved of it. Mark Twain

Motion in Three Dimensions

Velocity (constant acceleration) (m/s)

$\boldsymbol{v} = \boldsymbol{v}_0 + \boldsymbol{a}t = (vx, vy, vz)$

$v_x = v_{x0} + a_x t$

$v_y = v_{y0} + a_y t$

$v_z = v_{z0} + a_z t$

Displacement (constant acceleration) (m)

$\boldsymbol{s} = \boldsymbol{s}_0 + \boldsymbol{v}_0 t + ½\boldsymbol{a}t^2 = (x,y,z)$

$x = x_0 + v_{x0}t + ½a_x t^2$

$y = y_0 + v_{y0}t + ½a_y t^2$

$z = z_0 + v_{z0}t + ½a_z t^2$

Spherical coordinates (A, θ, ϕ)

$x = A \sin\theta \cos\phi$

$y = A \sin\theta \sin\phi$

$z = A \cos\theta$

$A = \sqrt{x^2 + y^2 + z^2}$

$\tan\theta = \dfrac{\sqrt{x^2 + y^2}}{z}$

$\tan\phi = y/x$

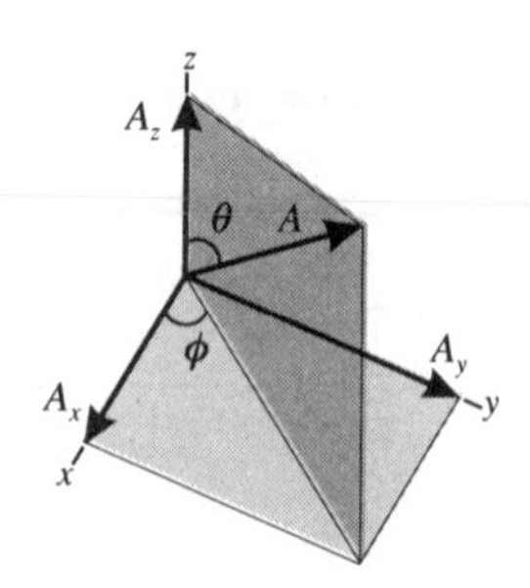

Fig 7.2.3 Vector in Three Dimensions

Circular Motion

Total Acceleration (m/s^2)

$\boldsymbol{a} = \boldsymbol{a}_c + \boldsymbol{a}_t$

Centripetal Acceleration (m/s^2)

$a_c = \dfrac{v^2}{r}$

$\boldsymbol{a}_c = -\dfrac{v^2}{r}\boldsymbol{r}$

Tangential Acceleration (m/s^2)

(related to the change in magnitude of velocity)

$a_t = \dfrac{dv}{dt}$

$\boldsymbol{a}_t = \dfrac{dv}{dt}\theta$

Rotational Motion of Rigid Bodies

A **rigid body** is an object or system of particles whose shape doesn't change. A rigid body will rotate about a fixed axis known as the **axis of rotation**.

Angular Displacement (m rad)

$s = r\theta$

Angular Position (rad)

θ (rad) = $(\pi/180°)\,\theta$ degrees

$\theta = s/r$

$*s$ = arclength

Angular Position with Constant Angular Acceleration

$\theta = \theta_0 + \omega_0 t + ½\,\alpha t^2$

$\theta = \theta_0 + ½(\omega + \omega_0)t$

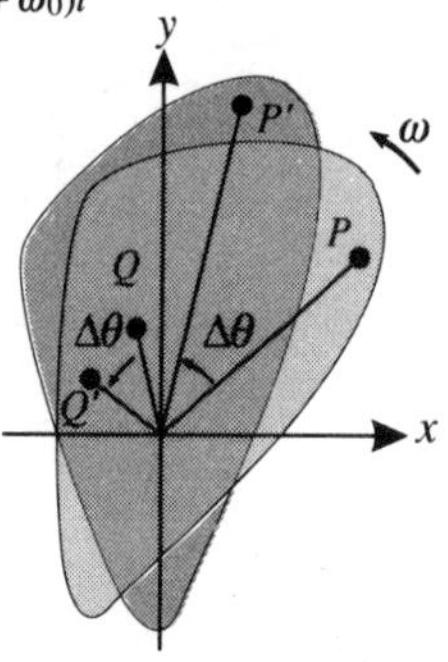

Fig 7.2.4 Rigid Body Rotation

Angular Velocity (rad/s)

Average Angular Velocity

$\bar{\omega} = \Delta\theta/\Delta t$

Instantaneous Angular Velocity

$\omega = \lim_{\Delta t \to 0} \Delta\theta/\Delta t = d\theta/dt$

Angular Velocity with Constant Angular Acceleration

$\omega = \omega_0 + \alpha t$

$\omega^2 = \omega_0^2 + 2\alpha\Delta\theta$

Angular Acceleration (rad/s^2)

Average Angular Acceleration

$\alpha = \Delta\omega/\Delta t$

Instantaneous Angular Acceleration

$\alpha = \lim_{\Delta t \to 0} \Delta\omega/\Delta t = d\omega/dt$

Other Useful Formulas

Center of Mass
$\boldsymbol{r}_c = (x_c, y_c, z_c)$

Center of Mass for a Group of Particles
$$x_c = \frac{\Sigma m_i x_i}{\Sigma m_i},\ y_c = \frac{\Sigma m_i y_i}{\Sigma m_i},\ z_c = \frac{\Sigma m_i z_i}{\Sigma m_i}$$

Center of Mass for a Rigid Body (m)
$$\boldsymbol{r}_c = \frac{1}{M}\int \boldsymbol{r}\,dm$$

Velocity of Center of Mass (m / s)
$$\boldsymbol{v}_c = \frac{d\boldsymbol{r}_c}{dt} = \frac{\Sigma m_i \boldsymbol{v}_i}{M}$$

Acceleration of Center of Mass (m / s^2)
$$\boldsymbol{a}_c = \frac{d\boldsymbol{v}_c}{dt} = \frac{\Sigma m_i \boldsymbol{a}_i}{M}$$

Moment of Inertia (kg m^2)
$$I = \int r^2\,dm$$
$$I = \Sigma m_i r_i^2$$

Angular Momentum (kg m^2 / s)
$L = mvr \sin\theta$
$L = \boldsymbol{r} \times \boldsymbol{p}$
$L = \boldsymbol{r} \times m\boldsymbol{v}$
$L = I\omega$

Torque Given to a Rigid Body (J)
$\tau = rF \sin\theta$
$\tau = \boldsymbol{r} \times \boldsymbol{F}$
$\tau = I\alpha$
$\tau = d\boldsymbol{L} / dt$
$\bar{\tau} = \Delta\boldsymbol{L} / \Delta t$

Power Given to a Rigid Body (W)
$P = \tau\omega$

Work Done by External Forces to Rotate a Rigid Body (J)
$$W = \int \tau\,d\theta = \tfrac{1}{2}I\omega^2 - \tfrac{1}{2}I\omega_0^2$$

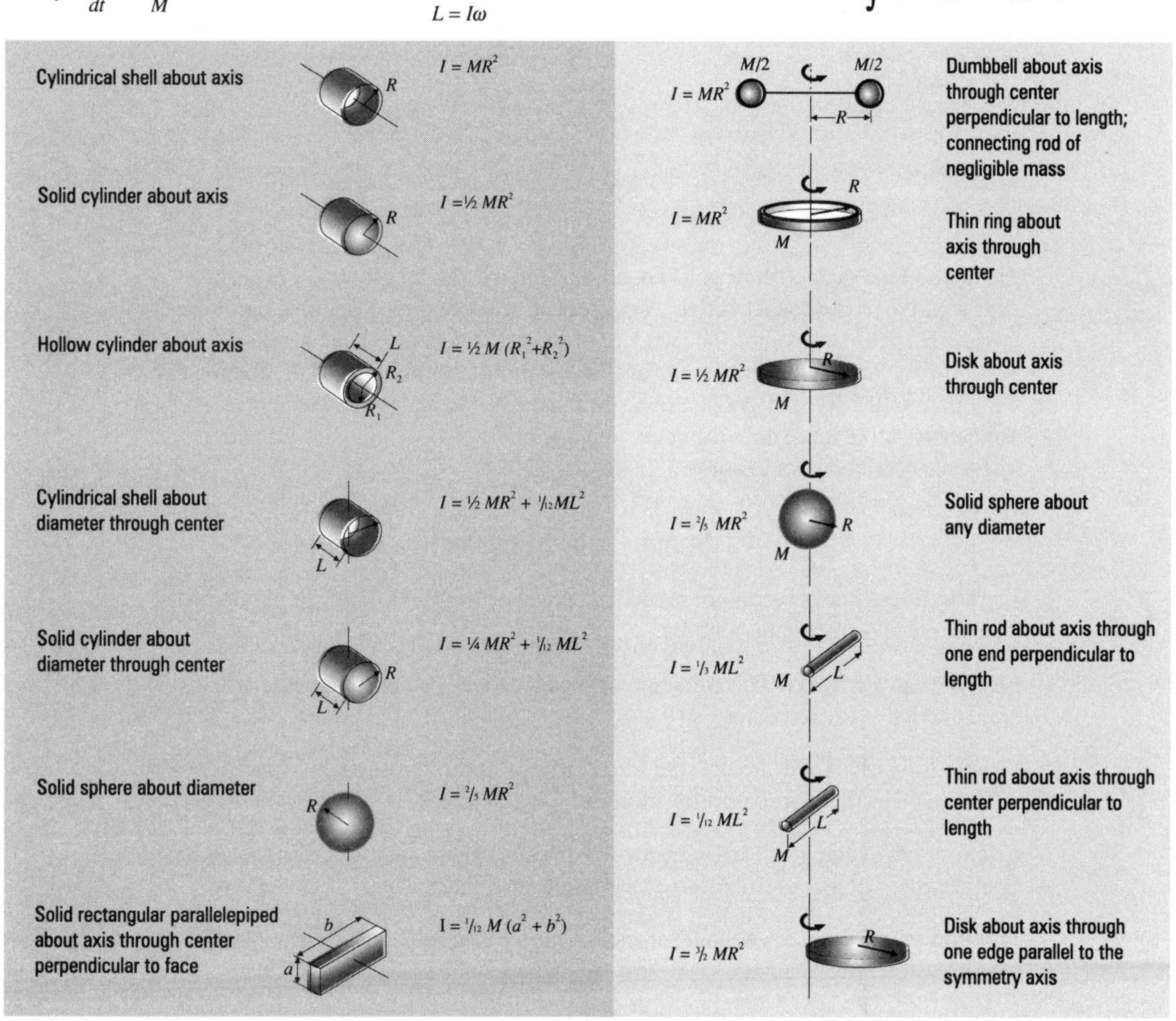

Fig 7.2.5 Moments of Inertia in Uniform Objects

7.3 Force

Symbols

ℓ = a vector whose magnitude is the length of the wire
ℓ_Y = length of wire
θ = unit vector
θ = angle
ρ = density of air
A = cross sectional area
a_c = centripetal acceleration
a_t = tangential acceleration
B = magnetic field vector
C = drag coefficient
C_Y = circumference of ring
E = electric field
F = force
F_{av} = time averaged force
F_c = centripetal acceleration
F_N = normal force
F_t = tangential acceleration
G = gravitational constant (6.67×10^{-11} N m^2/kg^2)
g = average acceleration due to gravity on Earth (9.81 m/s^2)
I = impulse
I_e = electric current
k_e = Coulomb's constant (9.0×10^9 N m^2/C^2)
k_s = spring constant
m = mass
M_e = mass of Earth
p = momentum
q = electric charge
r = unit vector
r = radius
r_{ab} = distance between two objects' gravity centers
r_e = radius of earth
t = time
v = velocity
w = weight

Think of a **force** as a push or pull. The unit of force is the **Newton**, N. Forces can be **conservative** or **nonconservative**. A **conservative force** does work on a particle moving between two points, independent of the path taken (see Mechanical Energy, Section 7.4). When the particle returns to the starting position (closed path), the total work is zero (the force of gravity, or a spring attached to a mass, for example). A **nonconservative force** does different amounts of work, depending on the path taken (the force of friction, for example).

Some Types of Forces

Electric force. The attraction or repulsion between electric charges. Similar charges repel while opposite charges attract.

Electromagnetic force. The attraction or repulsion between two charged particles moving relative to each other.

Fictitious forces. These are not true forces! They appear to exist in rotating frames of reference (centrifugal force, coriolis force).

Friction. A resistive force arising between objects that are sliding across each other. Friction arises from molecular forces in regions of contact.

Fundamental forces. These four forces in nature – electromagnetic, gravitational, strong nuclear, weak nuclear – can explain (at the atomic level) all other forces.

Ask yourself whether you are happy, and you cease to be so. John S. Mill

Gravitational force. The force of attraction between particles with mass.

Lorentz force. The force acting on a charged particle moving through a magnetic field.

Magnetic forces. Either the force of attraction or repulsion between magnetic poles, or the force arising from the motion of an electric charge.

Molecular forces. Attractive electromagnetic forces that bind atoms together in a molecule (see Section 5.5).

Normal force. A force that counterbalances the gravitational force. (It keeps you from falling through the floor and being sucked to the center of the Earth.)

Strong nuclear forces. The force responsible for holding an atom's nucleus together. The force is charge-independent, attractive at short range (10^{-15} m), and repulsive at very short range (4×10^{-16} m).

Tension. The force that a rope exerts on an object.

Weak nuclear forces. A short range nuclear force (10^{-15} m) causing instability in atoms; responsible for β decay (see Section 5.10).

Newton's Laws

Newton's First Law

A body remains at rest or in constant uniform motion in a straight line unless a force acts upon it. This resistance to changes in motion is known as **inertia**.

$$\text{If } \sum \boldsymbol{F} = 0, \text{ then } \boldsymbol{a} = 0$$

Newton's Second Law

A body accelerates in the direction of the resultant force upon it. The momentum of a body changes with time according to the resultant force upon it.

$$\boldsymbol{F} = m\boldsymbol{a}$$
$$\boldsymbol{F} = \Delta(mv)/\Delta t$$

Newton's Third Law

When a body applies a force upon a second body, the second body applies an equal and opposite force upon the first.

$$\boldsymbol{F}_{ab} = -\boldsymbol{F}_{ba}$$

Resultant Force

The **resultant force** is the sum of all forces acting upon a body. When a body receives two forces simultaneously it follows the diagonal of the parallelogram formed by the two force vectors. The time it takes for motion along the diagonal equals the time it would take to move along each force vector separately.

Formulas to Calculate the Resultant Force

Resultant force

$$\sum \boldsymbol{F} = \boldsymbol{F}_1 + \boldsymbol{F}_2 + \cdots + \boldsymbol{F}_n = m_1\boldsymbol{a}_1 + m_2\boldsymbol{a}_2 + \cdots \; m_n\boldsymbol{a}_n$$

In three dimensions

$$\sum F = \left(\sum F_x, \sum F_y, \sum F_z\right) = \left(\sum F_x\right)\mathbf{i} + \left(\sum F_y\right)\mathbf{j} + \left(\sum F_z\right)\mathbf{k}$$

$$\sum F_x = F_{x_1} + F_{x_2} + \cdots + F_{x_n} = m_1 a_{x_1} + m_2 a_{x_2} + \cdots + m_n a_{x_n}$$

$$\sum F_y = F_{y_1} + F_{y_2} + \cdots + F_{y_n} = m_1 a_{y_1} + m_2 a_{y_2} + \cdots + m_n a_{y_n}$$

$$\sum F_z = F_{z_1} + F_{z_2} + \cdots + F_{z_n} = m_1 a_{z_1} + m_2 a_{z_2} + \cdots + m_n a_{z_n}$$

The Force of Gravity and Weight

Weight is the force of **gravity** on a body. Gravitational force does not disappear when an object is at rest on a surface. Gravitational force is balanced by the **normal force, $\boldsymbol{F}_{\mathrm{N}}$**.

The force of gravity on a body

$\boldsymbol{F}_g = \boldsymbol{w}$

$\boldsymbol{F}_g = m\boldsymbol{g}$

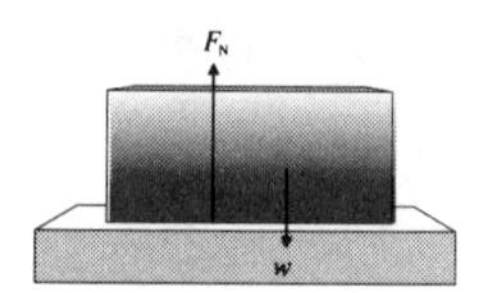

Fig. 7.3.1 Normal Force (F_{N}), Gravitational Force (w)

The Force of Friction

Friction is a force that resists motion between two bodies in contact. It is proportional to the normal force and depends on the materials in contact.

Static friction is the force that resists initiating motion.

μ_s = coefficient of static friction (depends on materials in contact)

$$F_s \leq \mu_s F_{\mathrm{N}}$$

Kinetic friction is the force that resists ongoing motion.

μ_k = coefficient of kinetic friction (depends on materials in contact)

$$F_k = \mu_k F_{\mathrm{N}}$$

Lawyers are like nuclear bombs: you only have them because the other side does and once you use them you can kiss everything goodbye.

The Free Body Diagram

Free body diagrams help you to picture the forces at work on each body in a physics problem. Here are a few tips on drawing free body diagrams, with examples below:

1) Isolate the body you want information on.
2) Draw it along the relevant geometrical axis (keep it simple).
3) Draw all force vectors acting on the body as arrows to size.
4) Label all forces.
5) Break these force vectors into their components (along axis).
6) Use the diagram to predict the answer, then work on solving the problem.

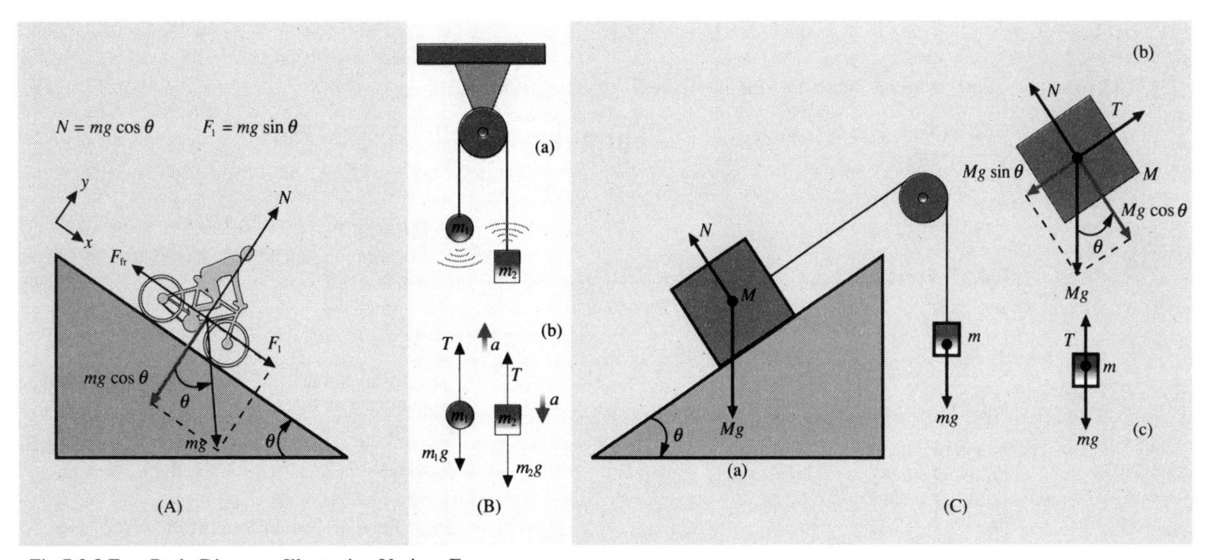

Fig 7.3.2 Free Body Diagrams Illustrating Various Forces

Women fall in love through their ears and men through their eyes. Woodrow Wyatt

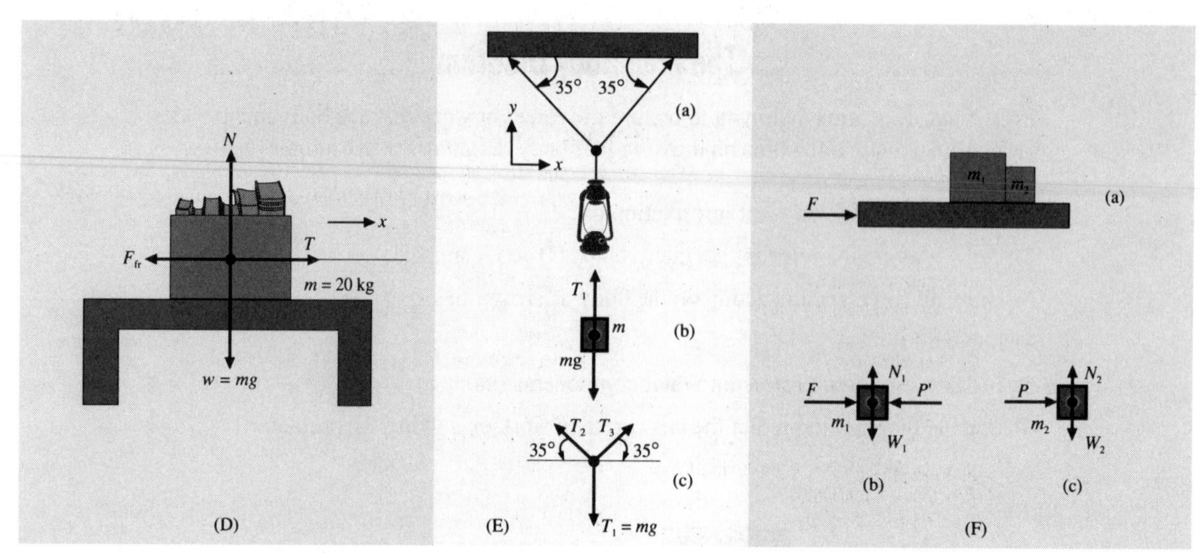

Fig. 7.3.2 (cont.) Free Body Diagrams Illustrating Various Forces

Formulas

Gravitational Forces

$$F_g = \frac{Gm_a m_b}{r_{ab}^2}$$

$$\boldsymbol{F}_g = \frac{-Gm_a m_b}{r_{ab}^2}\boldsymbol{r}_{ab}$$

Electric Force

$$F_e = \frac{k_e q_a q_b}{r_{ab}^2}$$

$$\boldsymbol{F}_e = k_e \frac{q_a q_b}{r_{ab}^2}\boldsymbol{r}_{ab}$$

$$\boldsymbol{F}_e = q\boldsymbol{E}$$

Magnetic Forces

$$F_m = qvB \sin\theta$$

$$F_m = I_e\, \ell\, B \sin\theta$$

$$\boldsymbol{F}_m = q\boldsymbol{v} \times \boldsymbol{B}$$

$$\boldsymbol{F}_m = I\boldsymbol{\ell} \times \boldsymbol{B}$$

$$\boldsymbol{F}_m = I_e \int d\boldsymbol{\ell} \times \boldsymbol{B}$$

Lorenz Force

$$\boldsymbol{F}_L = q\boldsymbol{E} + q\boldsymbol{v} \times \boldsymbol{B}$$

Hooke's Law

Forces given by a spring

$$F_s = -k_s \Delta x$$

Force and Momentum

$$\boldsymbol{F} = \frac{d\boldsymbol{p}}{dt} = m\frac{d\boldsymbol{v}}{dt}$$

Impulse of a Force

$$\boldsymbol{I} = \Delta\boldsymbol{p}$$

$$\boldsymbol{I} = \int \boldsymbol{F}\, dt$$

$$\boldsymbol{I} = \boldsymbol{F}_{av}\Delta t$$

Time Averaged Force

$$\boldsymbol{F}_{av} = \frac{1}{\Delta t}\int \boldsymbol{F}\, dt$$

Centripetal Force

Uniform circular motion directed toward the center

$$\boldsymbol{F}_c = m\boldsymbol{a}_c$$

$$\boldsymbol{F}_c = -\frac{mv^2}{r}\boldsymbol{r}$$

Tangential Force

Nonuniform motion tangent to the circular motion

$$\boldsymbol{F}_t = m\boldsymbol{a}_t$$

$$\boldsymbol{F}_t = m\frac{d|r|}{dt}\theta$$

Total Force in Nonuniform Motion

$$\boldsymbol{F} = \boldsymbol{F}_c + \boldsymbol{F}_t$$

Terminal Velocity in Air

$$v_T = \sqrt{\frac{2mg}{C\rho A}}$$

Earth's Escape Velocity

$$v_{esc} = \sqrt{\frac{2GM_e}{r_e}}$$

Air Drag

$$R_A = -\tfrac{1}{2}C\rho A v^2$$

Surface Tension

$$\gamma = F / 2C_y \text{ (ring)}$$

$$\gamma = F / 2\ell_y \text{ (wire)}$$

7.4 Work, Energy & Momentum

Symbols	
$'$ = final	$\mathbf{k}$ = unit vector in z axis
$_0$ = initial	k_e = Coulomb's constant (9.0 x 10^9 N m^2/C^2)
τ = torque	k_s = spring constant
θ = angle	m = mass
ω = angular velocity	M = mass of block
c = speed of light in a vacuum (3.00 x 10^8 m/s)	p = momentum
	P = power
E_k = kinetic energy	q = electric charge
E_{mec} = mechanical energy	R = resistance
E_p = potential energy	r = radius
E_{pg} = gravitational potential energy	r_{ab} = distance between a and b
g = acceleration due to gravity (9.81 m/s^2)	s = displacement
G = gravitational constant (6.67 x 10^{-11} N m^2/kg^2)	t = time
	V = voltage
h = height	v_b = speed of block and bullet
I = impulse	v = velocity
I_e = electric current	W = work
I_m = moment of inertia	W_c = work by conservative force
$\mathbf{i}$ = unit vector in x axis	W_{nc} = work by nonconservative force
$\mathbf{j}$ = unit vector in y axis	

Energy is a term that is not easy to define. Here, we'll define it as the ability to do **work**. Bear in mind, though, that not all energy is available to do work (some heat is not available). The total energy in the universe is constant. Energy is neither created nor destroyed. We say that energy is **conserved** because it can be transformed into other forms with the total energy remaining constant. The unit of energy is the joule, J. Common types of energy include chemical, electromagnetic, kinetic, mechanical, nuclear, potential and thermal.

Kinetic Energy

Kinetic energy is the energy of motion.

$$E_k = \tfrac{1}{2}mv^2$$

Kinetic Energy Relationships

$E_k = \tfrac{1}{2}mv^2$

$E_k = p^2/2m$

Rotational Kinetic Energy

$E_k = \tfrac{1}{2}I\omega^2$

Relativistic Kinetic Energy
(at speeds comparable to the speed of light)

$$E_k = mc^2\left(\frac{1}{\sqrt{1-(v/c)^2}} - 1\right)$$

Change in Kinetic Energy

$\Delta E_k = \tfrac{1}{2}mv^2 - \tfrac{1}{2}mv_o^2$

Potential Energy

Potential energy is stored energy.

Gravitational Potential Energy

$E_{pg} = mgh$ (assuming $E_{pg} = 0$, at $h = 0$)

$E_{pg} = -\frac{Gm_a m_b}{r_{ab}}$

Elastic Potential Energy

$E_{ps} = \frac{1}{2} k_s x^2$

Electric Potential Energy

$E_{pe} = \frac{k_e q_a q_b}{r_{ab}}$

Mechanical Energy

The total **mechanical energy** of a body is equal to the sum of the kinetic and potential energy.

$$E_{mec} = E_k + E_p$$

When **conservative forces** are involved, mechanical energy is conserved (see Section 7.3). This is an example of conservation of energy:

$$\Delta E_k + \Delta E_p = 0$$
$$\Delta E_k = -\Delta E_p$$

Conservation of mechanical energy in a free falling object:

$$\frac{1}{2}mv_0^2 + mgy_0 = \frac{1}{2}mv^2 + mgy$$

The total mechanical energy in a mass-spring system:

$$E_{mec} = \frac{1}{2}mv_0^2 + \frac{1}{2}k_s x_0^2 = \frac{1}{2}mv^2 + \frac{1}{2}k_s x^2$$

Work

The **work** done on an object is the product of the component of force in the direction of displacement and displacement.

Work - Constant Force

$W = (F \cos\theta)\, s$

$W = \boldsymbol{F} \cdot \boldsymbol{s}$ (dot product)

Work - Varying Force
(one dimension)

$W = \lim_{\Delta x \to 0} \sum F_x \Delta x = \int F_x dx$

Work - Variable Force
(more than one dimension)

$W = \int \boldsymbol{F} \cdot d\boldsymbol{s}$

$W = \int F_x dx\, \mathbf{i} + \int F_y dy\, \mathbf{j} + \int F_z dz\, \mathbf{k}$

Work and Rotational Motion

$W = \tau\theta$

Work Done by the Force of Gravity

$W_g = -\Delta E_{pg} = -\Delta mgh = mgh_0 - mgh$

Work-Energy Theorem

The **work-energy theorem** describes the relationship between work and mechanical energy. The work done by all conservative forces on a body equals the change in kinetic energy. Work done by all nonconservative forces on a body equals the total mechanical energy change.

$$W_c = \Delta E_k = -\Delta E_p$$
$$W_c = \frac{1}{2}mv^2 - \frac{1}{2}mv_0^2$$

$$\Delta E_k = W_{nc} + W_c$$
$$W_{nc} = \Delta E_{mec} = \Delta E_k + \Delta E_p$$

Work Done by a Spring

Conservative work done by a spring attached to a mass:

$$W_c = \tfrac{1}{2}k_s x_0^2 - \tfrac{1}{2}k_s x^2$$

Nonconservative forces acting on a spring:

$$W_{nc} = (\tfrac{1}{2}mv^2 + \tfrac{1}{2}kx^2) - (\tfrac{1}{2}mv_0^2 + \tfrac{1}{2}kx_0^2)$$

Power

Power is the rate of energy transfer per unit of time. The unit of power is the watt, W. (Note: In formulas below, W = work.)

$$P = \frac{dW}{dt} = \mathbf{F} \cdot \mathbf{v}$$

Average Power

$$\overline{P} = \frac{\Delta W}{\Delta t}$$

Instananeous Power

$$P = \lim_{\Delta t \to 0} \frac{\Delta W}{\Delta t}$$

$$P = \frac{dW}{dt}$$

Power Given to a Rigid Body

$$P = \tau\omega$$

Power Lost in a Resistor

$$P = I_e V$$

$$P = I_e^2 R$$

$$P = \frac{V^2}{R}$$

Momentum

Linear momentum is the product of mass and velocity for a body. The linear momentum for a system of bodies as they collide into one another is conserved.

$$\mathbf{p}_1 + \mathbf{p}_2 + \ldots + \mathbf{p}_n = \mathbf{p}_{1'} + \mathbf{p}_{2'} + \ldots + \mathbf{p}_{n'}$$

The total initial and final momentum in a system of bodies are equal.

Angular momentum is the cross-product of a body's position with respect to the axis of rotation and its linear momentum.

Linear Momentum

$$p = mv$$

Angular Momentum (L) of a Rigid Body

$$L = mvr \sin\theta$$

$$\mathbf{L} = \mathbf{r} \times \mathbf{p} = \mathbf{r} \times m\mathbf{v} = I_m\omega$$

$$\Delta L = \tau\Delta t$$

Impulse

Impulse is the change in momentum caused by a force. If one force is much greater, and acts for a brief period of time, it is often used for the purpose of **impulse approximation**.

$$\mathbf{I} = \int \mathbf{F}\, dt = \Delta \mathbf{p}$$

Impulse When Force Is Constant

$$\mathbf{I} = \Delta \mathbf{p}$$

$$\mathbf{I} = \mathbf{F}\Delta t$$

Time Averaged Force

Used as an aid in calculating impulse given by a force.

$$\overline{\mathbf{F}}_{av} = \frac{1}{\Delta t}\int \mathbf{F}\, dt$$

Eighty percent of pollution is caused by plants and trees. Ronald Reagan

Collisions

For any **collision**, the total momentum is conserved. This means that the momentum before the collision is equal to the momentum after the collision. There are three types of collisions: **elastic**, **inelastic,** and **perfectly inelastic**.

Elastic collisions. During elastic collisions, kinetic energy and momentum are conserved. For a two-object collision:

$$E_{k1} + E_{k2} = E_{k1}' + E_{k2}'$$
$$\tfrac{1}{2}m_1v_1^2 + \tfrac{1}{2}m_2v_2^2 = \tfrac{1}{2}m_1v_1'^2 + \tfrac{1}{2}m_2v_2'^2$$
$$m_1v_1 + m_2v_2 = m_1v'_1 + m_2v'_2$$

Inelastic collision. During inelastic collisions objects undergo deformation upon contact, with some sticking. In inelastic collisions momentum is conserved but kinetic energy is not. Although most collisions fall into this category, there are no easy formulas.

Perfectly inelastic collision. When two objects collide under these conditions they stick together. Momentum is conserved but kinetic energy is not. For a two-object collision:

$$E_{k1} + E_{k2} \neq E_k'$$
$$E_{k1} + E_{k2} = \tfrac{1}{2}m_1v_1^2 + \tfrac{1}{2}m_2v_2^2$$
$$E_k' = \tfrac{1}{2}(m_1 + m_2)v^{2\prime}$$
$$m_1v_1 + m_2v_2 = (m_1 + m_2)\, v'$$
$$v' = \frac{m_1v_1 + m_2v_2}{(m_1 + m_2)}$$

The Ballistic Pendulum

The **ballistic pendulum** is a tool used to measure the speed of a projectile (a bullet). It's an example of a perfectly inelastic collision.

Conservation of Momentum
$mv = (m + M)v_b$

Initial Speed of Bullet
$v = \frac{m + M}{m}\sqrt{2gh}$

Kinetic and Potential Energy of Block and Bullet (at the moment of collision)
$E_k = \tfrac{1}{2}(m + M)v_b^2,\ E_{p0} = 0$

Kinetic and Potential Energy of Block and Bullet (at maximum height)
$E_k' = 0\ ;\ E_p' = E_k$

Initial Speed of Block and Bullet
$v_b = \sqrt{2gh}$

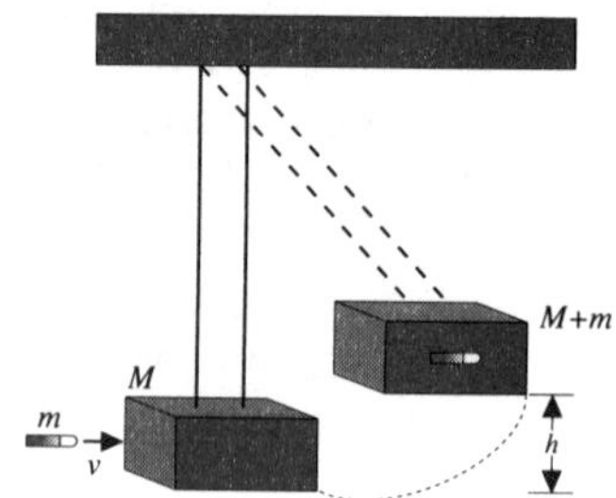

Fig. 7.4.1 Ballistic Pendulum

He who praises everybody praises nobody. Dr. Samuel Johnson

7.5 Waves & Periodic Motion

Waves carry energy from one place to another through a **medium**, in the form of a **disturbance**. For **transverse** waves, particles move perpendicular to the wave motion. In **longitudinal** waves, particles move parallel to the wave motion. When two waves move through the same medium, the resultant wave function is the sum of all wave functions. This is known as the **superposition principle. Constructive** interference occurs where the resultant wave function is greater than either of the waves. **Destructive** interference occurs where the resultant wave function is less than either of the waves. When waves strike a medium with a different density, part or all of the wave is **reflected**. When the new medium is denser, part of the wave is reflected and inverted. When the new medium is less dense, part of the wave is reflected but not inverted.

Symbols	
α = angular acceleration	k = wave number
δ = phase angle	k_s = spring constant
λ = wavelength	L = length
$\lambda_n = 2L/n$	m = mass
ω = angular frequency	n = integer
μ = mass per unit length	ϕ = phase constant
ρ = density	P = pressure
A = amplitude	π = 3.14
B = bulk modulus	T = period
d = distance from pivot point to center of mass	V = volume
f = frequency	v = phase velocity (wave speed)
f' = frequency heard	v_s = velocity of sound
F_T = tension	v_o = velocity of observer
g = acceleration due to gravity (9.81 m/s^2)	v_s = velocity of source
I = intensity	x = displacement
I_m = moment of inertia	

Traveling Waves

***Y* Axis Displacement of *X* Traveling Wave**

(Only valid if pulse does not change with time.)

$y = f(x - vt)$ (+ x direction)

$y = f(x + vt)$ (− x direction)

Velocity of Traveling Wave on a String

$v = \sqrt{F_T / \mu}$

Traveling Harmonic Waves

Wave number

$k = 2\pi/\lambda$

Velocity of Wave

$v = \lambda f$

$v = \omega/k$

Wavelength

$\lambda = vT$

Frequency

$f = 1/T$

Harmonic Wave Functions

$y(x, t) = A \sin 2\pi[(x/\lambda) - (t/T)]$

$y = A \sin k(x - vt)$ (+ x direction)

$y = A \sin (kx - \omega t)$ (+ x direction)

$y = A \sin (kx - \omega t - \phi)$ (+ x direction)

$y = A \sin (kx + \omega t + \phi)$ (− x direction)

Angular Frequency

$\omega = 2\pi/T$

$\omega = kv$

$\omega = 2\pi f$

Standing Harmonic Waves

String With Two Fixed Ends

Wave Function

$y = 2A_0 \sin(kx) \cos \omega t$

nodes at $x = n\lambda/2$

antinodes at $x = n\lambda/4$

Harmonics ($n = 1, 2, 3 \ldots$)

$\lambda_n = 2L/n$

$f_n = nv/2L$

Sound Waves in Pipes

Harmonics (n = 1, 2, 3, ...)

One Open End

$f_n = v_s/\lambda_n = (2n - 1)v_s/4L$

Two Open Ends

$f_n = v_s/\lambda_n = nv_s/2L$

Sound

Sound waves are longitudinal. They travel through a medium as high pressure **compressions** and low pressure **rarefactions**. The human ear can usually only detect such waves with frequencies between 20 and 20,000 Hz.

$$dP = B\, d\rho/\rho$$

Velocity of Sound

$v_s = \sqrt{B/\rho}$

Intensity of Sound Waves (W/m^2)

$I = (\Delta P)^2/2\rho\, v_s$

Decibel Intensity Level

$\beta = 10 \log I/I_0$

Shock Waves

$\sin\theta = v/v_s$

Mach Number

Mach = v/v_s

Bulk Modulus

$$B = \frac{(\Delta P)V}{-\Delta V}$$

Source of Sound	β (dB)
Nearby jet airplane	150
Jackhammer; machine gun	130
Siren; rock concert	120
Subway; power mower	100
Busy traffic	80
Vacuum cleaner	70
Normal conversation	50
Mosquito buzzing	40
Whisper	30
Rustling leaves	10
Threshold of hearing	0

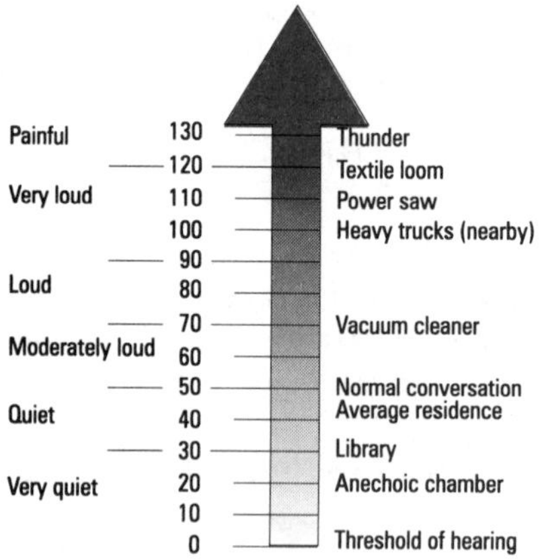

Fig 7.5.1 Decibel Intensity Levels

It is an equal failing to trust everybody, and to trust nobody. English proverb

Velocity of Sound in Various Media

Material	Density (g/cm^3)	Velocity (m/s)	Material	Density (g/cm^3)	Velocity (m/s)
Air (dry)	0.001	340	Iron	7.90	5120
Aluminum	2.70	5000	Water (sea)	1.03	1531
Brick	1.8	3650	Water (pure)	0.998	1500
Glass (pyrex)	2.32	5176	Wood (maple)	-	4110

Doppler Effect

Source Moving Relative to Observer

$f' = f / (1 \pm v_s/v)$

Sign convention:
(+) Source moves away from observer
(−) Source moves toward observer

Observer Moving Relative to Source

$f' = f(1 \pm v_O/v)$

Sign convention:
(+) Observer moves toward source
(−) Observer moves away from source

Periodic Motion

Simple Harmonic Motion

Displacement

$x = x_0 \cos 2\pi\, t/T$

Acceleration

$a_x = -a_0 \cos 2\pi\, t/T$

$a_x = -k_s\, x/m$

Velocity

$v_x = -v_0 \sin 2\pi\, t/T$

Energy

$E_p = \frac{1}{2} k_s x_0^2$

$E_p = \frac{1}{2} k_s A^2 \cos^2 (\omega t + \delta)$

$E_k = \frac{1}{2} m v^2$

$E_k = \frac{1}{2} m \omega^2 A^2 \sin^2 (\omega t + \delta)$

Total Energy $= \frac{1}{2} k_s A^2$

Hooke's Law

$F_s = -k_s x$

Harmonic Oscillator

$T = 2\pi \sqrt{m/k_s}$

$f = \left(\frac{1}{2\pi}\right)\sqrt{k_s / m}$

Simple Pendulum

$T = 2\pi \sqrt{L/g}$

$f = \left(\frac{1}{2\pi}\right)\sqrt{g / L}$

Physical Pendulum

$\theta \approx \sin\theta$

(valid for small angles in radians)

$\omega = \sqrt{(mgd / I_m)}$

$\alpha = \frac{-mgd}{I_m} \sin\theta$

$f = \frac{1}{T}$

$T = \frac{2\pi}{\omega} = \sqrt{\frac{I_m}{mgd}}$

About every month people shed an entire layer of skin.

7.6 Thermodynamics*

Symbols

$_0$ = initial value	n = number of moles
α = linear thermal expansion coefficient	P = pressure
β = volume thermal expansion coefficient	p = rate of heat loss by radiation
σ = Stefan's Constant (5.67×10^{-8} W/m^2 K^4)	q = convective heat transfer constant
A = surface area	Q = heat
c = specific heat	Q_c = lost heat
c_C = specific heat capacity of calorimeter	Q_H = absorbed heat
C_V = molar heat capacity at constant volume	R = gas constant (8.31 J/mol K or 0.0821 L atm/mol K)
C_p = molar heat capacity at constant pressure	R_V = R value
e = emissivity	S = entropy
H = rate of heat transfer	T = temperature
k_B = Boltzmann's constant (1.38×10^{-23} J/K)	T_c = temperature, cool
k_T = thermal conductivity	T_H = temperature, hot
L = length	T_{sur} = temperature of surroundings
L_H = latent heat	U = internal energy
L_T =thickness	V = volume
m = mass	W = work
m_c = mass of calorimeter	W_B = proportional probability of occurrence

Temperature. A number given to indicate the degree of warmth.
Internal energy (U). The energy a material has at a given temperature.
Heat. Energy being transferred as a result of temperature differences between materials.

Work and Heat

$$W = \int P\, dV$$

Processes and Their Relationships

Adiabatic process. No heat is transferred between a system and its surroundings.
Isothermal process. No change occurs in the temperature of the system.
Isochoric process (isovolumetric). The volume of the system does not change.
Isobaric process. No change occurs in pressure in the process.

Processes

Adiabatic Process ($Q = 0$)
(insulated from surroundings)
$\Delta U = -W$
Isothermal Process (ideal gas, $\Delta U = 0$)
$W = Q = nRT \ln (V/V_0)$
$W = nRT \ln (P_0/P)$
Isochoric Process (isovolumetric, $W = 0$)
$Q = \Delta U$
Isobaric Process (constant pressure)
$\Delta W = P\,\Delta V$

Thermal Expansion
(see Thermal Expansion Coefficients, Section 7.10)

Thermal Expansion (linear)
$\Delta L = L_0 \alpha \Delta T$
$L = L_0\, [1 + \alpha(T - T_0)]$
Thermal Expansion (volume)
$\Delta V = \beta\, V_0 \Delta T$
Heat of Transformation
(Latent Heat of Transformation)
$L_H = Q/m$

* Also see Section 5.9, Thermochemistry.

Laws of Thermodynamics

First Law of Thermodynamics

The change in internal energy is equal to the energy input minus the energy output. This is due to the law of conservation of energy.

$$dU = dQ - dW$$
$$\Delta U = Q - W$$

Thermal Efficiency of a Machine (ideal)

$$\text{Thermal efficiency} = \frac{W}{Q_H} = \frac{Q_H - Q_C}{Q_H}$$

Coefficient of Performance (ideal)

Refrigerator = $Q_C/W = T_C/T_H - T_C$

Heat pump = $Q_H/W = T_H/T_H - T_C$

Second Law of Thermodynamics

Each spontaneous chemical or physical change increases the entropy of the universe. The entropy in the universe constantly increases. Heat cannot by itself pass from a colder to a warmer body.

$$dS = dQ/T \text{ (reversible process)}$$
$$\Delta S = \Delta Q/T \text{ (reversible process)}$$

Change in Entropy in an Isothermal Expansion of an Ideal Gas

$\Delta S = nR \ln(V/V_0)$

$\Delta S = nR \ln(P_0/P)$

ΔS is negative (−) for compression and positive (+) for expansion.

Change in Entropy During Irreversible Heat Transfer Between Two Masses

$\Delta S = m_a c_a \ln(T'/T_a) + m_b c_b \ln(T'/T_b)$

Boltzmann's Calculation for Disorder

$S = k_B \ln W_B$

Third Law of Thermodynamics

A perfect crystalline substance at absolute zero (0 K) has a standard state entropy ($S°$) equal to zero.

Zeroth Law of Thermodynamics

If A and B are separate and each are in thermal equilibrium with C, then A, B and C are all in thermal equilibrium and are at the same temperature. This law is used to define temperature.

Heat Transfer

Heat conduction is the transfer of heat through materials in contact. It is the slowest form of heat transfer. **Convection** is the transfer of heat by movement of heated material. **Radiation** is the transfer of energy from one location to another by way of electromagnetic waves. It is the fastest form of heat transfer. Infrared radiation is associated with heat transfer.

Stefan-Boltzmann Law (radiant energy)

$p = \sigma e A T^4$

$p_{net} = \sigma e A(T^4 - T_{sur}^4)$

Law of Heat Conduction

$H = \Delta Q/\Delta t = k_T A\, \Delta T/L_T$

$H = -k_T A\, dT/dL$

$H = A(T_2 - T_1)/\Sigma(L_{Ti}/k_{Ti}) = A(T_2 - T_1)/\Sigma(R_{vi})$

(for a wall made of different materials)

Heat Transferred by Convection

$H \approx qA\Delta T$ (approximation)

Thermal Conductivities at 25°C			
Material	**Thermal conductivity k_T (W/m °C)**	**Material**	**Thermal conductivity k_T (W/m °C)**
Air	0.025	Glass	0.72-0.86
Aluminum	239	Silver	424
Asbestos	0.08	Styrofoam	0.01
Concrete	0.8	Water	0.60
Copper	399	Wood	0.08

Insulators

Insulators are poor conductors of heat. The ***R* value** is a measure of how effective an insulator is.

$$R_v = \frac{L_T}{k_T}$$

The Dewar Flask (Thermos Bottle)

The **Dewar flask** is a double-walled pyrex container. The space between its two silvered inner walls holds a vacuum. Because it minimizes heat transfer with the outside, the Dewar flask is used for storing hot or cold liquids over long periods of time.

Specific Heat Capacity

Heat Capacity is the amount of heat required to change the temperature of an object by 1°C (see Heat Capacity, Section 5.9).

Heat Capacity
$\Delta Q = m\,c\Delta T$

Determining Specific Heat Capacity in a Calorimeter
$\Delta Q = mc\Delta T + m_C c_C \Delta T$

Molar Heat Capacity (ideal gas)
$C_V = (1/n)\,dU/dT$ (constant volume)
$C_V = 3/2\,R$ (constant volume)
$C_P = (1/n)\,dU/dT + R$ (constant pressure)
$C_P = C_V + R = 5/2\,R$ (constant pressure)

Specific Heat at 25°C and Atmospheric Pressure			
Material	**Specific Heat (J/kg °C)**	**Material**	**Specific Heat (J/kg °C)**
Aluminum	899	Mercury	140
Copper	387	Silver	235
Ethanol	2400	Water (14.5-15.5)	4186
Glass	837	Ice (-5°C)	2090
Gold	129	Liquid (25°C)	4169
Iron	445	Steam (100°C)	2010
Lead	129	Wood	1700

7.7 Electricity & Magnetism

Symbols	
λ = linear charge density	F_e = electric force
σ = surface charge density	F_m = magnetic force
σ_c = conductivity constant	I = current
ρ = volume charge density	J = current density
ϵ_o = permittivity of free space (8.85 x 10^{-12} $C^2/N\ m^2$)	k_c = dielectric constant
Φ_c = net electric flux through a closed surface	k_e = Coulomb's constant (9.0 x 10^9 N m^2/C^2)
Φ_e = electric flux	k_m = Biot-Savart constant (10^{-7} T m/A)
Φ_m = magnetic flux	L = inductance
θ = angle	m = mass
ℓ = length	n = number of charges
$\mathscr{E}$ = electromotive force, emf (a battery)	N = number of turns
π = 3.14	p = resistivity
μ = magnetic dipole moment	P = power
μ_o = permeability of free space, (1.26 x 10^{-26} N/A^2)	Q = charge
ℓ = vector in the direction of the current	q = charge
δ = phase constant	q_{in} = charge inside
ω = angular frequency	q_o =test charge
$\varkappa_C$ = capacitive reactance	r = distance separating points
$\varkappa_L$ = inductive reactance	r =unit vector
a = radius	R = resistance
A = area	s = displacement
B = magnetic field	t = time
C = capacitance	T = temperature
d = distance	U = internal energy
e = 2.71	V = electric potential
E = electric field	v = velocity
f = frequency	v_d = drift velocity

Electric Charge

There are two types of **electric charges**: positive and negative. Opposite charges attract while like charges repel. Each electric charge is quantized (always found in multiples of a fundamental unit, 1.6 × 10^{-19} C), and is conserved (charge is not created or destroyed). Charge moves freely through a conductor under the influence of an electric field but does not move easily through an insulator.

Nonuniformly Distributed Charge Densities

Line $\lambda = dq/dx$

Surface $\sigma = dq/dA$

Volume $\rho = dq/dV$

Potential Energy in a Pair of Point Charges

$$E_p = \frac{k_e q_a q_b}{r_{ab}}$$

Uniformly Distributed Charge Densities

Line $\lambda = q/x$

Surface $\sigma = q/A$

Volume $\rho = q/V$

Coulomb's Law

Coulomb's Law

$$F_e = \frac{k_e q_a q_b}{r_{ab}^2} r_{ab}$$

Coulomb's Constant

$$k_e = \frac{1}{4\pi \epsilon_o} = 9.0 \times 10^9 \text{ Nm}^2/\text{C}^2$$

Elementary Charge

Charge of an Electron

$$e^- = 1.6 \times 10^{-19}\text{C}$$

Charge of a Proton

$$e^+ = 1.6 \times 10^{-19}\text{C}$$

Freedom is the last, best hope of the Earth. Abraham Lincoln

Electric Field

Every electric charge gives off an **electric field**. When another charge is placed in this electric field, it will experience an electric force.

$$E = \frac{F_e}{q_0} = \frac{k_e q}{r^2} \boldsymbol{r}$$

Electric field lines are used as a visual aid in understanding the electric field. Lines begin on positive charges and end on either negative charges or at infinity (for excess charge). The number of field lines are proportional to the magnitude of the charge. Field lines can't cross each other.

Field for a Group of Charges

$$E = k_e \sum \frac{q_i}{r_i^2} \boldsymbol{r}_i$$

Field for Continuous Charge Distribution

$$E = k_e \lim_{\Delta q_i \to 0} \sum \left(\frac{\Delta q_i}{r_i^2} \right) \boldsymbol{r}_i = k_e \int \frac{dq}{r^2} \boldsymbol{r}$$

Field for Sphere or Point

$$E = \frac{k_e q}{r^2} \boldsymbol{r}$$

Field for a Charged Plane

$$E = 2\pi k_e \sigma \ \boldsymbol{n}$$

Field Just Outside Sphere

$$E = 4\pi k_e \sigma \ \boldsymbol{r}$$

Field Along Charged Wire

$$E = \frac{2k_e \lambda}{r^2}$$

Electric Flux

Electric flux measures the strength of the electric field penetrating some substance.

Gauss's Law

$$\Phi_c = \oint E \bullet dA = \frac{q_{in}}{\epsilon_0}$$

Uniform Area at Angle θ to Field

$$\Phi_e = \boldsymbol{E} \bullet \boldsymbol{A} = EA \cos \theta$$

Electric Flux

$$\Phi_e = \int E \bullet dA \text{ (surface)}$$

Electric Potential

Electric potential is the work required per unit charge to move a positive test charge to an arbitrary point. The unit of electric potential is the volt (V).

Electric Potential

$$V = -\int E \bullet ds = \frac{dU}{q_0}$$

Potential of Many Charges

$$V = k_e \sum \frac{q_i}{r_i}$$

Potential Difference

$$\Delta V = \frac{\Delta U}{q_0}$$

Continuous Charge Distribution Potential

$$V = k_e \int \frac{dq}{r}$$

Capacitance

A **capacitor** stores charge. A capacitor's **capacitance** is determined by this ratio: the charge on either conductor to the magnitude of the potential difference between them. The unit of capacitance is the farad (F). A dielectric is often used to increase a capacitor's capacitance.

Symbol Used for Capacitors

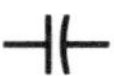

Capacitance

$C = \frac{Q}{V}$

Parallel Plate Capacitor

$C = \frac{\epsilon_0 A}{d}$

Energy Stored in a Capacitor

$U = \frac{1}{2}CV^2$

Parallel Plate Capacitor with Dielectric

$C = k_c \frac{\epsilon_0 A}{d}$

Dielectric Constant

$k_c = C$ (Dielectric) / C (Vacuum)

Dielectric Constants at 20-25°C

Material	Dielectric Constant k_c
Air (dry at 1 atm)	1.0006
Glass	5-10
Paper	3.5
Vacuum	1.0000
Water	78

Resistance

Resistance is a property that impedes the flow of current through a conductor.

Ohm's Law (constant resistance)

$$V = IR$$
$$J = \sigma_c E$$
$$R = V/I$$

Resistance (uniform conductor)

$R = \ell/\sigma_c A$

$R = p\,\ell/A$

Resistance (variations with temperature)

$R = R_0\,[1 + \alpha(T - T_0)]$

Resistivity (variations with temperature)

$p = p_0\,[1 + \alpha(T - T_0)]$

Symbol for a Resistor

Resistivity at 20°C

Material	Resistivity, p (Ω m)
Aluminum (conductor)	2.64×10^{-8}
Copper (conductor)	1.72×10^{-8}
Glass (insulator)	$10^7 - 10^{14}$
Silicon (semiconductor)	3000
Silver (conductor)	1.55×10^{-8}

Science is always wrong. It never solves a problem without creating ten more. George Bernard Shaw

Current

Electric current is the rate of flow of charge through a conductor. The unit of current is the ampere (A).

Average Current

$$I_{av} = \Delta Q/\Delta t = J A = nqv_d A$$

Instantaneous Current
$I = dQ/dt$

Current Density
$J = nqv_d = \sigma_c E$
$J = I/A$

Power Dissipated in a Conductor

Power is the rate at which charge loses energy as it passes through a conductor with resistance. The unit of power is the watt (W).

$$P = IV = I^2R = V^2/R$$

Direct Current Circuits

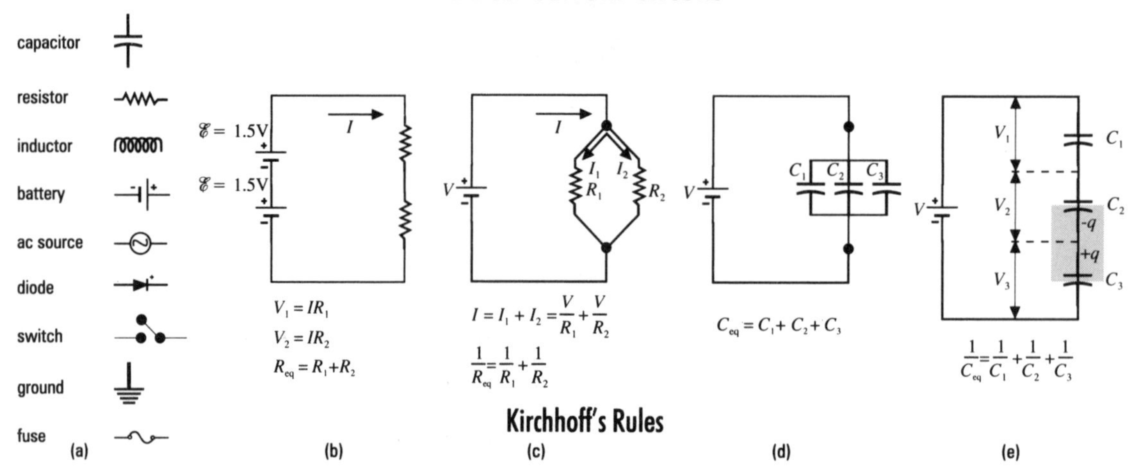

Fig 7.7.1 (a) Circuit Symbols; (b) Resistors in Series; (c) in Parallel; (d) Capacitors in Parallel; (e) in Series

Kirchhoff's Rules

1) **Junction rule.** The sum of currents entering a junction equals the sum of currents leaving the junction.
2) **Loop rule.** The sum of changes in potential around a closed conducting loop is zero.

Rules for Calculating Equivalent Resistance for Resistors in Series and in Parallel

Series
$V = IR_1 + IR_2 + \ldots + IR_n$
$I = I_1 = I_2 = \ldots = I_n$
$R_{eq} = R_1 + R_2 + \ldots + R_n$

Parallel
$I = I_1 + I_2 + \ldots + I_n$
$V = I_1R_1 = I_2R_2 = \ldots = I_nR_n$
$1/R_{eq} = 1/R_1 + 1/R_2 + \ldots + 1/R_n$

Rules for Calculating Equivalent Capacitance for Capacitors in Series and in Parallel

Capacitors in Series

$V = V_1 + V_2 + \ldots + V_n$

$Q = Q_1 = Q_2 = \ldots = Q_n$

$1/C_{eq} = 1/C_1 + 1/C_2 + \ldots + 1/C_n$

Capacitors in Parallel

$V = V_1 = V_2 = \ldots = V_n$

$Q = Q_1 + Q_2 + \ldots + Q_n$

$C_{eq} = C_1 + C_2 + \ldots + C_n$

The RC Circuit

Initial Current

$I_0 = \mathscr{E}/R$ (maximum)

Maximum Charge

$Q_{max} = C\mathscr{E}$

Current as a Function of Time

$I(t) = I_0\, e^{-t/RC}$

Charge as a Function of Time

$q(t) = Q_{max}[1 - e^{-t/RC}]$

Discharge as a Function of Time

$q(t) = Q_{max}\, e^{-t/RC}$

Magnetism

Magnetic Field

Biot – Savart Law

$$d\mathbf{B} = k_m \frac{I d\mathbf{s} \times \mathbf{r}}{r^2}$$

Ampère's Law

$$\oint \mathbf{B} \bullet d\mathbf{s} = \mu_0 I$$

Magnetic Field from a Long Wire

$$B = \frac{\mu_0 NI}{2\pi a}$$

Magnetic Field in a Solenoid

$$B = \frac{\mu_0 NI}{\ell}$$

Magnetic Flux Through a Surface

$\Phi_m = \int \mathbf{B} \bullet d\mathbf{A}$

$\Phi_m = BA \cos\theta$

Torque Exerted by Magnetic Field

$\tau = \mu \times \mathbf{B}$

$\tau = \mu B \sin\theta$

Magnetic Moment

$\mu = NIA$

Motion in a Magnetic Field

Lorentz Force

$\mathbf{F}_L = \mathbf{F}_e + \mathbf{F}_m$

$\mathbf{F}_L = q\mathbf{E} + q\mathbf{v} \times \mathbf{B}$

Magnetic Force on a Moving Charge

$F_m = qv_dB \sin\theta$

$\mathbf{F}_m = q\mathbf{v}_d \times \mathbf{B}$

Magnetic Force on Electric Current

$F_m = I\ell B \sin\theta$

$\mathbf{F}_m = I\ell \times \mathbf{B}$

Centripetal Acceleration

$a_c = qvB/m$

Radius of Circular Orbit

$r = mv/qB$

Angular Frequency

$\omega = qB/m$

Period

$T = 2\pi m/qB$

Faraday's Law of Induction

An electric current can be produced when the magnetic field changes. The total emf ($\mathscr{E}$) can be calculated by **Faraday's law of induction.**

$$\mathscr{E} = \oint \boldsymbol{E} \bullet d\boldsymbol{s} = \frac{d\Phi_m}{dt}$$

Inductance

Inductance is a measure of the opposition to change in current. The unit of inductance is the Henry (H). An **inductor** is a circuit element with a high inductance.

The Symbol for an Inductor

Inductance for a Coil with N Turns

$$L = \frac{N\Phi_m}{I}$$

Back emf Produced by an Inductor

$$\mathscr{E} = -\frac{L\,dI}{dt}$$

Energy Stored in an Inductor

$$U_m = ½LI^2$$

The RL Circuit
Current as a Function of Time

$$I(t) = \frac{\mathscr{E}}{R}(1 - e^{-RT/L})$$

The LC Circuit
Current as a Function of Time

$$I(t) = \frac{dQ}{dt} = \frac{-Q_{max}}{\sqrt{LC}} \sin\left(\frac{t}{\sqrt{LC}} + \delta\right)$$

Alternating Currents

Instantaneous Potential Difference Between Two Terminals

$$v = V_{max} \sin \omega t$$

Average Power

$$P_{av} = I_{rms} R = I_{rms} V_{rms}$$

Potential rms

$$V_{rms} = V_{max} / \sqrt{2}$$

Current rms

$$I_{rms} = I_{max} / \sqrt{2}$$

Capacitive Reactance

$$\chi_c = V_{max}/I_{max} = 1/2\pi fC$$

Inductive Reactance

$$\chi_L = V_{max}/I_{max} = 2\pi fL$$

RLC Series Circuits

Impedance

$$Z = \sqrt{R^2 + (\chi_L - \chi_c)^2}$$

Phase Angle ϕ

$$\tan \phi = (\chi_L - \chi_c)/R$$

Average Power

$$P = I_{rms}V_{rms} \cos \phi = I^2_{rms} R$$

Current rms

$$I_{rms} = V_{rms}/Z$$

Resonance Frequency when $\chi_L = \chi_c$

$$\omega_0 = \frac{1}{\sqrt{LC}}$$

Climb mountains to see the lowlands. Chinese proverb

7.8 Optics

Symbols	
θ_B = Brewster's angle	h_o = height of object
θ_c = critical angle	i = image distance
θ_i = angle of incidence	I = light intensity
θ_r = angle of reflection	I_i = incident intensity
δ = path difference	I_o = initial intensity
λ = wavelength	I_r = reflected intensity
λ_o = wavelength in a vacuum	L = distance from focal point to image
λ_n = wavelength in a medium	L_s = distance from screen
b = width of slit	m = order number
c = speed of light in a vacuum (3.0×10^8 m/s)	N = number of lines in grating
D = diameter	n = index of refraction
d = diameter of lens	o = object distance
d_p = distance between planes	P = power of a lens in diopters
d_s = separation between slits	r = radius
f = frequency	R = reflectance
f = focal length	R_c = radius of curvature
f_e = focal length of eye piece	t = thickness of the film
f_o = focal length of objective lens	v = velocity of light in some medium
h = Planck's constant (6.63×10^{-34} J s)	y = y axis
h_i = height of image	

Optics is the study of the properties of light. For geometric purposes we can assume that light travels in straight lines; this is known as **ray approximation**. When a light ray traveling through a medium encounters a second medium, part or all of it is **refracted**. The degree to which light is **reflected** is known as **reflectance**.

Reflection

The angle of incidence equals the angle of reflection.

Law of Reflection

$\theta_r = \theta_i$

Reflectance (normal incidence)

$R = I_r/I_i$

$R = [(n_2 - n_1)/(n_2 + n_1)]^2$

Critical Angle

$\sin \theta_c = n_2/n_1$

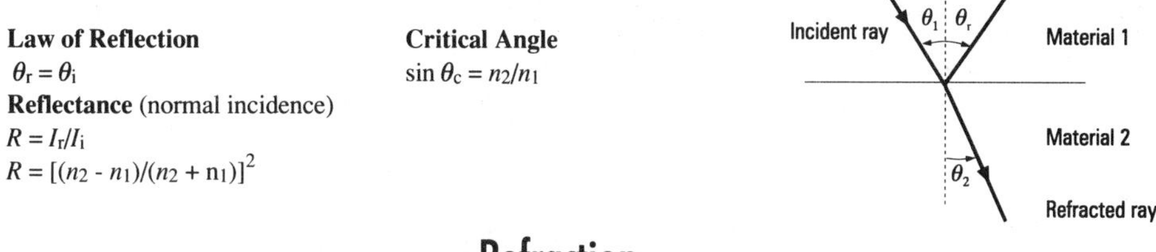

Fig 7.8.1 Light Refraction

Refraction

When light travels through a less dense to a more dense optical material the rays bend toward the normal.

Law of Refraction (Snell's law)

$$n_1 \sin \theta_1 = n_2 \sin \theta_2$$

Index of Refraction

$n = c/v$

$n = \lambda_0/\lambda_n$

$n_1/n_2 = \lambda_2/\lambda_1$

Refraction for a Spherical Surface

$n_1/o + n_2/i = (n_2 - n_1)/R$

Plane Refracting Surface

$i = -(n_2/n_1)\, o$

Information from Surfaces That Refract

Unit	Sign	Comment
o	+	Real object
o	–	Virtual object
i	+	Real image
i	–	Virtual image
r	+	Center of curvature in back
r	–	Center of curvature in front

Index of Refraction $\lambda = 589$ nm

Material	Index of refraction (*n*)
Air	1.0003
Carbon dioxide (CO_2)	1.0005
Diamond	2.417
Ethanol @ 20°C	1.361
Glass, flint	1.5-1.6
Glass, crown	1.6-1.8
Water —	
Ice @ 0°C	1.31
Liquid @ 25°C	1.3330

Formulas for Lenses

Magnification

Angular Magnification

$M = \theta'/\theta$

Lateral/Linear Magnification

$M = -i/o = h_i/h_o$

Magnifying Lens

$M = 25$ cm/f (image at infinity)

Telescope

$M = -f_o/f_e$

Compound Microscope

$M = -L\ 25\text{cm}/f_o f_e$

Other Formulas

f-Number

f-number $= f/d$

Lens Power

$P = 1/f$

Thin Lens Equation in Gaussian Form

$1/f = 1/i + 1/o$

Lens Maker's Equation

$1/f = (n-1)[(1/R_{c1}) - (1/R_{c2})]$

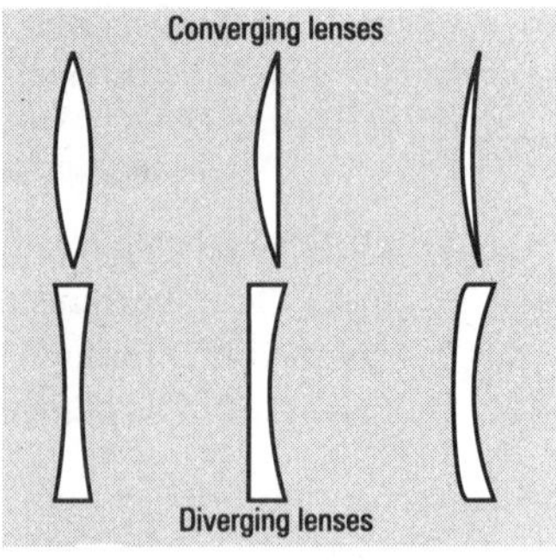

Fig 7.8.2 Converging, Diverging Lenses

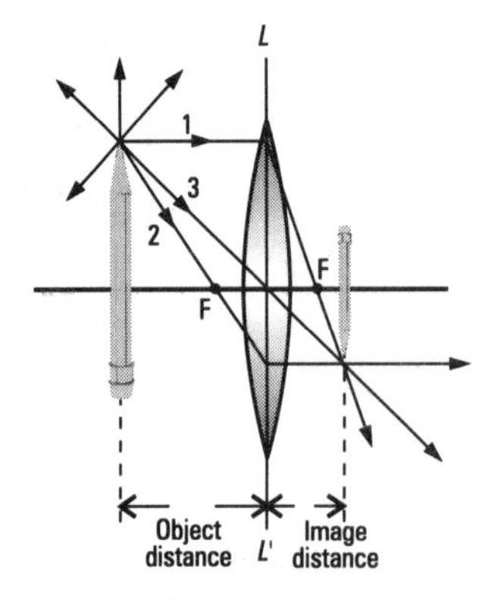

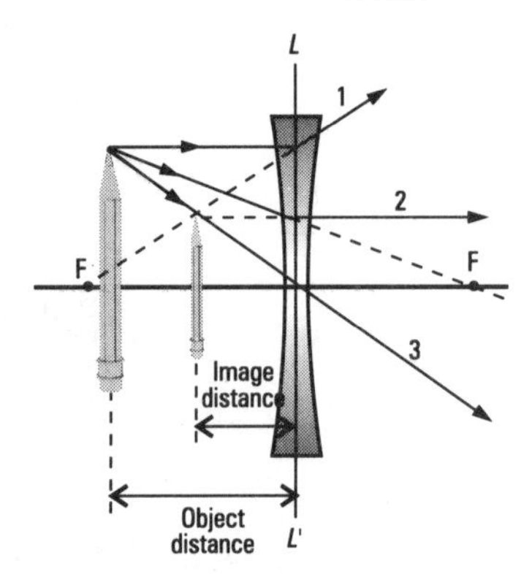

Fig. 7.8.3 Differences Between Object and Image Viewed Through Converging and Diverging Lenses

Mirrors

When light rays actually intersect at the image, we call this a **real image** (image can be projected onto a screen). A **virtual image** is formed where light rays appear to converge (image cannot be projected onto a screen).

Focal Length for a Concave Spherical Mirror
$f = r/2$

Mirror Equation
(paraxial rays)
$1/f = 1/i + 1/o$

Information From Mirrors

Unit	Sign	Comment
o	+	Real object formed in front of mirror
o	–	Virtual object formed behind mirror
i	+	Real object formed in front of mirror
i	–	Virtual object formed behind mirror
M	+	Image erect
M	–	Image inverted
both *f* and *r*	+	Concave mirror (center of curve in front)
both *f* and *r*	–	Convex mirror (center of curve behind)

Formulas

Energy of a Photon
$E = h f$

Double Slit Interference
$m = 0, 1, 2, 3, ...$

Constructive Interference (maxima)
$\delta = d_s \sin\theta = m\lambda$

Destructive Interference
$\delta = d_s \sin\theta = (m + 1/2)\lambda$

Bright Fringe
$y_{bright} = \lambda L_s/d_s$

Dark Fringe
$y_{dark} = \lambda L_s(m + ½)/d_s$

Average Light Intensity
$I_{av} = I_0 \cos^2 (\pi d_s (\sin\theta)/\lambda)$
$I_{av} = I_0 \cos^2 (\pi\ d_s y/\lambda L_s)$

Thin Film Interference

Reflection (maxima)
$2nt = (m + 1/2)\lambda \quad m = 0, 1, 2, 3, \ldots$

Reflection (minima)
$2nt = m\lambda \quad m = 0, 1, 2, 3, \ldots$

Diffraction
Limiting angle of resolution for circular aperture
$\sin\theta_m = 1.22\ \lambda/D$

Single Slit Diffraction
$m\lambda = b \sin\theta \quad m = 1, 2, 3, \ldots$ (minima)

Multiple Slit Diffraction
Multiple slit diffraction grating condition for maximum interference
$d_s \sin\theta = m\lambda \quad m = 0, 1, 2, 3, \ldots$ (grating maximum)

Resolving Power for Diffraction Grating
$R_p = Nm$

Bragg's Law
$2d_p \sin\theta = m\lambda$

Polarized Light Intensity
$I = I_0 \cos^2\theta$

Brewster's Angle
(for maximum polarization, by reflection)
$\tan\theta_B = n_2/n_1$

A thing is not vulgar merely because it is common. William Hazlitt

7.9 The Electromagnetic Spectrum

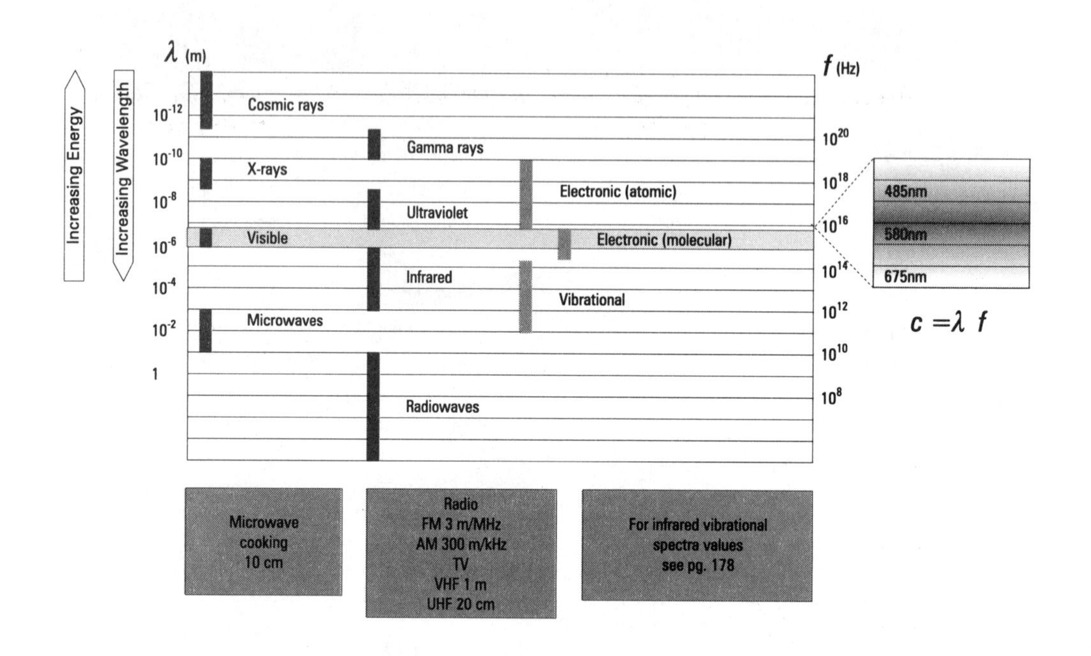

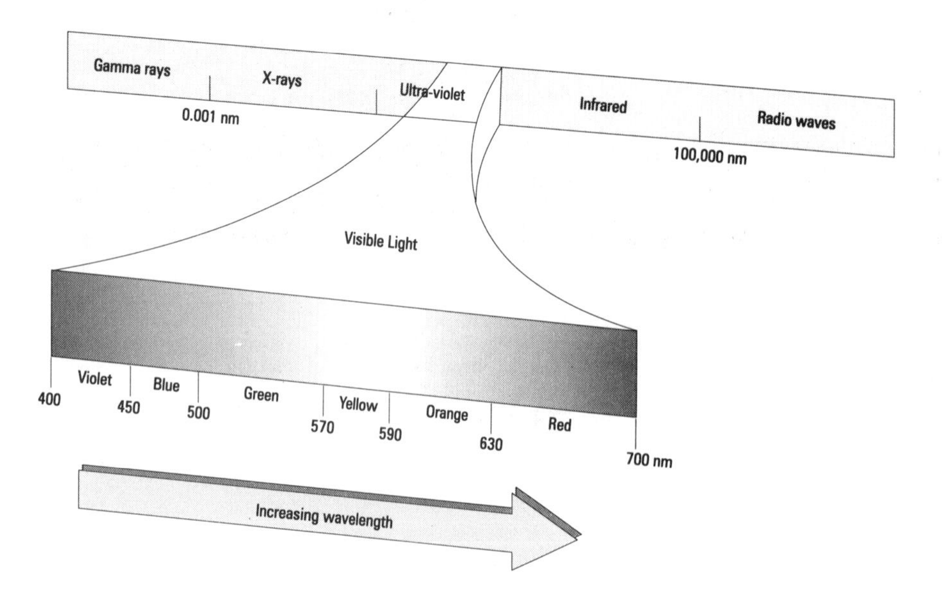

Fig. 7.9.1 The Electromagnetic Spectrum

7.10 Physical Data

Acceleration Due to Gravity Above the Earth's Surface

Altitude (m)	Acceleration due to gravity (m/s^2)	Altitude (m)	Acceleration due to gravity (m/s^2)
0	9.80665	10 000	9.7759
500	9.8051	20 000	9.7452
1000	9.8036	30 000	9.7147
1500	9.8020	40 000	9.6844
2000	9.8005	50 000	9.6542
2500	9.7989	60 000	9.6241
3000	9.7974	70 000	9.5942
3500	9.7959	80 000	9.5644
4000	9.7943	90 000	9.5348
4500	9.7928	100 000	9.5052
5000	9.7912	200 000	9.2175
5500	9.7897	300 000	8.9427
6000	9.7882	400 000	8.6799
6500	9.7866	500 000	8.4286
7000	9.7851	600 000	8.1880
7500	9.7836	700 000	7.9576
8500	9.7805	800 000	7.7368
9000	9.7789	900 000	7.5250
9500	9.7774	1 000 000	7.3218

Densities of Common Solids

Solid	Density (g/cm^3)	Solid	Density (g/cm^3)
Asphalt	1.1-1.5	Diamond	3.01-3.52
Bone	1.7-2.0	Glass (common)	2.4-2.8
Brick	1.4-2.2	Glue	1.27
Cardboard	0.69	Ice	0.917
Cement (set)	2.7-3.0	Ivory	1.83-1.92
Chalk	1.9-2.8	Leather (dry)	0.86
Clay	1.8-2.6	Paper	0.7-1.15
Cork	0.22-0.26	Rubber	0.9-1.2

No gain without pain. Franklin

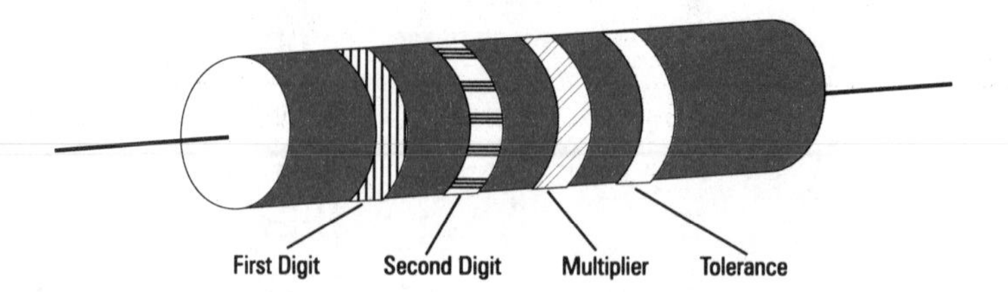

Fig. 7.10.1 Resistor Color Codes

Resistor Color Codes

Color	Digit	Multiplier	Tolerance
Gold		0.1	5%
Silver		0.01	10%
Colorless			20%
Black	0	1	
Brown	1	10	
Red	2	100	
Orange	3	1000	
Yellow	4	10^4	
Green	5	10^5	
Blue	6	10^6	
Violet	7	10^7	
Gray	8	10^8	
White	9	10^9	

Solar System Data

Body	Mass (kg)	Radius (km)	Density (g/cm^3)	Period of rotation (d)	Period of revolution about sun (d)	Acceleration of gravity (m/s^2)
Sun	1.99×10^{30}	696 000	-	-	-	270
Mercury	3.30×10^{23}	2439	5.43	58.6	88.0	3.6
Venus	4.87×10^{24}	6052	5.24	-243	225	8.9
Earth	5.98×10^{24}	6378	5.517	0.997	365.25	9.81
Mars	6.42×10^{23}	3393	3.94	1.03	688	3.8
Jupiter	1.90×10^{27}	71 398	1.33	0.414	4330	26
Saturn	5.69×10^{26}	56 800	0.70	0.438	10800	11
Uranus	8.66×10^{25}	25 400	1.30	-0.65	30600	11
Neptune	1.03×10^{26}	24 300	1.76	0.768	60400	14
Pluto	1.5×10^{22}	1500	1.1	-6.39	90500	0.44
Moon	7.35×10^{22}	1738	3.34	27.32	-	1.6

Thermal Expansion Coefficients Between 20 and 25°C

Material	Linear coefficient α (°C^{-1})	Volume coefficient β (°C^{-1})
Air	-	3.7×10^{-3}
Aluminum	2.4×10^{-5}	7.2×10^{-5}
Brass	1.9×10^{-5}	5.7×10^{-5}
Concrete	1.2×10^{-5}	-
Copper	1.7×10^{-5}	5.1×10^{-5}
Gasoline	-	9.6×10^{-4}
Glass (ordinary)	9.0×10^{-6}	2.7×10^{-5}
Glass (pyrex)	3.1×10^{-6}	9.0×10^{-6}
Invar	8.0×10^{-7}	2.1×10^{-6}
Lead	2.9×10^{-5}	8.7×10^{-5}
Mercury	-	1.8×10^{-4}
Steel	1.2×10^{-5}	3.6×10^{-5}

Viscosities for Various Liquids

Liquid	Viscosity (10^3 N s/m^2)	Liquid	Viscosity (10^3 N s/m^2)
Acetic acid	1.31	Pentane	0.24
Acetone	0.30	Petroleum ether	0.30
Benzene	0.65	1-Propanol	2.26
Carbon tetrachloride	0.97	2-Propanol	2.86
Chloroform	0.58	Pyridine	0.95
Cyclohexane	0.98	Tetrahydrofuran (THF)	0.55
Dioxane	1.44	Toluene	0.59
Diethyl ether	0.24	Trichloroethylene	0.95
Hexane	0.31	Water	1.00
Methanol	0.55	Whole blood	2.7
Methylene chloride	0.45	o-Xylene	0.81

MATHEMATICS

8

8.1 Translating Mathematics

Symbols Used in Mathematics

{a,b,c} = the set of elements including a,b,c
$\varnothing$ = empty set
$=$ = equal to
$\equiv$ = is defined as
$\approx$ = approximately equal to
$\neq$ = not equal to
$>$ = greater than
$\gg$ = much greater than
$\geq$ = greater than or equal to
$<$ = less than
$\ll$ = much less than
$\leq$ = less than or equal to
$+$ = addition
$-$ = subtraction
$\pm$ = plus or minus
$\times$ = multiplication
$\div$ = division
/ = division
a^n = *a* raised to the power of *n*, or *a* multiplied by itself *n* times (if *n* is a positive integer)
$\sqrt{a}$ = the square root of *a*
$\sqrt[n]{a}$ = the *n*th root of *a*
Δa = the change in *a* (final - initial)
$\perp$ = perpendicular (90°)
$\in$ = an element of a set
$\notin$ = not an element of a set
θ = angle
∞ = infinity
° = degrees
$\int a\,dt$ = integral of *a* with respect to *t*
$\subseteq$ = a subset of
$\cup$ = union
$\cap$ = intersection
α = significance level for a statistical test
μ = population mean
$\pi = 3.141592654...$
σ = standard deviation for a population
Σ = the sum of, summation sign
a = the vector quantity **a**
$|\mathbf{a}|$ = the magnitude of vector **a**
$\mathbf{a} \bullet \mathbf{b}$ = the dot product (scalar product) of vectors **a** and **b**
$\mathbf{a} \times \mathbf{b}$ = the cross product of vectors **a** and **b**
$\mathbf{AB}$ = the vector from point *A* to point *B*
A = matrix **A**
$\mathbf{A}^t$ = the transpose of matrix **A**
$\mathbf{A}^{-1}$ = the inverse of matrix **A**
$|\mathbf{A}|$ = the determinant of matrix **A**
c = circumference
$\mathbb{C}$ = set of complex numbers of the form $a + bi, i = \sqrt{-1}$
$\text{cis}\,\theta = \cos\theta + i\sin\theta$
d = depth, diameter
d.f. = degrees of freedom
$\frac{dx}{dt}$ = derivative of *x* with respect to *t*
$\frac{\partial x}{\partial y}$ = partial derivative of *x* with respect to *y*
det **A** = the determinant of matrix **A**
$e = 2.718281828...$
i = imaginary unit, $i = \sqrt{-1}$
l = length
$\lim_{\Delta t \to 0} \frac{\Delta x}{\Delta t}$ = limit of Δx as Δt approaches 0
log *a* = logarithm of *a* (base 10)
ln *a* = natural logarithm of *a* (base e)
$\mathbb{N}$ = whole natural numbers (any whole positive number including zero)
$n!$ = *n* factorial $(n)(n-1)(n-2)...(2)(1)$
${}_nC_r$ = number of combinations (order irrelevant)
${}_nP_r$ = number of permutations (order important)
$\mathbb{Q}$ = rational numbers; positive or negative numbers that can be expressed as a fraction
$\overline{\mathbb{Q}}$ = irrational numbers; cannot be expressed as a fraction
r = radius
R = radius
$\mathbb{R}$ = real numbers including rational and irrational numbers
$\mathbb{R}^2$ = two-dimensional real number space
$\mathbb{R}^3$ = three-dimensional real number space
$\mathbb{R}^n$ = *n*-dimensional real number space
s = sample standard deviation
$T:\mathbb{R}^n \rightarrow \mathbb{R}^m$ = transformation from $\mathbb{R}^n$ to $\mathbb{R}^m$
w = width
$\bar{x}$ = sample mean
$\mathbb{Z}$ = Integers {...,-2,-1,0,1,2,...}
z = standard score
z = complex number $(a + bi)$
$|z|$ = modulus or absolute value of *z*
$\bar{z}$ = conjugate of *z*

Quick Tests for Whole Number Divisibility

Divisible by	...if these conditions are met
1	All numbers are divisible by 1 (1, 13, 35).
2	The last digit is even or equal to zero (18, 50, 106).
3	The sum of all individual digits is divisible by 3 (15, 21, 81).
4	The number formed by the last two digits is divisible by 4 or is 00 (524, 900, 1036).
5	The last digit is 5 or 0 (10, 25, 250).
6	Both 2 and 3 can be shown to be factors (18, 36, 72).
7	We don't have any quick tests for divisibility by 7.
8	The number formed by the last three digits is divisible by 8 (1064, 5128, 8016).
9	The sum of all the individual digits is divisible by 9 (18, 81, 4419).
10	The last digit is a zero (10, 220, 61720).

8.2 Geometry & Algebra

Geometrical Formulas

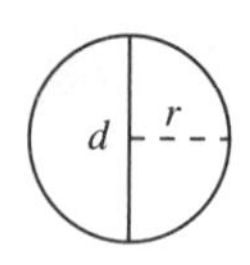

Circle
Area $= \pi r^2$
Circumference $= 2\pi r = \pi d$

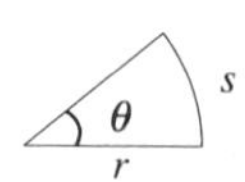

Circle (slice)
Area $= \dfrac{\theta r^2}{2}$
Arclength $= r\theta$

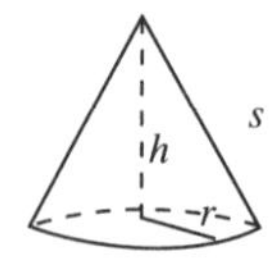

Cone (right circular)
Volume $= 1/3\ \pi r^2 h$
Total surface area $= \pi r^2 + \pi r s$
Lateral surface area $= \pi r s$
Slant height $= s = \sqrt{r^2 + h^2}$

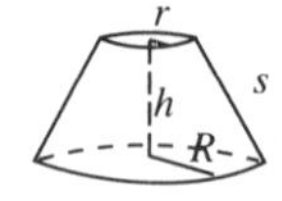

Cone (frustrum of right circular)
Volume $= \dfrac{\pi\left(r^2 + rR + R^2\right)h}{3}$
Lateral surface area $= \pi s(R + r)$

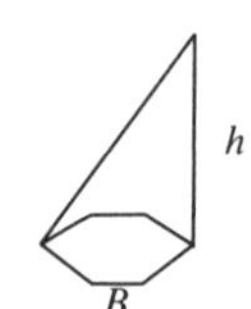

Cone (all, pyramids)
Volume $= 1/3$ (base area) h
$V = 1/3\ Bh$

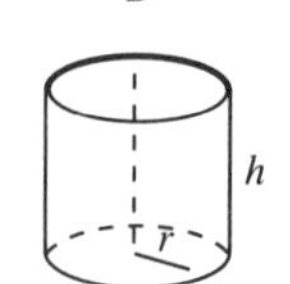

Cylinder (right circular)
Volume $= \pi r^2 h$
Total surface area $= 2\pi r^2 + 2\pi r h$
Lateral surface area $= 2\pi r h$

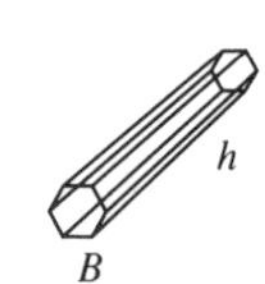

Cylinder (or prism)
Volume = (area of base*) h
$V = Bh$
*base and top must be parallel

Ellipse
Area $= \pi ab$
Circumference $\approx 2\pi\sqrt{\dfrac{a^2 + b^2}{2}}$

Parallelogram
Area $= bh$

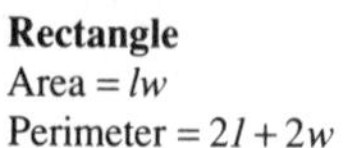

Rectangle
Area $= lw$
Perimeter $= 2l + 2w$

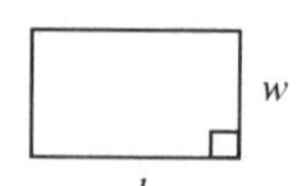

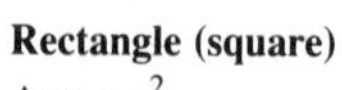

Rectangle (square)
Area $= s^2$
Perimeter $= 4s$

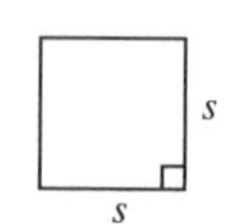

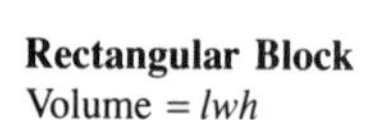

Rectangular Block
Volume $= lwh$

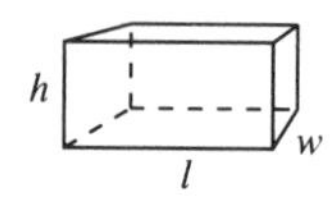

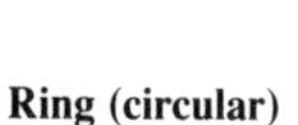

Ring (circular)

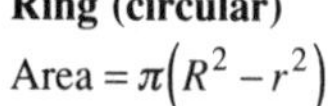

Area $= \pi\left(R^2 - r^2\right)$

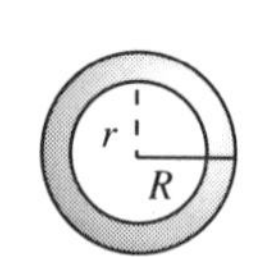

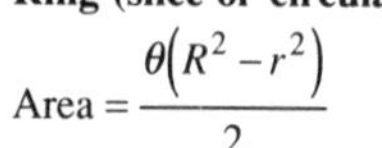

Ring (slice of circular)
Area $= \dfrac{\theta\left(R^2 - r^2\right)}{2}$

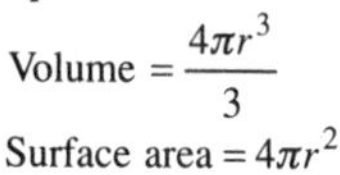

Sphere
Volume $= \dfrac{4\pi r^3}{3}$
Surface area $= 4\pi r^2$

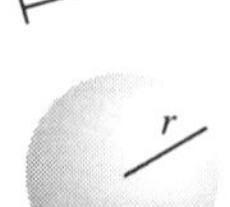

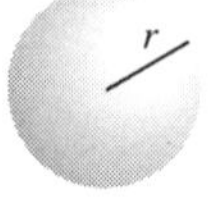

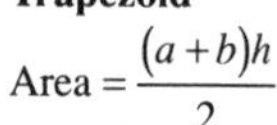

Trapezoid
Area $= \dfrac{(a + b)h}{2}$

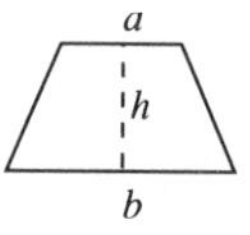

Triangle
Area $= \dfrac{bh}{2}$
$h = a \sin\theta$
$c^2 = a^2 + b^2 - 2ab\cos\theta$
(θ in radians)

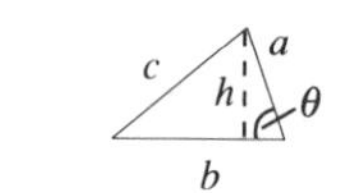

Triangular Wedge
Area of upper face
$=$ (Area of base) $\sec\theta$
(θ in radians)

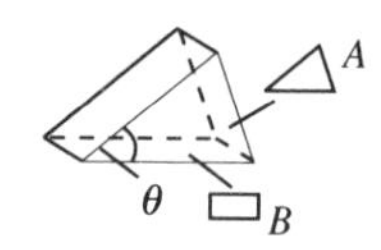

Minds are like parachutes — they only function when open. Thomas Dewar

Algebra

Arithmetic Operations

Commutative Law

$ab = ba$

$a + b = b + a$

Associative Law

$(ab)c = a(bc)$

$(a+b)+c = a+(b+c)$

Distributive Law

$a(b+c) = ab + ac$

$-(b+c) = -b - c$

Fractions

$$\frac{a+c}{b} = \frac{a}{b} + \frac{c}{b}$$

$$\frac{a}{b} + \frac{c}{d} = \frac{ad+bc}{bd}$$

$$\frac{ab+ac}{a} = b + c$$

$$\frac{a-b}{c-d} = \frac{b-a}{d-c}$$

$$\frac{-a}{b} = -\frac{a}{b} = \frac{a}{-b}$$

$$\frac{a}{b}\frac{c}{d} = \frac{ac}{bd}$$

$$\frac{(a/b)}{c/d} = \frac{a}{b}\frac{d}{c}$$

$$\frac{a}{(b/c)} = \frac{ac}{b}$$

$$\frac{a/b}{c} = \frac{a}{bc}$$

$$a\frac{b}{c} = \frac{ab}{c}$$

Exponents and Radicals

$a^0 = 1 \ (a \neq 0)$

$a^1 = a$

$(ab)^x = a^x b^x$

$a^x a^y = a^{x+y}$

$\frac{a^x}{a^y} = a^{x-y} \ (a \neq 0)$

$(a^x)^y = a^{xy}$

$a^{m/n} = \sqrt[n]{a^m} = (\sqrt[n]{a})^m$

$\left(\frac{a}{b}\right)^x = \frac{a^x}{b^x}$

$a^{-x} = \frac{1}{a^x} \ (a \neq 0)$

$\sqrt[n]{ab} = \sqrt[n]{a}\sqrt[n]{b}$

$\sqrt{\frac{a}{b}} = \frac{\sqrt{a}}{\sqrt{b}}$

$\sqrt[n]{\frac{a}{b}} = \frac{\sqrt[n]{a}}{\sqrt[n]{b}}$

$\sqrt[n]{\sqrt[m]{a}} = \sqrt[m]{\sqrt[n]{a}} = \sqrt[nm]{a}$

$\sqrt{a} = a^{1/2}$

$\sqrt[n]{a} = a^{1/n}$

$\sqrt{ab} = \sqrt{a}\sqrt{b}$

If $x = \sqrt[n]{a}$ then $x^n = a$

Logarithms

$\log_a uv = \log_a u + \log_a v$

$\log_a u/v = \log_a u - \log_a v$

$\log_a u^n = n \log_a u$

$\ln ax = \ln a + \ln x$

$\ln a/x = \ln a - \ln x$

$\ln l/x = -\ln x$

$\ln x^n = n \ln x$

The Binomial Theorem (*n is positive)

$(a+b)^1 = a + b$

$(a+b)^2 = a^2 + 2ab + b^2$

$(a+b)^3 = a^3 + 3a^2b + 3ab^2 + b^3$

$$(a+b)^n = a^n + na^{n-1}b + \frac{n(n-1)}{2!}a^{n-2}b^2 + \cdots + nab^{n-1} + b^n$$

$(a-b)^1 = a - b$

$(a-b)^2 = a^2 - 2ab + b^2$

$(a-b)^3 = a^3 - 3a^2b + 3ab^2 - b^3$

$$(a-b)^n = a^n - na^{n-1}b + \frac{n(n-1)}{2!}a^{n-2}b^2 - \cdots + na(-b)^{n-1} + b^n$$

The Difference of Positive Like Integer Powers

$$a^n - b^n = (a-b)\left(a^{n-1} + a^{n-2}b + a^{n-3}b^2 + \cdots + ab^{n-2} + b^{n-1}\right) \text{ when } n > 1$$

Special Factors

$a^2 - b^2 = (a-b)(a+b)$

$a^3 + b^3 = (a+b)(a^2 - ab + b^2)$

$a^3 - b^3 = (a-b)(a^2 + ab + b^2)$

$a^4 - b^4 = (a^2 - b^2)(a^2 + b^2)$

Completing the Square

$$ax^2 + bx + c = a\left(x + \frac{b}{2a}\right)^2 + \left(c - \frac{b^2}{4a}\right)$$

The Quadratic Equation

For $ax^2 + bx + c = 0, \; a \neq 0$

$$x = \frac{-b \pm \sqrt{b^2 - 4ac}}{2a}$$

Equations for Planes

Cartesian Equation for a Plane

$Ax + By + Cz + D = 0$

Vector Equation for a Plane

$r = r_0 + sa + tb$

Parametric Equation for a Plane

$x = x_0 + sa_1 + tb_1$

$y = y_0 + sa_2 + tb_2$

$z = z_0 + sa_3 + tb_3$

Graph Equations

Straight Line

$y = mx + b$

$(y_2 - y_1) = m(x_2 - x_1)$

$m = \dfrac{y_2 - y_1}{x_2 - x_1} = \dfrac{\Delta y}{\Delta x} = \tan\theta$

Parabola

$y = ax^2 + b$

Quadratic Function

$y = ax^2 + bx + c$

Circle

$x^2 + y^2 = r^2$ (center (0, 0))

$(x-h)^2 + (y-k)^2 = r^2$, (with center (h, k))

Ellipse

$\dfrac{x^2}{a^2} + \dfrac{y^2}{b^2} = 1$ (center (0, 0))

$\dfrac{(x-h)^2}{a^2} + \dfrac{(y-k)^2}{b^2} = 1$ (with center (h, k))

Exponential

$y = a^x$

Logarithmic

$y = \log_a x$

It is not the love of money which is the root of all evil—it is greed. Gordon Coleman

Vectors

Vector: Quantity with both magnitude and direction.

Properties of Vectors

$a+b=b+a$
$a+(b+c)=(a+b)+c$
$a+0=a$

$1a=a$
$(n)a=na$
$m(na)=(mn)a=mna$

$m(a+b)=ma+mb$
$(m+n)a=ma+na$

Unit Vector: b is a unit vector in the a direction. $b=\frac{a}{|a|}$

Dot Product Properties (Scalar Product)

$a \bullet b=|a||b|\cos\theta$
$\cos\theta=\frac{a \bullet b}{|a||b|}$
$a \bullet b=0$ when $a \perp b$

$a \bullet b=b \bullet a$
$a \bullet (b+c)=a \bullet b+a \bullet c$
$k(a \bullet b)=(ka)\bullet b=a \bullet (kb)$
$a \bullet a=|a|^2$

Cross Product Properties

$a\times b=-(b\times a)$
$k(a\times b)=ka\times b=a\times kb$
$a\times(b+c)=(a\times b)+(a\times c)$

$|a\times b|=|a||b|\sin\theta,\ a,\ b,\neq 0$
$\sin\theta=\frac{|a\times b|}{|a||b|}$
$a\times b=0$ if a and b are linearly dependent

Cartesian Coordinate System Vectors

	Formula	R^2 (two dimensions)	R^3 (three dimensions)	R^n (n dimensions)
Vector notations	a	(a_1,a_2)	(a_1,a_2,a_3)	$(a_1,a_2,\cdots,a_n)$
	a	(x,y)	(x,y,z)	
	a	$x\hat{i}+y\hat{j}$	$x\hat{i}+y\hat{j}+z\hat{k}$	
Magnitude	$\lvert a\rvert$	$\sqrt{a_1^2+a_2^2}$	$\sqrt{a_1^2+a_2^2+a_3^2}$	$\sqrt{a_1^2+a_2^2+\cdots+a_n^2}$
	$\lvert a\rvert$	$\sqrt{x^2+y^2}$	$\sqrt{x^2+y^2+z^2}$	
Scalar multiplication	ka	(ka_1,ka_2)	(ka_1,ka_2,ka_3)	$(ka_1,ka_2,\cdots,ka_n)$
Vector addition	$a+b$	(a_1+b_1,a_2+b_2)	$(a_1+b_1,a_2+b_2,a_3+b_3)$	$(a_1+b_1,a_2+b_2,\cdots,a_n+b_n)$
	$a+b$	(x_1+x_2,y_1+y_2)	$(x_1+x_2,y_1+y_2,z_1+z_2)$	
	$a+b$	$(a_1+b_1)\hat{i}+(a_2+b_2)\hat{j}$	$(a_1+b_1)\hat{i}+(a_2+b_2)\hat{j}+(a_3+b_3)\hat{k}$	
Vector joining points A and B	AB	(b_1-a_1,b_2-a_2)	$(b_1-a_1,b_2-a_2,b_3-a_3)$	$(b_1-a_1,b_2-a_2,\cdots,b_n-a_n)$
Dot product	$a \bullet b$	$a_1b_1+a_2b_2$	$a_1b_1+a_2b_2+a_3b_3$	$a_1b_1+a_2b_2+\ldots+a_nb_n$
Angle between vectors	$\cos\theta=\frac{a\bullet b}{\lvert a\rvert\lvert b\rvert}$	$\frac{a_1b_1+a_2b_3}{\sqrt{a_1^2+a_2^2}\sqrt{b_1^2+b_2^2}}$	$\frac{a_1b_1+a_2b_2+a_3b_3}{\sqrt{a_1^2+a_2^2+a_3^2}\sqrt{b_1^2+b_2^2+b_3^2}}$	$\frac{a_1b_1+a_2b_2+\cdots+a_nb_n}{\sqrt{a_1^2+a_2^2+\cdots+a_n^2}\sqrt{b_1^2+b_2^2+\cdots+b_n^2}}$
Vector projection of a on b	$\left(\frac{a\bullet b}{b\bullet b}\right)b$	$\left(\frac{a_1b_1+a_2b_2}{b_1b_1+b_2b_2}\right)(b_1,b_2)$	$\left(\frac{a_1b_1+a_2b_2+a_3b_3}{b_1b_1+b_2b_2+b_3b_3}\right)(b_1,b_2,b_3)$	$\left(\frac{a_1b_1+a_2b_2+\cdots+a_nb_n}{b_1b_1+b_2b_2+\cdots+b_nb_n}\right)(b_1,b_2,\cdots b_n)$
Scalar projection of a on b	$\frac{a\bullet b}{\lvert b\rvert}$	$\frac{a_1b_1+a_2b_2}{\sqrt{b_1^2+b_2^2}}$	$\frac{a_1b_1+a_2b_2+a_3b_3}{\sqrt{b_1^2+b_2^2+b_3^3}}$	$\frac{a_1b_1+a_2b_2+\cdots+a_nb_n}{\sqrt{b_1^2+b_2^2+\cdots+b_n^2}}$
Cross product	$a\times b$		$(a_2b_3-a_3b_2,a_3b_1-a_1b_3,a_1b_2-a_2b_1)$	
Magnitude of the cross product	$\lvert a\times b\rvert=\lvert a\rvert\lvert b\rvert\sin\theta$		$\sqrt{a_1^2+a_2^2+a_3^2}\sqrt{b_1^2+b_2^2+b_3^2}\sin\theta$	

Matrices

Properties of Matrices

$A+B=B+A$
$A+(B+C)=(A+B)+C$
$A+(-A)=0$
$IA=A$
$IA=AI=A$

$n(mA)=(nm)A$
$n(A+B)=nA+nB$
$(n+m)A=nA+mA$
$A(BC)=(AB)C$
$A(B+C)=AB+AC$
$(A+B)C=(AC+BC)$

Matrix Addition (matrices must be the same size)

$$A+B=\begin{bmatrix} a_{11} & a_{12} \\ a_{21} & a_{22} \end{bmatrix}+\begin{bmatrix} b_{11} & b_{12} \\ b_{21} & b_{22} \end{bmatrix}\begin{bmatrix} a_{11}+b_{11} & a_{12}+b_{12} \\ a_{12}+b_{12} & a_{22}+b_{22} \end{bmatrix}$$

Scalar Multiplication

$$kA=k\begin{bmatrix} a_{11} & a_{12} \\ a_{21} & a_{22} \end{bmatrix}=\begin{bmatrix} ka_{11} & ka_{12} \\ ka_{21} & ka_{22} \end{bmatrix}$$

Matrix Multiplication

The number of columns in the first matrix must equal the number of rows in the second matrix.

$$AB=\begin{bmatrix} a_{11} & a_{12} \\ a_{21} & a_{22} \end{bmatrix}\begin{bmatrix} b_{11} & b_{12} \\ b_{21} & b_{22} \end{bmatrix}=\begin{bmatrix} a_{11}b_{11}+a_{12}b_{21} & a_{11}b_{12}+a_{12}b_{22} \\ a_{21}b_{11}+a_{22}b_{21} & a_{21}b_{12}+a_{22}b_{22} \end{bmatrix}$$

$$AX=\begin{bmatrix} a_{11} & a_{12} \\ a_{21} & a_{22} \end{bmatrix}\begin{bmatrix} x \\ y \end{bmatrix}=\begin{bmatrix} a_{11}x+a_{12}y \\ a_{21}x+a_{22}y \end{bmatrix}$$

$$AI=\begin{bmatrix} a_{11} & a_{12} \\ a_{21} & a_{22} \end{bmatrix}\begin{bmatrix} 1 & 0 \\ 0 & 1 \end{bmatrix}=\begin{bmatrix} a_{11} & a_{12} \\ a_{21} & a_{22} \end{bmatrix}$$

Determinant of a 2 × 2 Matrix

$$\det A=\begin{vmatrix} a_{11} & a_{12} \\ a_{21} & a_{22} \end{vmatrix}=a_{11}a_{22}-a_{21}a_{12}$$

Cramer's Rule (used to solve unknowns using determinants)

For the following equations:

$$a_1x+b_1y=c_1$$
$$a_2x+b_2y=c_2$$

How it works: $D=\begin{vmatrix} a_1 & b_1 \\ a_2 & b_2 \end{vmatrix}$; $D_1=\begin{vmatrix} c_1 & b_1 \\ c_2 & b_2 \end{vmatrix}$; $D_2=\begin{vmatrix} a_1 & c_1 \\ a_2 & c_2 \end{vmatrix}$

The solution: $x=\frac{D_1}{D}$, $y=\frac{D_2}{D}$, provided $D\neq 0$

2 × 2 Matrix Transformations

Matrix	Transformation	Matrix	Transformation
$\begin{bmatrix}1 & 0\\0 & 1\end{bmatrix}$	Identity Matrix (I)	$\begin{bmatrix}0 & 1\\1 & 0\end{bmatrix}$	Reflection about line $x = y$
$\begin{bmatrix}0 & 0\\0 & 1\end{bmatrix}$	Projection onto y axis	$\begin{bmatrix}-1 & 0\\0 & -1\end{bmatrix}$	Reflection about origin
$\begin{bmatrix}1 & 0\\0 & 0\end{bmatrix}$	Projection onto x axis	$\begin{bmatrix}1 & b\\0 & 1\end{bmatrix}$	Horizontal shear
$\begin{bmatrix}a & 0\\0 & a\end{bmatrix}$	Magnification or dilation	$\begin{bmatrix}a & b\\c & d\end{bmatrix}$	Arbitrary transformation
$\begin{bmatrix}a & 0\\0 & b\end{bmatrix}$	Stretch and / or compression in two directions	$\begin{bmatrix}\cos\theta & -\sin\theta\\\sin\theta & \cos\theta\end{bmatrix}$	Rotation matrix

Complex Numbers

Complex Numbers

$$z = a + bi$$
$$i^2 = -1$$

Complex Number Arithmetic

$$(a+bi)+(c+di)=(a+c)+(b+d)i$$
$$(a+bi)-(c+di)=(a-c)+(b-d)i$$

Complex Conjugate Properties

$$\bar{z} = a - bi$$
$$\overline{z_1 + z_2} = \bar{z} + \bar{z}_2$$
$$z\bar{z} = a^2 + b^2$$
$$\overline{z_1 z_2} = \bar{z}_1 \bar{z}_2$$

Polar Coordinates

$$a = (r, \theta)$$
$$x = r\cos\theta$$
$$y = r\sin\theta$$
$$r = \sqrt{x^2 + y^2}$$
$$\theta = \arctan\frac{y}{x}, \ or\ \pi + \arctan\frac{y}{x}$$

The Polar Form of a Complex Number

$$z = r(\cos\theta + i\sin\theta)$$
$$= r\ cis\theta, (r \le 0)$$
$$= re^{i\theta}$$

The Polar Complex Conjugate

$$\bar{z} = r\ cis(-\theta)$$

Inverse of a Complex Number

$$z^{-1} = \frac{\bar{z}}{z^2} = \frac{a-bi}{a^2+b^2} = \frac{a}{a^2+b^2} - \frac{b}{a^2+b^2}i$$

Complex Number Multiplication

$$(a+bi)(c+di) = (ac-bd)+(ad+bc)i$$

Complex Modulus (absolute value) Properties

$$|z| = |a+bi| = \sqrt{a^2+b^2}$$
$$|z| = |\bar{z}|$$
$$z\bar{z} = |z|^2$$
$$|z_1 z_2| = |z_1||z_2|$$

Polar Complex Multiplication and Division

(where $z_1 = r_1\ cis\ \theta_1$, $z_2 = r_2$ cis θ_2)

$$z_1 z_2 = r_1 r_2\ cis\ (\theta_1 + \theta_2)$$
$$\frac{z_1}{z_2} = \frac{r_1}{r_2} cis\ (\theta_1 - \theta_2)$$

de Moivre's Theorem

$$(r\ cis\theta)^n = r^n\ cis\ (n\theta)$$
$$(r\ cis\theta)^{-n} = r^{-n}\ cis\ (-n\theta)$$
$$\left(re^{i\theta}\right)^n = r^n\ e^{in\theta}$$

Complex Roots

$$z_k =^n \sqrt{r}\ cis\left(\frac{\theta + 2k\pi}{n}\right), (k = 0, 1, 2, \ldots, n-1)$$

Nothing is really work unless you would rather be doing something else. J.M. Barrie

8.3 Trigonometry

Common Angle Conversions

$1° = (\pi/180)$ rad $= 0.017453292$ rad 1 rad $= (180/\pi)° = 57.2957795°$

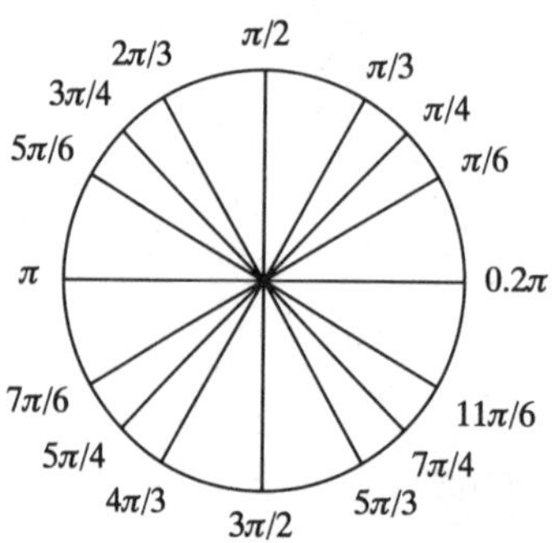

Fig. 8.3.1

$1/2 = 0.5$
$1/\sqrt{2} = 0.70106$
$2/\sqrt{3} = 1.15470$
$1/\sqrt{3} = 0.57735$
$\sqrt{3}/2 = 0.86603$
$\sqrt{3} = 1.73205$

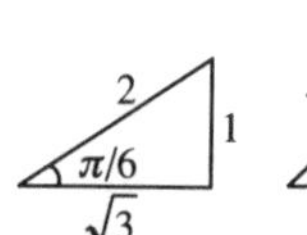

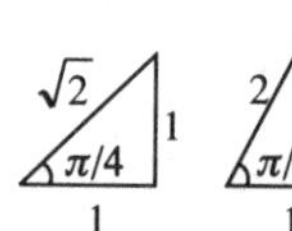

2 √3 π/3 1

Fig. 8.3.2 Reference Triangles

Common Angle Conversions

Degrees (θ)	Radians (θ)	sin θ	csc θ (1/sinθ)	cos θ	sec θ (1/cosθ)	tan θ	cot θ (1/tanθ)
0°	0	0	–	1	1	0	–
30°	$\pi/6$	$1/2$	2	$\sqrt{3}/2$	$2/\sqrt{3}$	$1/\sqrt{3}$	$\sqrt{3}$
45°	$\pi/4$	$1/\sqrt{2}$	$\sqrt{2}$	$1/\sqrt{2}$	$\sqrt{2}$	1	1
60°	$\pi/3$	$\sqrt{3}/2$	$2/\sqrt{3}$	$1/2$	2	$\sqrt{3}$	$1/\sqrt{3}$
90°	$\pi/2$	1	1	0	–	–	0
120°	$2\pi/3$	$\sqrt{3}/2$	$2/\sqrt{3}$	$-1/2$	-2	$-\sqrt{3}$	$-1/\sqrt{3}$
135°	$3\pi/4$	$1/\sqrt{2}$	$\sqrt{2}$	$-1/\sqrt{2}$	$-\sqrt{2}$	-1	-1
150°	$5\pi/6$	$1/2$	2	$-\sqrt{3}/2$	$-2/\sqrt{3}$	$-1/\sqrt{3}$	$\sqrt{3}$
180°	π	0	–	-1	-1	0	–
210°	$7\pi/6$	$-1/2$	-2	$\sqrt{3}/2$	$-2/\sqrt{3}$	$1/\sqrt{3}$	$\sqrt{3}$
225°	$5\pi/4$	$-1/\sqrt{2}$	$-\sqrt{2}$	$-1/\sqrt{2}$	$-\sqrt{2}$	1	1
240°	$4\pi/3$	$-\sqrt{3}/2$	$-2/\sqrt{3}$	$-1/2$	-2	$\sqrt{3}$	$1/\sqrt{3}$
270°	$3\pi/2$	-1	-1	0	–	–	0
300°	$5\pi/3$	$-\sqrt{3}/2$	$-2/\sqrt{3}$	$1/2$	2	$-\sqrt{3}$	$-1/\sqrt{3}$
315°	$7\pi/4$	$-1\sqrt{2}$	$-\sqrt{2}$	$1/\sqrt{2}$	$\sqrt{2}$	-1	-1
330°	$11\pi/6$	$-1/2$	-2	$\sqrt{3/2}$	$2/\sqrt{3}$	$-1/\sqrt{3}$	$-\sqrt{3}$
360°	2π	0	–	1	1	0	–

Thought is free. William Shakespeare

Fundamental Trigonometric Identities

$r^2 = x^2 + y^2$

$\frac{y^2}{r^2} + \frac{x^2}{r^2} = \frac{r^2}{r^2} = 1$

$\sin\theta = \frac{y}{r} = \frac{\text{opposite}}{\text{hypotenuse}} = \frac{1}{\csc\theta}$

$\cos\theta = \frac{x}{r} = \frac{\text{adjacent}}{\text{hypotenuse}} = \frac{1}{\sec\theta}$

$\tan\theta = \frac{y}{x} = \frac{\text{opposite}}{\text{adjacent}} = \frac{1}{\cot\theta} = \frac{\sin\theta}{\cos\theta}$

$\sin(-\theta) = -\sin\theta \quad \tan(-\theta) = -\tan\theta \quad \sin\left(\frac{\pi}{2} - \theta\right) = \cos\theta$

$\cos(-\theta) = \cos\theta \quad \cos\left(\frac{\pi}{2} - \theta\right) = \sin\theta \quad \tan\left(\frac{\pi}{2} - \theta\right) = \cot\theta$

$\sec^2\theta = 1 + \tan^2\theta$

$\csc^2\theta = 1 + \cot^2\theta$

$\sin^2\theta + \cos^2\theta = 1$

Addition Formulas

$\sin(\theta + \Phi) = \sin\theta\cos\Phi + \cos\theta\sin\Phi$

$\cos(\theta + \Phi) = \cos\theta\cos\Phi - \sin\theta\sin\Phi$

$\tan(\theta + \Phi) = \frac{\tan\theta + \tan\Phi}{1 - \tan\theta\tan\Phi}$

Subtraction Formulas

$\sin(\theta - \Phi) = \sin\theta\cos\Phi - \cos\theta\sin\Phi$

$\cos(\theta - \Phi) = \cos\theta\cos\Phi + \sin\theta\sin\Phi$

$\tan(\theta - \Phi) = \frac{\tan\theta - \tan\Phi}{1 + \tan\theta\tan\Phi}$

Product Formulas

$\sin\theta\cos\Phi = \tfrac{1}{2}[\sin(\theta + \Phi) + \sin(\theta - \Phi)]$

$\cos\theta\cos\Phi = \tfrac{1}{2}[\cos(\theta + \Phi) + \cos(\theta - \Phi)]$

$\sin\theta\sin\Phi = \tfrac{1}{2}[\cos(\theta - \Phi) - \cos(\theta + \Phi)]$

Double-Angle Formulas

$\sin 2\theta = 2\sin\theta\cos\theta$

$\cos 2\theta = \cos^2\theta - \sin^2\theta = 2\cos^2\theta - 1 = 1 - 2\sin^2\theta$

$\tan 2\theta = \frac{2\tan\theta}{1 - \tan^2\theta}$

Power Reducing Formulas

$\sin^2\theta = \frac{1 - \cos 2\theta}{2}$

$\cos^2\theta = \frac{1 + \cos 2\theta}{2}$

$\tan^2\theta = \frac{1 - \cos 2\theta}{1 + \cos 2\theta}$

Law of Cosines for All Triangles

$a^2 = b^2 + c^2 - 2bc\cos A$

$b^2 = a^2 + c^2 - 2ac\cos B$

$c^2 = a^2 + b^2 - 2ab\cos C$

Law of Sines

$\frac{\sin A}{a} = \frac{\sin B}{b} = \frac{\sin C}{c}$

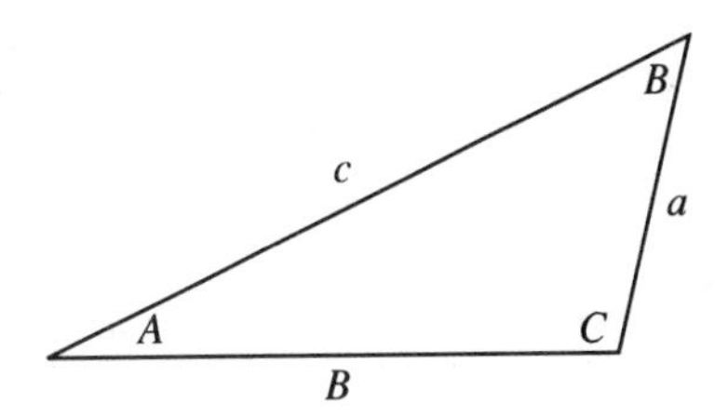

Fig. 8.3.3

Boys will be boys, and so will a lot of middle aged men. F. McKinney

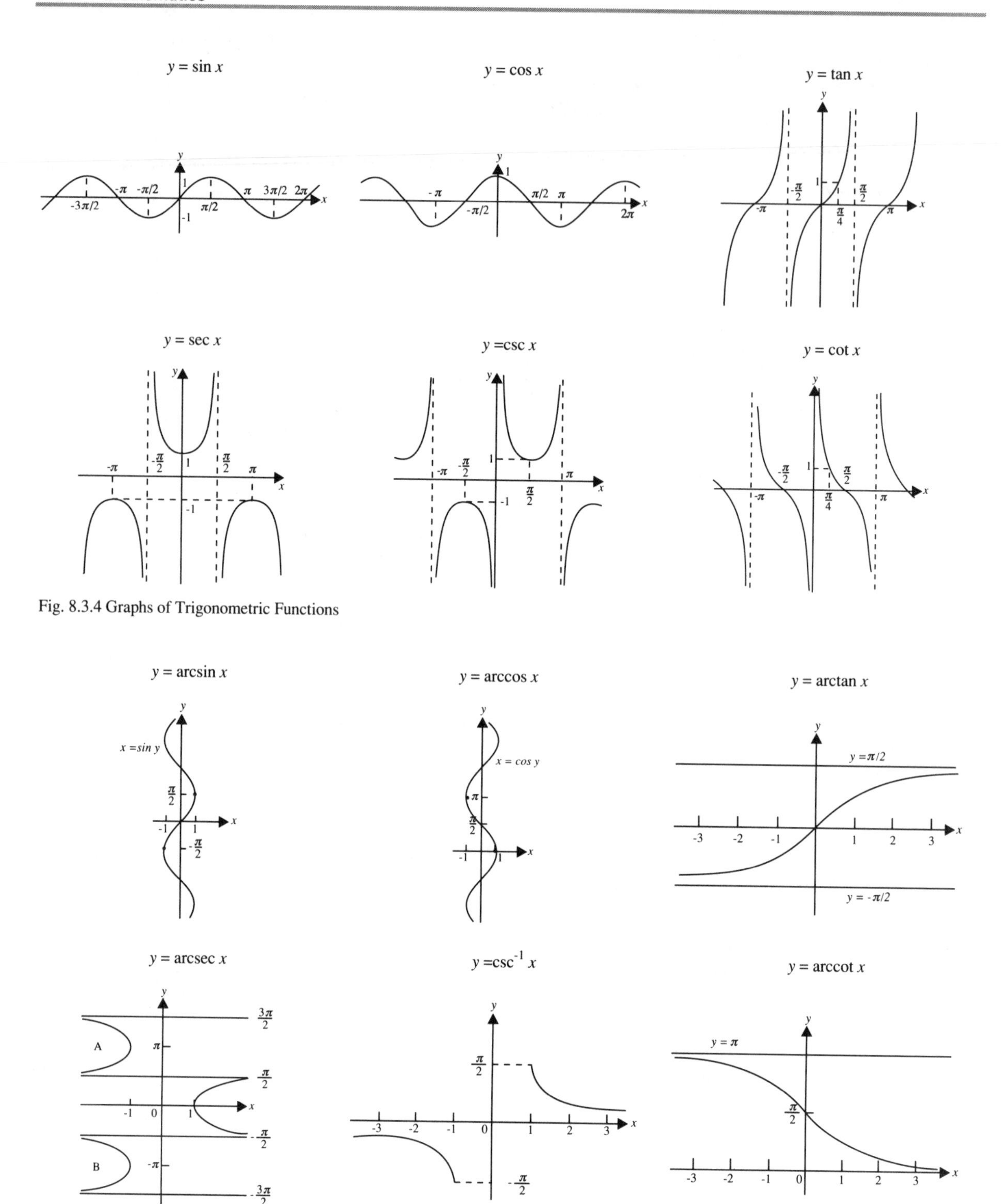

Fig. 8.3.4 Graphs of Trigonometric Functions

Fig. 8.3.5 Graphs of Inverse Trigonometric Functions

Hyperbolic Functions and Identities

$$\cosh x = \frac{e^x + e^{-x}}{2}$$

$$\sinh x = \frac{e^x - e^{-x}}{2}$$

$$\tanh x = \frac{e^x - e^{-x}}{e^x + e^{-x}}$$

$$\coth x = \frac{e^x + e^{-x}}{e^x - e^{-x}}$$

$$\text{sech } x = \frac{2}{e^x + e^{-x}}$$

$$\text{csch } x = \frac{2}{e^x - e^{-x}}$$

$$\sinh 2x = 2 \sinh x \cosh x$$

$$\cosh 2x = \cosh^2 x + \sinh^2 x$$

$$\cosh^2 x = \frac{\cosh 2x + 1}{2}$$

$$\sinh^2 x = \frac{\cosh 2x - 1}{2}$$

$$\cosh^2 x - \sinh^2 x = 1$$

$$\tanh^2 x = 1 - \text{sech}^2 x$$

$$\coth^2 x = 1 + \text{csch}^2 x$$

Derivatives

$$f'(x) = y' = \frac{dy}{dx} = \lim_{\Delta x \to 0} \frac{\Delta y}{\Delta x} = \lim_{\Delta x \to 0} \frac{y(x + \Delta x) - y\Delta}{\Delta x} = \lim_{\Delta x \to 0} \frac{f(x + \Delta x) - f(x)}{\Delta x}$$

Derivative Rules

Constant Rule

$$\frac{d(cu)}{dx} = c\frac{du}{dx}$$

$$\frac{d}{dx}(cu) = cu'$$

Sum Rule

$$\frac{d}{dx}(u + v) = \frac{du}{dx} + \frac{dv}{dx}$$

$$\frac{d}{dx}(u + v) = u' + v'$$

Difference Rule

$$\frac{d}{dx}(u - v) = \frac{du}{dx} - \frac{dv}{dx}$$

$$\frac{d}{dx}(u - v) = u' - v'$$

Product Rule

$$\frac{d}{dx}(uv) = u\frac{dv}{dx} + v\frac{du}{dx}$$

$$\frac{d}{dx}(uv) = uv' + vu'$$

Quotient Rule

$$\frac{d}{dx}\left(\frac{u}{v}\right) = \frac{v\frac{du}{dx} - u\frac{dv}{dx}}{v^2}$$

$$\frac{d}{dx}\left(\frac{u}{v}\right) = \frac{vu' - uv'}{v^2}$$

Power Rule

$$\frac{d}{dx}(x^n) = nx^{n-1}$$

Chain Rule

$$\frac{dy}{dx} = \frac{dy}{du} \cdot \frac{du}{dx}$$

Second Derivative

$$\frac{d^2 y}{dx^2} = \frac{d}{dx}\left(\frac{dy}{dx}\right)$$

Common Derivatives List

Involving Constants

$$\frac{d}{dx}[cu] = cu'$$
$$\frac{d}{dx}[c] = 0$$

Involving [x]

$$\frac{d}{dx}[x] = 1$$

Involving Sum or Difference

$$\frac{d}{dx}[u \pm v] = u' \pm v'$$

Involving e^u, ln and Absolute Values

$$\frac{d}{dx}[|u|] = \frac{u}{|u|}(u'),\ u \neq 0$$
$$\frac{d}{dx}[\ln u] = \frac{u'}{u}$$
$$\frac{d}{dx}[e^u] = e^u u'$$

Involving Inverse Trigometric Functions

$$\frac{d}{dx}[\arcsin u] = \frac{u'}{\sqrt{1-u^2}} \quad \left(\frac{-\pi}{2} \leq \arcsin u \leq \frac{\pi}{2}\right)$$
$$\frac{d}{dx}[\arccos u] = \frac{-u'}{\sqrt{1-u^2}} \quad (0 \leq \arccos u \leq \pi)$$
$$\frac{d}{dx}[\arctan u] = \frac{u'}{1+u^2} \quad \left(\frac{-\pi}{2} < \arctan u < \frac{\pi}{2}\right)$$
$$\frac{d}{dx}[\text{arccot } u] = \frac{-u'}{1+u^2} \quad (0 \leq \arctan u \leq \pi)$$
$$\frac{d}{dx}[\text{arcsec } u] = \frac{u'}{|u|\sqrt{u^2-1}} \quad \left(\begin{array}{l} 0 \leq \text{arcsec } u \leq \pi \\ \text{arcsec } u \neq \frac{\pi}{2} \end{array}\right)$$
$$\frac{d}{dx}[\text{arccsc } u] = \frac{-u'}{|u|\sqrt{u^2-1}} \quad \left(\begin{array}{l} -\frac{\pi}{2} \leq \text{arccsc } u \leq \frac{\pi}{2} \\ \text{arccsc } u \neq 0 \end{array}\right)$$

Involving Product

$$\frac{d}{dx}[uv] = uv' + vu'$$

Involving Quotient

$$\frac{d}{dx}\left[\frac{u}{v}\right] = \frac{vu' - uv'}{v^2}$$

Involving Powers

$$\frac{d}{dx}[u^n] = nu^{n-1}u'$$

Involving Trigonometric Functions

$$\frac{d}{dx}[\sin u] = (\cos u)u'$$
$$\frac{d}{dx}[\cos u] = -(\sin u)\, u'$$
$$\frac{d}{dx}[\tan u] = (\sec^2 u)\, u'$$
$$\frac{d}{dx}[\cot u] = -(\csc^2 u)u'$$
$$\frac{d}{dx}[\sec u] = (\sec u \tan u)u'$$
$$\frac{d}{dx}[\csc u] = -(\csc u \cot u)u'$$

Hyperbolic Functions

$$\frac{d}{dx}(\sinh u) = (\cosh u)\, u'$$
$$\frac{d}{dx}(\cosh u) = (\sinh u)\, u'$$
$$\frac{d}{dx}(\tanh u) = (\text{sech}^2\, u)\, u'$$
$$\frac{d}{dx}(\coth u) = (-\text{csch}^2\, u)\, u'$$
$$\frac{d}{dx}(\text{sech } u) = (-\text{sech } u \tanh u)\, u'$$
$$\frac{d}{dx}(\text{csch } u) = (-\text{csch } u \coth u)\, u'$$

Techniques for Integration

The Educated Guess (use this when you don't know what you're doing)
1) Determine a reasonable answer.
2) Compare its derivative with the problem.
3) Modify answer and try again by making improvements.

The Substitution Method

For integrals of the type $\int f(g(x))\, g'(x)\, dx$:

(f and g' are continuous functions)
1) Substitute by letting $u = g(x)$ and $du = g'(x)\, dx$.
2) This will give $\int f(u)\, du$.
3) Integrate with respect to u, then replace u with $g(x)$ to obtain the answer.

Trigonometric Substitutions (using trigonometric identities to help)

$a^2 - u^2$ is replaced by $a^2 \cos^2 \theta$ using $u = a \sin \theta$.

$a^2 + u^2$ is replaced by $a^2 \sec^2 \theta$ using $u = a \tan \theta$.

$u^2 - a^2$ is replaced by $a^2 \tan^2 \theta$ using $u = a \sec \theta$.

Integration by Parts

$$\int u\, dv = u\, v - \int v\, du$$

Hints on choosing u and dv:
1) u should be something that becomes simpler when differentiated.
2) dv should be something with a simple integral.

Trigonometric Integrals

1) For $\int \sin mx \cos nx\, dx$ use $\int ½[\sin (m - n)x + \sin (m + n)x]$.
2) For $\int \sin mx \sin nx\, dx$ use $\int ½[\cos (m - n)x - \cos (m + n)x]$.
3) For $\int \cos mx \cos nx\, dx$ use $\int ½[\cos (m - n)x + \cos (m + n)x]$.
4) For $\int \sin^m x \cos^n x\, dx$ the substitution depends on m and n.
   a) If both m and n are even, then use the half-angle identities:
      $\sin^2 x = ½[1 - \cos 2x]$ and $\cos^2 x = ½[1 + \cos 2x]$
   b) If the power of cosine is odd ($n = 2k + 1$), replace the $\cos^{2k+1} x$ with $\cos x\,(1 - \sin^2 x)^{2k}$.
      Let $u = \sin x$, $du = \cos x\, dx$.

Success is the best revenge. French proverb

c) If the power of sine is odd ($m = 2k + 1$), replace the $\sin^{2k+1} x$ with $\sin x$ $(1 - \cos^2 x)^{2k}$.
Let $u = \cos x$, $du = (-\sin x)\, dx$

5) For $\int \tan^m x \sec^n x\, dx$ the substitution depends on m and n.

a) If the power of secant is even ($n = 2k$), replace $\sec^{2k} x$ with $\sec^2 x\,(1 + \tan^2 x)^{k-1}$.
Let $u = \tan x$, $du = \tan x\, dx$.

b) If the power of tangent is odd ($n = 2k + 1$), remove a factor of $\sec x \tan x$, and replace the remaining $\tan^2 x$ with $\sec^2 x - 1$.
Let $u = \sec x$, $du = \sec x \tan x\, dx$.

Partial Fractions for Integrating Rational Functions

Use this technique to convert a complicated rational function to a sum of simpler fractions. This makes it easier to integrate.

$$x - a\text{, use } \frac{A}{x-a}$$

$$(x-a)^k\text{, use } \frac{A_1}{x-a} + \frac{A_2}{(x-a)^2} + \ldots + \frac{A_k}{(x-a)^k}$$

$$\frac{1}{(x-a)(x-b)}\text{, use } \frac{A}{x-a} + \frac{B}{x-b}$$

$$(ax^2 + bx + c)\text{, use } \frac{Ax+B}{(ax^2+bx+c)}$$

$$(ax^2+bx+c)^k\text{, use } \frac{A_1x+B_1}{(ax^2+bx+c)} + \frac{A_2x+B_2}{(ax^2+bx+c)^2} + \ldots + \frac{A_kx+B_k}{(ax^2+bx+c)^k}$$

Rationalizing Substitutions

If the integral contains an expression like $\sqrt[n]{g(x)}$, then let $u = \sqrt[n]{g(x)}$, or $u^n = g(x)$.

Weierstrass Substitution

To convert any function of sin (x) and cos (x) into a rational function, use the following substitutions:

$$\cos(x) = \frac{1-t^2}{1+t^2} \qquad t = \tan\left(\frac{x}{2}\right)$$

$$\sin(x) = \frac{2t}{1+t^2} \qquad dx = \frac{2}{1+t^2}\, dt$$

Common Integrals

Quick List

$$\int u^n \, du = \frac{u^{n+1}}{n+1} + C, n \neq -1$$

$$\int \frac{1}{u} \, du = \ln |u| + C$$

$$\int \frac{1}{\sin x \cos x} \, dx = \ln |\tan x| + C$$

$$\int \frac{1}{1+e^x} \, dx = x - \ln\left(1+e^x\right) + C$$

$$\int x \ln x \, dx = \frac{x^2\left[(2 \ln x) - 1\right]}{4} + C$$

$$\int \cos u \, du = \sin u + C$$

$$\int \sin u \, du = -\cos u + C$$

$$\int \sin ax \, dx = -\frac{1}{a} \cos ax + C$$

$$\int \cos ax \, dx = \frac{1}{a} \sin ax + C$$

$$\int \tan ax \, dx = \frac{1}{a} \ln |\sec ax| + C$$

$$\int \frac{dx}{1+\sin ax} = -\frac{1}{a} \tan\left(\frac{\pi}{4} - \frac{ax}{2}\right) + C$$

$$\int \frac{dx}{1-\sin ax} = \frac{1}{a} \tan\left(\frac{\pi}{4} + \frac{ax}{2}\right) + C$$

$$\int \frac{dx}{1+\cos ax} = \frac{1}{a} \tan\frac{ax}{2} + C$$

$$\int \frac{dx}{1-\cos ax} = -\frac{1}{a} \cot\frac{ax}{2} + C$$

$$\int x \sin ax \, dx = \frac{1}{a^2} \sin ax - \frac{x}{a} \cos ax + C$$

$$\int x \cos ax \, dx = \frac{1}{a^2} \cos ax + \frac{x}{a} \sin ax + C$$

$$\int \cot ax \, dx = \frac{1}{a} \ln |\sin ax| + C$$

$$\int \sec ax \, dx = \frac{1}{a} \ln |\sec ax + \tan ax| + C$$

$$\int \csc ax \, dx = -\frac{1}{a} \ln |\csc ax + \cot ax| + C$$

$$\int \arcsin ax \, dx = x \arcsin ax + \frac{1}{a} \sqrt{1-a^2x^2} + C$$

$$\int \arccos ax \, dx = x \arccos ax - \frac{1}{a} \sqrt{1-a^2x^2} + C$$

$$\int \arctan ax \, dx = x \arctan ax - \frac{1}{2a} \ln\left(1+a^2x^2\right) + C$$

$$\int e^{ax} \, dx = \frac{1}{a} e^{ax} + C$$

$$\int \ln ax \, dx = x \ln ax - x + C$$

$$\int \frac{du}{\sqrt{1-u^2}} = \arcsin u + C, \text{ when } u^2 < 1$$

$$\int \frac{du}{1+u^2} = \arctan u + C$$

$$\int \frac{du}{u\sqrt{u^2-1}} = \arccos \left|\frac{1}{u}\right|, \text{ when } u^2 > 1$$

$$\int u \, du = uv - \int v \, du$$

$$\int a^u \, du = \frac{a^u}{\ln a} + C, a \neq 1, a > 0$$

Involving u

$$\int u^n\,du = \frac{u^{n+1}}{n+1} + C,\ n \neq -1$$

$$\int \frac{1}{u}\,du = \ln|u| + C$$

$$\int u\,dv = uv - \int v\,du$$

$$\int a^u\,du = \frac{a^u}{\ln a} + C,\ a \neq 1,\ a > 0$$

Involving $ax + b$ and $a + bx$

$$\int (ax+b)^n\,dx = \frac{(ax+b)^{n+1}}{a(n+1)} + C,\ n \neq -1$$

$$\int \frac{1}{ax+b}\,dx = \frac{1}{a}\ln|ax+b| + C$$

$$\int \frac{x}{(ax+b)^2}\,dx = \frac{1}{a^2}\left[\ln|ax+b| + \frac{b}{ax+b}\right] + C$$

$$\int \frac{dx}{x(ax+b)} = \frac{1}{b}\ln\left|\frac{x}{ax+b}\right| + C$$

$$\int \frac{x}{(a+bx)^n}\,dx = \frac{1}{b^2}\left[\frac{-1}{(n-2)(a+bx)^{n-2}} + \frac{a}{(n-1)(a+bx)^{n-1}}\right] + C,\ n \neq 1, 2$$

$$\int \frac{x^2}{a+bx}\,dx = \frac{1}{b^3}\left[-\frac{bx}{2}(2a-bx) + a^2\ln|a+bx|\right] + C$$

$$\int \frac{x^2}{(a+bx)^2}\,dx = \frac{1}{b^3}\left[a+bx - \frac{a^2}{a+bx} - 2a\ln|a+bx|\right] + C$$

$$\int \frac{x^2}{(a+bx)^n}\,dx = \frac{1}{b^3}\left[\frac{-1}{(n-3)(a+bx)^{n-3}} + \frac{2a}{(n-2)(a+bx)^{n-2}} - \frac{a^2}{(n-1)(a+bx)^{n-1}}\right] + C,\ n \neq 1, 2, 3$$

$$\int \frac{1}{x(a+bx)}\,dx = \frac{1}{a}\ln\left|\frac{x}{a+bx}\right| + C$$

$$\int \frac{dx}{x(a+bx)^2} = \frac{1}{a}\left[\frac{1}{a+bx} + \frac{1}{a}\ln\left|\frac{x}{a+bx}\right|\right] + C$$

$$\int \frac{1}{x^2(a+bx)}\,dx = -\frac{1}{a}\left[\frac{1}{x} + \frac{b}{a}\ln\left|\frac{x}{a+bx}\right|\right] + C$$

$$\int \frac{x}{ax+b}\,dx = \frac{x}{a} - \frac{b}{a^2}\ln|ax+b| + C$$

Involving $a^2 \pm x^2$

$$\int \frac{dx}{a^2+x^2} = \frac{1}{a}\arctan\frac{x}{a} + C$$

$$\int \frac{dx}{(a^2+x^2)^2} = \frac{x}{2a^2(a^2+x^2)} + \frac{1}{2a^3}\arctan\frac{x}{a} + C$$

$$\int \frac{dx}{a^2-x^2} = \frac{1}{2a}\ln\left|\frac{x+a}{x-a}\right| + C$$

$$\int \frac{dx}{(a^2-x^2)^2} = \frac{x}{2a^2(a^2-x^2)} + \frac{1}{2a^2}\int \frac{dx}{a^2-x^2} + C$$

Involving $\sqrt{a^2 \pm x^2}$

$$\int \frac{dx}{\sqrt{a^2+x^2}} = \operatorname{arcsinh}\frac{x}{a} + C = \ln\left|x+\sqrt{a^2+x^2}\right| + C$$

$$\int \sqrt{a^2+x^2}\,dx = \frac{x}{2}\sqrt{a^2+x^2} + \frac{a^2}{2}\operatorname{arcsinh}\frac{x}{a} + C$$

$$\int x^2\sqrt{a^2+x^2}\,dx = \frac{x\left(a^2+2x^2\right)\sqrt{a^2+x^2}}{8} - \frac{a^4}{8}\operatorname{arcsinh}\frac{x}{a} + C$$

$$\int \frac{\sqrt{a^2+x^2}}{x}\,dx = \sqrt{a^2+x^2} - a\operatorname{arcsinh}\left|\frac{a}{x}\right| + C$$

$$\int \frac{\sqrt{a^2+x^2}}{x^2}\,dx = \operatorname{arcsinh}\frac{x}{a} - \frac{\sqrt{a^2+x^2}}{x} + C$$

$$\int \frac{x^2}{\sqrt{a^2+x^2}}\,dx = -\frac{a^2}{2}\operatorname{arcsinh}^{-1}\frac{x}{a} + \frac{x\sqrt{a^2+x^2}}{2} + C$$

$$\int \frac{dx}{x\sqrt{a^2+x^2}} = -\frac{1}{a}\ln\left|\frac{a+\sqrt{a^2+x^2}}{x}\right| + C$$

$$\int \frac{dx}{x^2\sqrt{a^2+x^2}} = -\frac{\sqrt{a^2+x^2}}{a^2x} + C$$

$$\int \frac{dx}{\sqrt{a^2-x^2}} = \arcsin\frac{x}{a} + C$$

$$\int \sqrt{a^2-x^2}\,dx = \frac{x}{2}\sqrt{a^2-x^2} + \frac{a^2}{2}\arcsin\frac{x}{a} + C$$

$$\int x^2\sqrt{a^2-x^2}\,dx = \frac{a^4}{8}\arcsin\frac{x}{a} - \frac{1}{8}x\sqrt{a^2-x^2}\left(a^2-2x^2\right) + C$$

$$\int \frac{\sqrt{a^2-x^2}}{x}\,dx = \sqrt{a^2-x^2} - a\ln\left|\frac{a+\sqrt{a^2-x^2}}{x}\right| + C$$

$$\int \frac{\sqrt{a^2-x^2}}{x^2}\,dx = -\arcsin\frac{x}{a} - \frac{\sqrt{a^2-x^2}}{x} + C$$

$$\int \frac{x^2}{\sqrt{a^2-x^2}}\,dx = \frac{a^2}{2}\arcsin\frac{x}{a} - \frac{1}{2}x\sqrt{a^2-x^2} + C$$

$$\int \frac{dx}{x\sqrt{a^2-x^2}} = -\frac{1}{a}\ln\left|\frac{a+\sqrt{a^2-x^2}}{x}\right| + C$$

$$\int \frac{dx}{x^2\sqrt{a^2-x^2}} = -\frac{\sqrt{a^2-x^2}}{a^2x} + C$$

$$\int \frac{dx}{\sqrt{x^2-a^2}} = \operatorname{arccosh}\frac{x}{a} + C = \ln\left|x+\sqrt{x^2-a^2}\right| + C$$

$$\int \sqrt{x^2-a^2}\,dx = \frac{x}{2}\sqrt{x^2-a^2} - \frac{a^2}{2}\operatorname{arccosh}\frac{x}{a} + C$$

$$\int x^2\sqrt{x^2-a^2}\,dx = \frac{x}{8}\left(2x^2-a^2\right)\sqrt{x^2-a^2} - \frac{a^4}{8}\operatorname{arccosh}\frac{x}{a} + C$$

$$\int \frac{\sqrt{x^2-a^2}}{x}\,dx = \sqrt{x^2-a^2} - a\operatorname{arcsec}\left|\frac{x}{a}\right| + C$$

$$\int \frac{\sqrt{x^2-a^2}}{x^2}\,dx = \operatorname{arcosh}\frac{x}{a} - \frac{\sqrt{x^2-a^2}}{x} + C$$

$$\int \frac{x^2}{\sqrt{x^2-a^2}}\,dx = \frac{a^2}{2}\operatorname{arccosh}\frac{x}{a} + \frac{x}{2}\sqrt{x^2-a^2} + C$$

$$\int \frac{dx}{x\sqrt{x^2-a^2}} = \frac{1}{a}\operatorname{arcsec}\left|\frac{x}{a}\right| + C = \frac{1}{a}\arccos\left|\frac{a}{x}\right| + C$$

$$\int \frac{dx}{x^2\sqrt{x^2-a^2}} = \frac{\sqrt{x^2-a^2}}{a^2x} + C$$

Involving $\sqrt{ax+b}$

$$\int \frac{\sqrt{ax+b}}{x}\,dx = 2\sqrt{ax+b} + b\int \frac{dx}{x\sqrt{ax+b}}$$

$$\int \frac{dx}{x\sqrt{ax+b}} = \frac{2}{\sqrt{-b}}\arctan\sqrt{\frac{ax+b}{-b}} + C,\ \text{if } b<0$$

$$\int \frac{dx}{x\sqrt{ax+b}} = \frac{1}{\sqrt{b}}\ln\left|\frac{\sqrt{ax+b}-\sqrt{b}}{\sqrt{ax+b}+\sqrt{b}}\right| + C,\ \text{if } b>0$$

$$\int \frac{1}{x^n\sqrt{a+bx}}\,dx = \frac{-1}{a(n-1)}\left[\frac{\sqrt{a+bx}}{x^{n-1}} + \frac{(2n-3)b}{2}\int \frac{1}{x^{n-1}\sqrt{a+bx}}\,dx\right],\ n\neq 1$$

$$\int \frac{\sqrt{a+bx}}{x^n}\,dx = \frac{-1}{a(n-1)}\left[\frac{(a+bx)^{3/2}}{x^{n-1}} + \frac{(2n-5)b}{2}\int \frac{\sqrt{a+bx}}{x^{n-1}}\,dx\right],\ n\neq 1$$

An injury is much sooner forgotten than an insult. Lord Chesterfield

Involving sin, cos, tan

$$\int \cos u\,du = \sin u + C$$

$$\int \sin u\,du = -\cos u + C$$

$$\int \sin ax\,dx = -\frac{1}{a}\cos ax + C$$

$$\int \cos ax\,dx = \frac{1}{a}\sin ax + C$$

$$\int \sin^2 ax\,dx = \frac{x}{2} - \frac{\sin 2ax}{4a} + C$$

$$\int \cos^2 ax\,dx = \frac{x}{2} + \frac{\sin 2ax}{4a} + C$$

$$\int \sin^n ax\,dx = \frac{-\sin^{n-1} ax \cos ax}{na} + \frac{n-1}{n}\int \sin^{n-2} ax\,dx$$

$$\int \cos^n ax\,dx = \frac{\cos^{n-1} ax \sin ax}{na} + \frac{n-1}{n}\int \cos^{n-2} ax\,dx$$

$$\int \frac{dx}{1+\sin ax} = -\frac{1}{a}\tan\left(\frac{\pi}{4} - \frac{ax}{2}\right) + C$$

$$\int \frac{dx}{1-\sin ax} = \frac{1}{a}\tan\left(\frac{\pi}{4} + \frac{ax}{2}\right) + C$$

$$\int \frac{dx}{1+\cos ax} = \frac{1}{a}\tan\frac{ax}{2} + C$$

$$\int \frac{dx}{1-\cos ax} = -\frac{1}{a}\cot\frac{ax}{2} + C$$

$$\int x \sin ax\,dx = \frac{1}{a^2}\sin ax - \frac{x}{a}\cos ax + C$$

$$\int x \cos ax\,dx = \frac{1}{a^2}\cos ax + \frac{x}{a}\sin ax + C$$

$$\int x^n \sin ax\,dx = -\frac{x^n}{a}\cos ax + \frac{n}{a}\int x^{n-1}\cos ax\,dx$$

$$\int x^n \cos ax\,dx = \frac{x^n}{a}\sin ax - \frac{n}{a}\int x^{n-1}\sin ax\,dx$$

$$\int \tan ax\,dx = \frac{1}{a}\ln|\sec ax| + C$$

$$\int \tan^2 ax\,dx = \frac{1}{a}\tan ax - x + C$$

$$\int \tan^n ax\,dx = \frac{\tan^{n-1} ax}{a(n-1)} - \int \tan^{n-2} ax\,dx,\ n \neq 1$$

$$\int \frac{1}{\sin x \cos x}\,dx = \ln|\tan x| + C$$

Involving cot, sec, csc

$$\int \cot ax\,dx = \frac{1}{a}\ln|\sin ax| + C$$

$$\int \cot^2 ax\,dx = -\frac{1}{a}\cot ax - x + C$$

$$\int \cot^n ax\,dx = \frac{-\cot^{n-1} ax}{a(n-1)} - \int \cot^{n-2} ax\,dx + C,\ n \neq 1$$

$$\int \sec ax\,dx = \frac{1}{a}\ln|\sec ax + \tan ax| + C$$

$$\int \csc ax\,dx = -\frac{1}{a}\ln|\csc ax + \cot ax| + C$$

$$\int \sec^2 ax\,dx = \frac{1}{a}\tan ax + C$$

$$\int \csc^2 ax\,dx = -\frac{1}{a}\cot ax + C$$

$$\int \sec^n ax\,dx = \frac{\sec^{n-2} ax \tan ax}{a(n-1)} + \frac{n-2}{n-1}\int \sec^{n-2} ax\,dx,\ n \neq 1$$

$$\int \csc^n ax\,dx = \frac{-\csc^{n-2} ax \cot ax}{a(n-1)} + \frac{n-2}{n-1}\int \csc^{n-2} ax\,dx,\ n \neq 1$$

Involving Hyperbolic Functions

$$\int \sinh ax\,dx = \frac{1}{a}\cosh ax + C$$

$$\int \cosh ax\,dx = \frac{1}{a}\sinh ax + C$$

$$\int \operatorname{sech} ax\,dx = \frac{1}{a}\arcsin(\tan ax) + C$$

$$\int \operatorname{csch} ax\,dx = \frac{1}{a}\ln\left|\tanh\frac{ax}{2}\right| + C$$

$$\int \operatorname{sech}^2 ax\,dx = \frac{1}{a}\tanh ax + C$$

$$\int \operatorname{csch}^2 ax\,dx = -\frac{1}{a}\coth ax + C$$

Involving Inverse Trigonometric Functions

$$\int \arcsin ax\,dx = x \arcsin ax + \frac{1}{a}\sqrt{1-a^2x^2} + C$$

$$\int \arccos ax\,dx = x \arccos ax - \frac{1}{a}\sqrt{1-a^2x^2} + C$$

$$\int \arctan ax\,dx = x \arctan ax - \frac{1}{2a}\ln\left(1+a^2x^2\right) + C$$

$$\int \operatorname{arccot} x\,dx = x \operatorname{arccot} x - \ln\left|x + \sqrt{1+x^2}\right| + C$$

$$\int \operatorname{arcsec} x\,dx = x \operatorname{arcsec} x - \ln\left|x + \sqrt{x^2-1}\right| + C$$

$$\int \operatorname{arccsc} x\,dx = x \operatorname{arccsc} x + \ln\left|x + \sqrt{x^2-1}\right| + C$$

Nature never breaks her own laws. Leonardo Da Vinci

Involving e^x

$$\int e^{ax}\,dx = \frac{1}{a}e^{ax} + C$$

$$\int xe^{ax}\,dx = \frac{e^{ax}}{a^2}(ax-1) + C$$

$$\int x^n e^{ax}\,dx = \frac{1}{a}x^n e^{ax} - \frac{n}{a}\int x^{n-1}e^{ax}\,dx$$

$$\int e^{ax}\sin bx\,dx = \frac{e^{ax}}{a^2+b^2}(a\sin bx - b\cos bx) + C$$

$$\int e^{ax}\cos bx\,dx = \frac{e^{ax}}{a^2+b^2}(a\cos bx + b\sin bx) + C$$

$$\int \frac{1}{1+e^x}\,dx = x - \ln\left(1+e^x\right) + C$$

Involving $\ln x$

$$\int \ln ax\,dx = x\ln ax - x + C$$

$$\int x^n(\ln ax)^m\,dx = \frac{x^{n+1}(\ln ax)^m}{n+1} - \frac{m}{n+1}\int x^n(\ln ax)^{m-1}\,dx,\ n \neq -1$$

$$\int \frac{(\ln ax)^m}{x}\,dx = \frac{(\ln ax)^{m+1}}{m+1} + C,\ m \neq -1$$

$$\int \frac{dx}{x\ln ax} = \ln\left|\ln ax\right| + C$$

When your work speaks for itself, don't interrupt. Henry Kaiser

8.5 Statistics

Common Statistical Formulas

Population Formulas

Population Mean

$$\mu = \frac{\Sigma\, x_i}{N}$$

Population Variance

$$\sigma^2 = \frac{\Sigma(x_i - \mu)^2}{N}$$

Population Standard Deviation

$$\sigma = \sqrt{\sigma^2} = \frac{\Sigma(x_i - \mu)^2}{N}$$

Sample Formulas

Sample Mean

$$\bar{x} = \frac{\Sigma\, x_i}{n}$$

Sample Variance

$$s^2 = \frac{\Sigma(x_i - \bar{x})^2}{n-1} = \frac{n(\Sigma\, x_i^2) - (\Sigma\, x_i)^2}{n(n-1)}$$

$$s = \sqrt{\frac{\Sigma(x_i - \bar{x})^2}{n-1}} = \sqrt{\frac{n(\Sigma\, x_i^2) - (\Sigma\, x_i)^2}{n(n-1)}}$$

Confidence Intervals for the Mean

95% Confidence Interval

$$\bar{x} \pm z_{\alpha/2} \frac{\sigma}{\sqrt{n}} = \bar{x} \pm 1.96 \frac{\sigma}{\sqrt{n}}$$

99% Confidence Interval

$$\bar{x} \pm z_{\alpha/2} \frac{\sigma}{\sqrt{n}} = \bar{x} \pm 2.575 \frac{\sigma}{\sqrt{n}}$$

Estimate for σ

$$\sigma \approx s$$

* The sample must be larger or equal to 30.

Test Statistics

Large Sample (when $n \geq 30$)

$$z = \frac{\bar{x} - u_0}{\sigma / \sqrt{n}} \approx \frac{\bar{x} - u_0}{s / \sqrt{n}}$$

Small Sample (when $n < 30$)

$$t = \frac{\bar{x} - u_0}{s / \sqrt{n}}$$

Regression Line Formulas

$$\hat{y} = \hat{\beta}_0 + \hat{\beta}_1 x$$

$$S_{xy} = \Sigma(x_i - \bar{x})(y_i - \bar{y}) = \Sigma x_i y_i - \frac{(\Sigma x_i)(\Sigma y_i)}{n}$$

$$S_{xx} = \Sigma(x_i - \bar{x})^2 = \Sigma\left(x_i^2\right) - \frac{(\Sigma x_i)^2}{n}$$

$$S_{yy} = \Sigma(y_i - \bar{y})^2 = \Sigma\left(y_i^2\right) - \frac{(\Sigma y_i)^2}{n}$$

$$\hat{\beta}_1 = \frac{S_{xy}}{S_{xx}}$$

$$\hat{\beta}_0 = \bar{y} - \hat{\beta}_1 \bar{x}$$

Correlation Coefficient

$$r = \frac{S_{xy}}{\sqrt{S_{xx}\sqrt{S_{yy}}}}$$

Permutations and Combinations

$$P_r^n = \frac{n!}{(n-r)!} \text{ (order important)}$$

$$C_r^n = \frac{n!}{r!(n-r)!} = \frac{P_r^n}{r!} \text{ (order not important)}$$

Success is not measured by how many failures you achieve. David B. Fogel

The Large Sample Statistical Test

Assumes n were randomly selected and the n population is greater than or equal to 30. To test a hypothesis about a population mean (use normal curve area table on page 280):

1) **The Null Hypothesis.** H_0: $\mu = \mu_0$

2) **Alternative Hypothesis.** This is the hypothesis that you want to support.

One - Tailed Tests	**Two - Tailed Tests**
a) H_a: $\mu > \mu_0$ or,	H_a: $\mu \neq \mu_0$
b) H_a: $\mu < \mu_0$	

3) **The Test Statistic.** This is a value that you get from a sample. It's the "decision maker."

$$z = \frac{\bar{x} - \mu_0}{\sigma / \sqrt{n}} \approx \frac{\bar{x} - \mu_0}{s / \sqrt{n}}$$

$$\bar{x} = \sum x_i / n$$

$$s = \sqrt{\frac{n\left(\sum x_i\right) - \left(\sum x_i\right)^2}{n(n-1)}}$$

4) **The Rejection Region.** The value that will cause the rejection of the null hypothesis.

One - Tailed Tests	**Two - Tailed Tests**
a) $z > z_\alpha$ or,	$z > z_{\alpha/2}$ or $z < -z_{\alpha/2}$
b) $z < -z_\alpha$, when b) is the alternative hypothesis.	

5) **Conclusion:** When your test statistic is in the rejection region, then you reject the null hypothesis (H_0) and support the alternative hypothesis (H_a).

The Small Sample Statistical Test

The small sample statistical test is used when your sample size is less than 30. It's the same as above except the test statistic and rejection region are different (use critical values table on page 281).

Test Statistic:

$$t = \frac{\bar{x} - \mu_0}{s / \sqrt{n}}$$

Rejection Region:

One - Tailed Test	**Two - Tailed Test**
a) $t > t_\alpha$ or,	a) $t > t_{\alpha/2}$ or,
b) $t < -t_\alpha$ when b) is the alternative hypothesis	b) $t < -t_{\alpha/2}$

Science is the great antidote to the poison of enthusiasm and superstition. Adam Smith

Normal Curve Areas

To determine z_α or $z_{\alpha/2}$, you need to know how significant your test is going to be.

To be 90% confident of your test, let $\alpha = 0.1$ for a one-tailed test and $\alpha/2 = 0.05$ for a two-tailed test. For 95% and 99% confidence, let $\alpha = 0.05, 0.01$, and $\alpha/2 = 0.025, 0.005$ respectively.

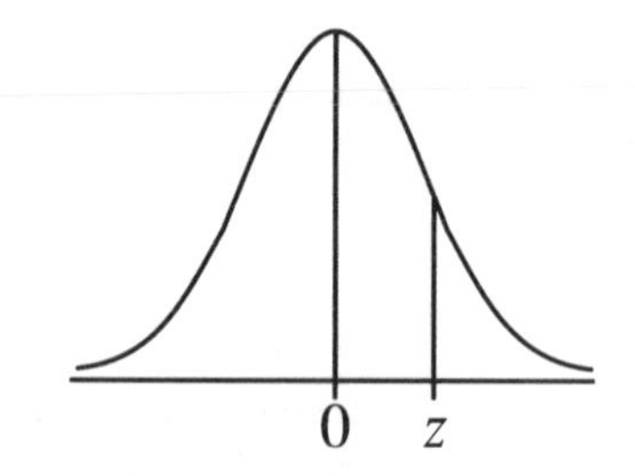

z - distribution table

The entries below give the areas under the standard normal curve between z = 0 and the given positive z value specified by entries on the right and top

z	0.00	0.01	0.02	0.03	0.04	0.05	0.06	0.07	0.08	0.09
0.0	0.0000	0.0040	0.0080	0.0120	0.0160	0.0199	0.0239	0.0279	0.0319	0.0359
0.1	0.0398	0.0438	0.0478	0.0517	0.0557	0.0596	0.0636	0.0675	0.0714	0.0753
0.2	0.0793	0.0832	0.0871	0.0910	0.0948	0.0987	0.1026	0.1064	0.1103	0.1141
0.3	0.1179	0.1217	0.1255	0.1293	0.1331	0.1368	0.1406	0.1443	0.1480	0.1517
0.4	0.1554	0.1591	0.1628	0.1664	0.1700	0.1736	0.1772	0.1808	0.1844	0.1879
0.5	0.1915	0.1950	0.1985	0.2019	0.2054	0.2088	0.2123	0.2157	0.2190	0.2224
0.6	0.2257	0.2291	0.2324	0.2357	0.2389	0.2422	0.2454	0.2486	0.2517	0.2549
0.7	0.2580	0.2611	0.2642	0.2673	0.2704	0.2734	0.2764	0.2794	0.2823	0.2852
0.8	0.2881	0.2910	0.2939	0.2967	0.2995	0.3023	0.3051	0.3078	0.3106	0.3133
0.9	0.3159	0.3186	0.3212	0.3238	0.3264	0.3289	0.3315	0.3340	0.3365	0.3389
1.0	0.3413	0.3438	0.3461	0.3485	0.3508	0.3531	0.3554	0.3577	0.3599	0.3621
1.1	0.3643	0.3665	0.3686	0.3708	0.3729	0.3749	0.3770	0.3790	0.3810	0.3830
1.2	0.3849	0.3869	0.3888	0.3907	0.3925	0.3944	0.3962	0.3980	0.3997	0.4015
1.3	0.4032	0.4049	0.4066	0.4082	0.4099	0.4115	0.4131	0.4147	0.4162	0.4177
1.4	0.4192	0.4207	0.4222	0.4236	0.4251	0.4265	0.4279	0.4292	0.4306	0.4319
1.5	0.4332	0.4345	0.4357	0.4370	0.4382	0.4394	0.4406	0.4418	0.4429	0.4441
1.6	0.4452	0.4463	0.4474	0.4484	0.4495	0.4505	0.4515	0.4525	0.4535	0.4545
1.7	0.4554	0.4564	0.4573	0.4582	0.4591	0.4599	0.4608	0.4616	0.4625	0.4633
1.8	0.4641	0.4649	0.4656	0.4664	0.4671	0.4678	0.4686	0.4693	0.4699	0.4706
1.9	0.4713	0.4719	0.4726	0.4732	0.4738	0.4744	0.4750	0.4756	0.4761	0.4767
2.0	0.4772	0.4778	0.4783	0.4788	0.4793	0.4798	0.4803	0.4808	0.4812	0.4817
2.1	0.4821	0.4826	0.4830	0.4834	0.4838	0.4842	0.4846	0.4850	0.4854	0.4857
2.2	0.4861	0.4864	0.4868	0.4871	0.4875	0.4878	0.4881	0.4884	0.4887	0.4890
2.3	0.4893	0.4896	0.4898	0.4901	0.4904	0.4906	0.4909	0.4911	0.4913	0.4916
2.4	0.4918	0.4920	0.4922	0.4925	0.4927	0.4929	0.4931	0.4932	0.4934	0.4936
2.5	0.4938	0.4940	0.4941	0.4943	0.4945	0.4946	0.4948	0.4949	0.4951	0.4952
2.6	0.4953	0.4955	0.4956	0.4957	0.4959	0.4960	0.4961	0.4962	0.4963	0.4964
2.7	0.4965	0.4966	0.4967	0.4968	0.4969	0.4970	0.4971	0.4972	0.4973	0.4974
2.8	0.4974	0.4975	0.4976	0.4977	0.4977	0.4978	0.4979	0.4979	0.4980	0.4981
2.9	0.4981	0.4982	0.4982	0.4983	0.4984	0.4984	0.4985	0.4985	0.4986	0.4986
3.0	0.4987	0.4987	0.4987	0.4988	0.4988	0.4989	0.4989	0.4989	0.4990	0.4990